Subcellular Biochemistry

Volume 81

Series editor
J. Robin Harris
University of Mainz, Mainz, Germany

The book series SUBCELLULAR BIOCHEMISTRY is a renowned and well recognized forum for disseminating advances of emerging topics in Cell Biology and related subjects. All volumes are edited by established scientists and the individual chapters are written by experts on the relevant topic. The individual chapters of each volume are fully citable and indexed in Medline/Pubmed to ensure maximum visibility of the work.

More information about this series at http://www.springer.com/series/6515

Mary Ann Asson-Batres • Cecile Rochette-Egly
Editors

The Biochemistry of Retinoid Signaling II

The Physiology of Vitamin A - Uptake, Transport, Metabolism and Signaling

Springer

Editors
Mary Ann Asson-Batres
Tennessee State University
Nashville, TN, USA

Cecile Rochette-Egly
Functional Genomics & Cancer
Institut de Genetique et de Department
Functional Genomics & Cancer
Illkirch, France

ISSN 0306-0225
Subcellular Biochemistry
ISBN 978-94-024-0943-7   ISBN 978-94-024-0945-1 (eBook)
DOI 10.1007/978-94-024-0945-1

Library of Congress Control Number: 2016957852

Printed on acid-free paper

This Springer imprint is published by Springer Nature
The registered company is Springer Science+Business Media B.V.
The registered company address is: Van Godewijckstraat 30, 3311 GX Dordrecht, The Netherlands

Preface

Vitamin A and its active derivative, retinoic acid (RA), have been known for a century to be essential for all steps of life, from embryo to adult. Over this time span, the biochemical properties of vitamin A and vitamin A derivatives and their mechanisms of action at the subcellular, cellular, and systems level have been deciphered by countless groups of researchers working in laboratories around the world. The field continues to produce novel insights into the actions of vitamin A, which have turned out to be surprisingly diverse and elegantly controlled by an amazing complexity of biosynthetic and regulatory processes that are beautifully orchestrated to execute its distribution and conversion to physiologically active derivatives. To update and synthesize this spectacular array of disparate findings, the editorial staff of Springer invited us to produce a multivolume book series to recapitulate all the historical and recent discoveries that define vitamin A biology. This volume represents the second in that series.

The first volume was published in 2014. It covered topics related to the structure and biochemistry of nuclear retinoic acid receptors, their interactions with retinoic acid, and their role as transcription factors as well as signaling moieties in the cytoplasm, regulating the expression of target genes and intermediates involved in cell growth, differentiation, development, and organogenesis.

In this second volume, topics cover vitamin A uptake, transport, and storage, the enzymatic conversion of vitamin A into retinoic acid; the formation and regeneration of the retinaldehyde chromophore that facilitates vision; and, along a very different vein, processes where vitamin A is active as a molecule in its own right.

Volume II is divided into nine chapters, each contributed by an author, whose area of expertise has had an impact on his/her respective field. All chapters follow a similar format which begins with a general introduction to the area, followed by descriptions of (a) the earliest findings and history of the area, (b) key findings that contributed to the development of the field, and (c) research that defines our current understanding of the area. Each chapter concludes with the author's perspective on the relevance and future directions for the field.

Chapter 1 covers the nomenclature and chemistry of provitamin A carotenoids and vitamin A derivatives, named retinoids. Chapters 2, 3, and 4 focus on vitamin A uptake, transport, storage, and mobilization. Chapters 5, 6, and 7 address the mechanism of synthesis of RA from vitamin A, its shuttling into the nucleus, and its catabolism. Chapter 8 presents and discusses recent findings highlighting a novel mechanism of vitamin A action, distinct from that mediated by retinoic acid, wherein vitamin A acts directly to activate kinase signalosomes that are involved in lipid metabolism and energy homeostasis. Chapter 9 provides an up-to-date summation of foundational and new data describing the well-known and essential role of vitamin A in vision.

We thank all of the authors for their efforts and enthusiasm in preparing this volume. They comprehensively reviewed the literature and provided stimulating ideas for the future. We also thank and acknowledge Meran Owen for his invitation to put together this Retinoic Acid Signaling book series; his assistant, Tanja van Gaans for her help and assistance in bringing the volume to completion; and Springer Publishing for its support of this project.

It is our hope that this second volume, along with the first, will serve as a solid introduction for all interested readers and as a strong reference for all current and future scientists working in the field of vitamin A biology.

Nashville, TN, USA — Mary Ann Asson-Batres
Illkirch, France — Cecile Rochette-Egly
November 24, 2016

In Memoriam

Proteins that bind vitamin A and its derivatives facilitate the many actions of retinoids. The initial recognition of this concept grew out of research carried out by DeWitt Goodman and colleagues at Columbia University, who identified retinol-binding protein (RBP) and who proposed that it was the major transporter of vitamin A in the bloodstream [2], and Frank Chytil and colleagues at Vanderbilt University, who identified a "*macromolecular fraction... capable of binding [3H] retinol in vitro and which differs from the serum component*" [1]. This early body of work has driven much of the research that is highlighted in the chapters that comprise this volume.

David Ong

David Ong joined Frank Chytil's laboratory in late 1973 and proceeded to purify the "macromolecular component" that Frank and his teammates had identified. This molecule turned out to be the first cellular retinol-binding protein (CRBP) [3]. The discovery of CRBP stimulated a flurry of research by Chytil, Ong, and others that subsequently led to the identification and characterization of all of the retinoid-binding proteins that we now know are the transporters and mediators of vitamin A's storage, metabolism, and downstream actions. Fittingly, Dave and Frank shared the Osborne and Mendel Award from the American Institute of Nutrition (now the ASN) in 1983 for their revolutionary work identifying and characterizing the cellular retinoid-binding proteins.

Frank Chytil passed away on July 7, 2014, and Dave Ong passed away just one short year later, on April 25, 2015. We invited Dave to contribute a chapter describing the history and biology of the cellular retinol-binding proteins and his and Frank's contributions to this field, but he declined in favor of enjoying his newly entered state of retirement, offering instead to serve as a consultant to Joe Napoli, the author of Chap. 2. Despite his many insightful comments and generosity in sharing firsthand information, Dave felt he hadn't done enough to warrant coauthorship on Joe's chapter and asked simply to be acknowledged.

Noa Noy

Just before publication, we were very sorry to learn that Noa Noy became gravely ill and, quite unexpectedly, passed away on October 18th, 2016. Noa was a truly innovative researcher in the field of retinoids and **her work** spanned two important areas of inquiry as detailed here in Chaps. 3 and 7. She was a very good friend and colleague, and she will be missed by all the 'retinoids' community.

References

1. Bashor MM, Toft DO, Chytil F (1973) *In-vitro* binding of retinol to rat tissue components. Proc Natl Acad Sci USA 70:3483–3487
2. Kanai M, Raz A, Goodman DS (1968) Retinol-binding protein: the transport protein for vitamin A in human plasma. J Clin Invest 47:2025–2044
3. Ong DE, Chytil F (1978) Cellular retinol-binding protein from rat liver. Purification and characterization. J Biol Chem 253:828–832

Contents

Chapter 1
Carotenoids and Retinoids: Nomenclature, Chemistry, and Analysis

Earl H. Harrison and Robert W. Curley, Jr.

Abstract Carotenoids are polyenes synthesized in plants and certain microorganisms and are pigments used by plants and animals in various physiological processes. Some of the over 600 known carotenoids are capable of metabolic conversion to the essential nutrient vitamin A (retinol) in higher animals. Vitamin A also gives rise to a number of other metabolites which, along with their analogs, are known as retinoids. To facilitate discussion about these important molecules, a nomenclature is required to identify specific substances. The generally accepted rules for naming these important molecules have been agreed to by various Commissions of the International Union of Pure and Applied Chemistry and International Union of Biochemistry. These naming conventions are explained along with comparisons to more systematic naming rules that apply for these organic chemicals. Identification of the carotenoids and retinoids has been advanced by their chemical syntheses, and here, both classical and modern methods for synthesis of these molecules, as well as their analogs, are described. Because of their importance in biological systems, sensitive methods for the detection and quantification of these compounds from various sources have been essential. Early analyses that relied on liquid adsorption and partition chromatography have given way to high-performance liquid chromatography (HPLC) coupled with various detection methods. The development of HPLC coupled to mass spectrometry, particularly LC/MS-MS with Multiple Reaction Monitoring, has resulted in the greatest sensitivity and specificity in these analyses.

Keywords Carotenoid • Retinoid • Nomenclature • Vitamin A chemistry • Retinoid structure

E.H. Harrison (✉)
Department of Human Sciences, The Ohio State University, Columbus, OH 43210, USA
e-mail: Harrison.304@osu.edu

R.W. Curley, Jr.
College of Pharmacy, The Ohio State University, Columbus, OH 43210, USA
e-mail: Curley.1@osu.edu

M.A. Asson-Batres, C. Rochette-Egly (eds.), *The Biochemistry of Retinoid Signaling II*, Subcellular Biochemistry 81, DOI 10.1007/978-94-024-0945-1_1

Abbreviations

CTCL	cutaneous T-cell lymphoma
HPLC	high performance liquid chromatography
HWE	Horner-Wadsworth-Emmons modification
IUB	International Union of Biochemistry
IUPAC	International Union of Pure and Applied Chemistry
LC/MS	liquid chromatography/mass spectrometry
LC/MS-MS	liquid chromatography/mass spectrometry-mass spectrometry
MRM	multiple reaction monitoring
RAL	retinal
RA	retinoic acid
RAR	retinoic acid receptor
RXR	retinoid X receptor
ROL	retinol
TTNPB	tetrahydro-tetramethyl-napthalenyl-propenyl-benzoic acid
UV	ultraviolet

Introduction

Carotenoids are synthesized in plants and in certain microorganisms such as some bacteria, algae, and fungi. They are a group of pigments that are widespread in nature and responsible for the yellow/orange/red/purple colors of many fruits, flowers, birds, insects, and marine animals. Over 600 carotenoids have been isolated from natural sources and new ones continue to be discovered or synthesized. All carotenoids are derived from the basic linear structure of lycopene that contains 40 carbon atoms and an extended system of 13 conjugated double bonds. Carotenoids derive from this parent structure by cyclization at one or two ends of the chain and by dehydrogenation and/or oxidation.

The carotenoids and retinoids are all biosynthesized beginning with activated forms (pyrophosphates) of the five carbon molecule "isoprene" (see Fig. 1.1). The product of "head-to-tail" condensing of two of these units produces a "monoterpene". When three of these isoprene units are combined, the important sesquiterpene relay compound farnesyl pyrophosphate is produced. Using different enzymes and pathways, this sesquiterpene is converted into, for example, the sterols, the dolichols and other triterpenes. Four of these units combine in a similar manner to produce the diterpene geranylgeranyl pyrophosphate which is the precursor of quinones such as ubiquinone, the carotenoids, and other diterpenes [24].

Carotenoids are fundamentally important in the evolution and ecology of many taxa. Because they absorb light in some part of the visible spectrum, carotenoids are colored and carotenoid-based coloration is used by both plants and animals as

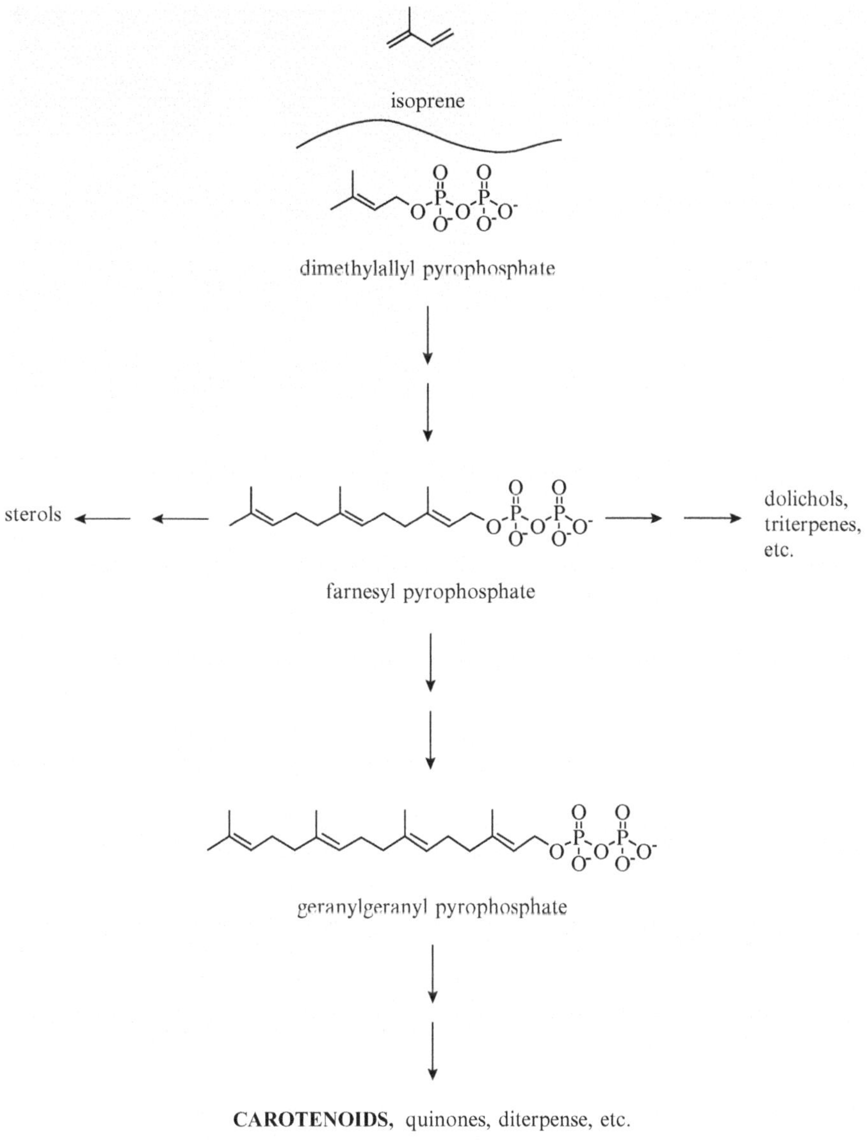

Fig. 1.1 Biosynthesis of carotenoids and other terpenoids

attractants (e.g. of pollinators for plants and of mates for animals) and to communicate fitness [2] (Fig. 1.2).

One of the principal functions of the carotenoids in the plant and animal kingdoms is as antioxidants. Singlet oxygen is a high energy reactive form of molecular oxygen which can be produced from ground state triplet oxygen by light-induced photosensitization as well as other chemical reactions. Other damaging reactive

Fig. 1.2 Carotenoids and mate choice. Male American goldfinch, *Carduelis tristis*, an avian species in which females choose mates based on variation in the extent of carotenoid-based plumage coloration

oxygen species include peroxyl and hydroxyl radicals. The carotenoids can protect cellular components from the damaging reactions of photo-oxidation and reactive oxygen species by multiple mechanisms including: 1) their very large molar absorption coefficient for light that allows them to protect directly against photo-oxidation; 2) their ability to quench directly highly reactive singlet oxygen; and 3) their loss of protons in response to interactions with reactive species that produces a much less reactive radical center in the carotenoid molecule which is stabilized by the polyene network [26]. Indeed, all three of these mechanisms are involved in the role that carotenoids play in the fundamental process of photosynthesis (see [3]) and in their putative roles as antioxidants in human health (see [40]).

Some of the carotenoids are metabolically converted to the essential nutrient vitamin A or retinol [18]. Vitamin A is an essential vitamin for higher animals, including humans. The vitamin is needed for normal embryogenesis and development and for vision, immunity, reproduction, and the maintenance of differentiated epithelial tissues. Vitamin A is the generic term used for all naturally occurring compounds containing a β-ionone ring, other than the carotenoids, that exhibit qualitatively the biological activity of retinol. In [34] Sporn et al. coined the term "retinoid" to refer to the natural and synthetic chemical derivatives of retinol and retinoic acid, regardless of whether they have vitamin A activity.

Retinol is a C20 isoprenoid (or a diterpenoid). It can be derived from metabolic conversion of some dietary carotenoids, which are C40 isoprenoids (or a tetraterpenoid). Such carotenoids are termed provitamin A carotenoids. In order to exhibit a provitamin A activity, the carotenoid molecule must have at least one unsubstituted β-ionone ring and the correct number and position of methyl groups in the polyene chain.

In this chapter we discuss the nomenclature and classification of retinoids and carotenoids as well as the chemical synthesis and quantitative analysis of these two classes of compounds.

Retinoid/Carotenoid Nomenclature and Classification

Retinoids

Karrer [23] established the structure of the dietary component in fat ("fat soluble A", now "vitamin A"), that McCollum and Davis had first discovered was essential for growth in mammals [31]. Soon after, Wald isolated a substance from frog and mammal eyes that he called "retinene" [38], and Morton suggested that the substance that Wald had isolated from eyes was the aldehyde of vitamin A which he called "retinaldehyde" [32]. In 1960, the International Union of Pure and Applied Chemistry (IUPAC) published recommendations on the nomenclature of the vitamins and proposed that the parent retinoids should be known as retinol (ROL; **1**; Fig. 1.3), retinal (RAL; **2**; Fig. 1.3) and retinoic acid (RA; **3**; Fig. 1.3) [7]. These names recapitulated the importance of these substances for vision in the retina and also made use of the suffixes normally used in organic chemistry to indicate one is dealing with an alcohol ("-ol"), aldehyde ("-al"), or carboxylic acid ("-oic acid") oxidation state at the polar terminus of the molecule. The accepted numbering scheme for the positions in these molecules is shown in Fig. 1.3. It should be noted that when there might be confusion between the molecule retinal and the adjective retinal (pertaining to the retina), that the use of retinaldehyde is still recommended.

The names for various vitamin A metabolites, and the accepted numbering convention for the positions within the molecules differ substantially from the systematic names and numbering system that would be used if one were to apply the accepted rules of nomenclature for organic chemicals. For example, using the latter nomenclature, RA with its carboxylic acid carbon as position 1 would be: (all-*E*)-3,7-dimethyl-9-(2,6,6-trimethyl-1-cyclohexen-1-yl)-2,4,6,8-nonatetraenoic acid.

In the early 1980s the Joint Commission on Biochemical Nomenclature of the IUPAC-International Union of Biochemistry (IUB) issued the recommendations on the nomenclature of retinoids. At that time [6], retinoids were defined as compounds composed of 4 isoprene units joined head-to-tail such that the products were monocyclic compounds with 5 conjugated double bonds and a functional group at the terminus of the acyclic portion of the molecules.

Following the Commision's recommendations, if the functional group at the 15-position is changed, the remainder of the hydrocarbon is referred to as the "retinyl" radical and the new functional group identified, for example there are esters of retinol known as retinyl acetate (**4** Fig. 1.3) and retinyl palmitate (**5**; Fig. 1.3). Changes to the state of hydrogenation of the parent structure are denoted by indicating the position(s) involved and "hydro" for addition of hydrogen or "dehydro" for removal of hydrogen. For example, the ROL derivative originally termed "vitamin A_2" would be 3,4-didehydroretinol (**6**; Fig. 1.3).

In the parent retinoids, all of the double bonds are in a *trans*, or *E*, configuration. Changes from this starting stereochemistry are described by using the lowest numbered carbon in the double bond effected and identifying it as now being *cis*, or

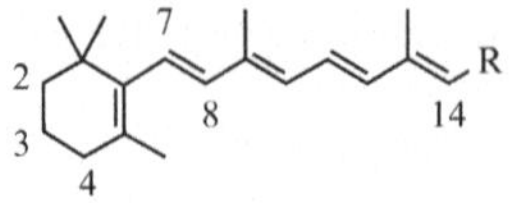

1 R = CH_2OH
2 R = CHO
3 R = COOH
4 R = CH_2OOCCH_3
5 R = $CH_2OOC(CH_2)_{14}CH_3$

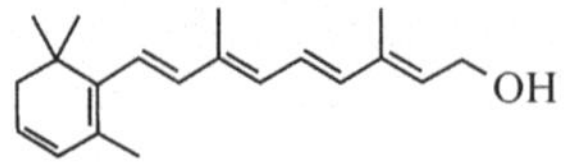

6

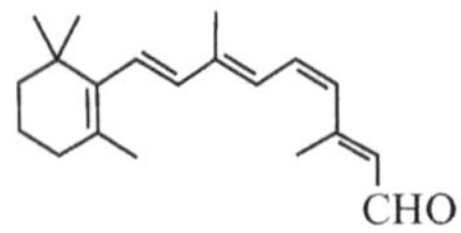

7

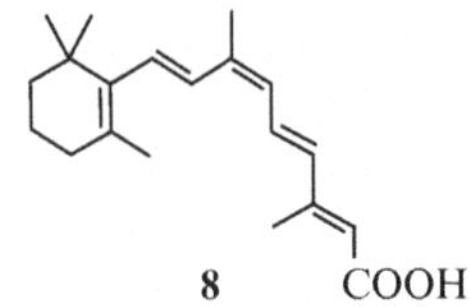

8

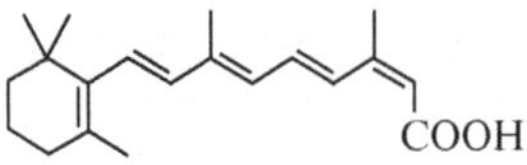

9

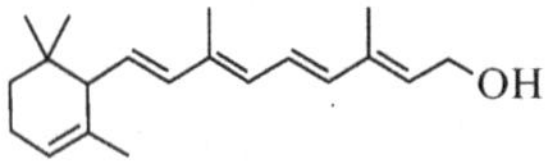

10

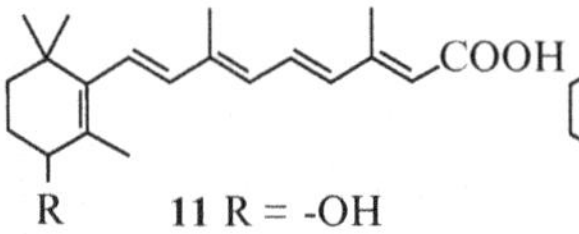

11 R = -OH
12 R = -O

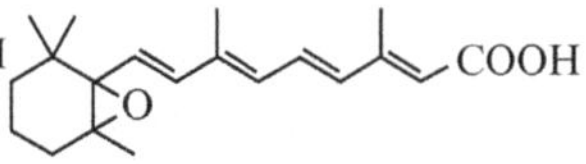

13

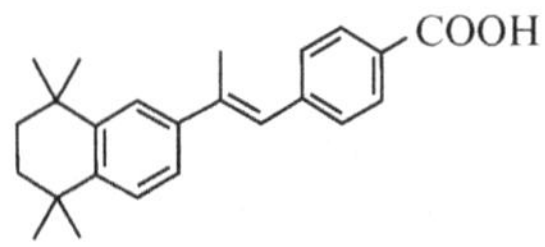

14

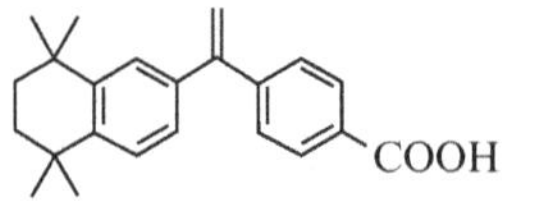

38

Fig. 1.3 Structure of retinoids; **1** = retinol, **2** = retinal, **3** = retinoic acid, **4** = retinyl acetate, **5** = retinyl palmitate, **6** = 3,4-didehydroretinol, **7** = 11-*cis*-retinal, **8** = 9-*cis*-retinoic acid, **9** = 13-cis-retinoic acid, **10** = α-retinol, **11** = 4-hydroxyretinoic acid, **12** = 4-oxoretinoic acid, **13** = 5,6-epoxyretinoic acid, **14** = TTNPB, **38** = bexarotene

Z in configuration. Important examples would be 11-*cis*-retinal (**7**; Fig. 1.3) and 9-*cis*- (**8**; Fig. 1.3) and 13-*cis*-retinoic acid (**9**; Fig. 1.3). If a double bond moves its position from that of the parent molecules, a more complicated naming situation arises. As an illustration, the form of the vitamin with the non-conjugated ring double bond known as α-retinol would also be named 4,5-didehydro-5,6-dihydroretinol (**10**; Fig. 1.3) and results in the parent retinol (**1**; Fig. 1.3) occasionally being referred to as β-retinol. Many of the important retinoid metabolites are the result of oxidation of the parent molecule at other than the 15-position. In these cases, as expected, the position(s) and type of substitution is identified. For example 4-hydroxy- (**11**; Fig. 1.3) and 4-oxoretinoic acid (**12**; Fig. 1.3) as well as 5,6-epoxyretinoic acid (**13**; Fig. 1.3) are known, although the latter metabolite would formally be named 5,6-epoxy-5,6-dihydroretinoic acid.

In efforts to increase the selectivity or potency and/or reduce the toxicity of pharmacologic doses of RA, many synthetic analogs of the parent molecule have been prepared. Those that are closely related to the parent retinoid structure have names that also resemble many of the modified structures above. However, the greatest success in preparing useful RA analogs have resulted from structures which bear limited resemblance to the classical retinoid parent molecules and thus have very different names than those described above. For example, one of the first very potent RA-like agonists to be synthesized was the stilbene known by the acronym TTNPB (**14**; Fig. 1.3) [29]. This compound is named systematically as 4-[(1*E*)-2-(5,6,7,8-tetrahydro-5,5,8,8-tetramethyl-2-naphthalenyl)-1-propen-1-yl]benzoic acid, making it easy to understand why the acronym TTNPB is popularly used to refer to this molecule.

A nomenclature for retinoids is essential to ensure that when we speak or write about a specific compound, there is no ambiguity about which chemical is being referred to. The nomenclature and classification of the retinoids has become standardized and well accepted. However, it is possible that the range of structure types that get classified as "retinoids" may continue to expand, particularly as the exploration of the non-genomic actions of retinoids are further elucidated.

Carotenoids

Carotenoids are 40-carbon organic pigments that are commonly biosynthesized in plants and also by some bacteria and fungi. Hydrocarbon forms of carotenoids are known as "carotenes" while oxygenated forms are termed "xanthophylls". As a class, these molecules are "tetraterpenes", that is they are made up of 4 monoterpene units which is equivalent to the joining of 8 isoprene units in a manner such that the arrangement of the isoprene units is reversed at the center of the molecule. The structure of the important orange plant carotenoid β-carotene (**15**; Fig. 1.4), commonly found in carrot (*Daucus carota*), was elucidated by Karrer and coworkers in 1930 [22]. They also suggested based on the chemical similarity, that central cleavage of **15** might be the source of two molecules of vitamin A.

Because there are many different carotenoid structures found in nature, there have been numerous Commissions that have attempted to systematize the nomenclature of carotenoids with updates occurring frequently as new structures were determined which forced modification to the rules. However, the rules for nomenclature of carotenoids have been reasonably stable since the joint IUPAC-IUB Commission approved rules in 1974 [5]. According to these rules, all carotenoids should be named using the stem "carotene" with the balance of the name indicating

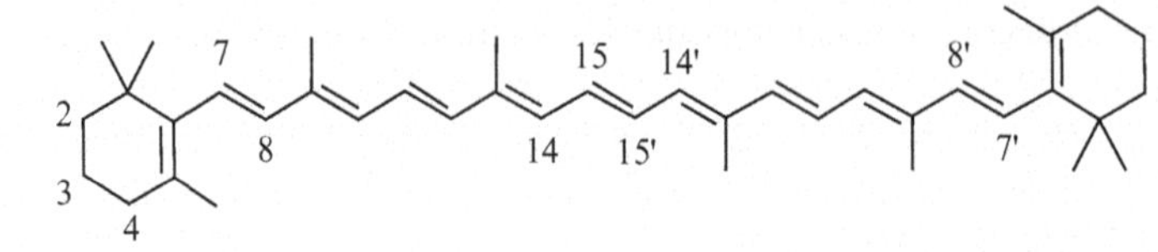

15

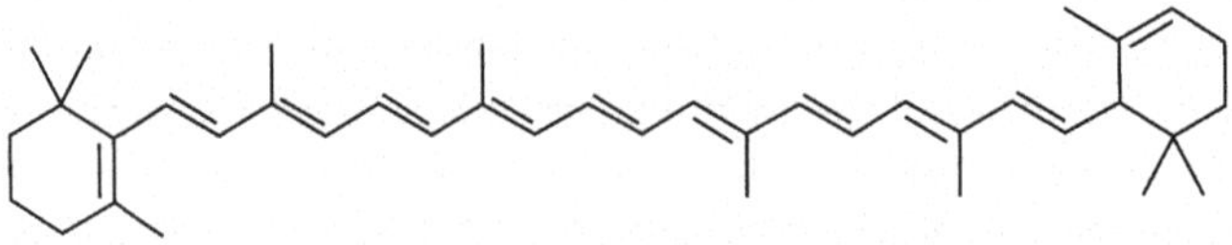

16

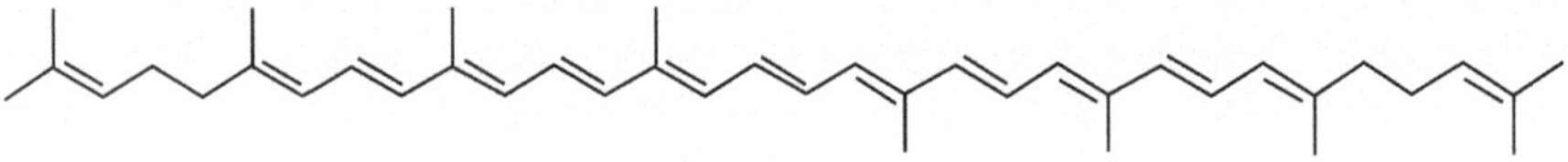

17

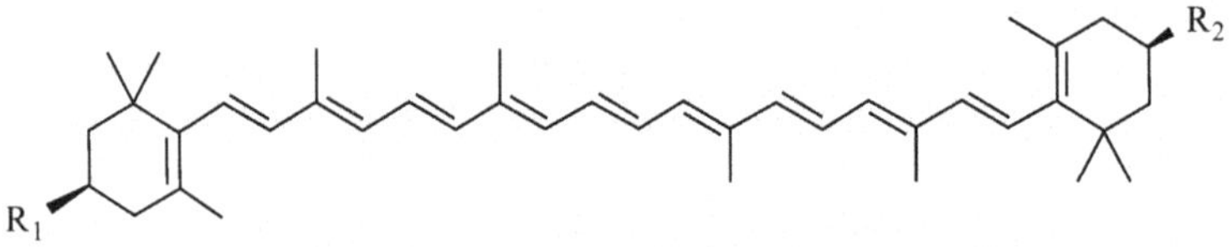

18 R_1 = OH, R_2 = H
20 R_1 = OH, R_2 = OH

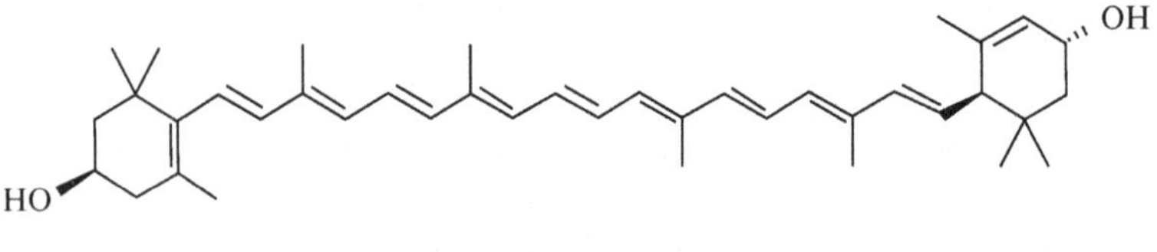

19

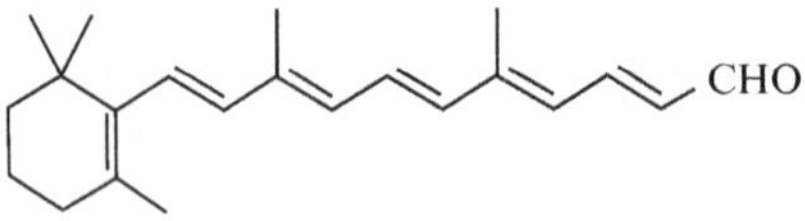

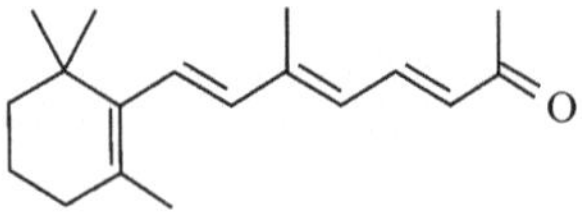

21 **22**

Fig. 1.4 Structure of carotenoids; **15** = β-carotene, **16** = α-carotene, **17** = lycopene, **18** = β-cryptoxanthin, **19** = lutein, **20** = zeaxanthin, **21** = β-apo-14'-carotenal, **22** = β-apo-13-carotenone

the nature of the end group(s), state of hydrogenation, oxygenation, etc. The basic numbering of these molecules is shown on the structure of β-carotene in Fig. 1.4 and reflects the symmetry of the molecule through a plane between the 15 and 15' carbon atoms. If both end groups are a cyclohexene ring with double bonds in the 5,6-position, the end group prefix is "β", as in β-carotene, which should technically be called β,β-carotene because it has two ring end groups. If the double bonds are instead in the 4,5-position, the prefix "ε" is used. The carotenoid known by the trivial name α-carotene (**16**; Fig. 1.4) would thus correctly be β,ε-carotene. If the terminus of the carotenoid is acyclic, the accepted prefix is "ψ" and thus the red tomato (*Solanum lycopersicum*) pigment lycopene (**17**; Fig. 1.4) would be formally known as ψ,ψ-carotene.

Since both β-carotene and α-carotene can be centrally cleaved enzymatically to yield at least one vitamin A molecule, as the aldehyde precursor retinal (**2**; Fig. 1.3), they are referred to as having provitamin A activity. Xanthophylls with only one oxygenated terminal ring, such as the one known by the trivial name β-cryptoxanthin (**18**; Fig. 1.4), can also have provitamin A activity and would correctly be called (3*R*)-β,β-caroten-3-ol. Other major xanthophylls that do not have provitamin A activity would include lutein (**19**; Fig. 1.4) and zeaxanthin (**20**; Fig. 1.4). Of course both of these carotenoids would be alternatively named as carotenes applying the rules for carotenoid nomenclature summarized here. Similar to the accepted naming of retinoids, β-carotene's name would be much different and quite a bit more complicated if the systematic rules for nomenclature of organic chemicals were applied: (all-*E*)-1,1'-(3,7,12,16-tetramethyl-1,3,5,7,9,11,13,15,17-octadecanonaene-1,18-diyl)bis[2,6,6-trimethylcyclohexene]!

When the carbon skeleton of a carotenoid has been shortened by removal of fragments the nomenclature rules prescribe the prefix "apo" preceded by its position (locant) to indicate that all of the carbon atoms beyond the carbon at the locant have been replaced by hydrogen atoms. Of course retinal (**2**; Fig. 1.3) would represent a very important apo-carotenoid derived from β-carotene but is rarely referred to by other than its accepted name. However, cleavage of a carotenoid such as β-carotene at other than the central double bond results in products without common names. The products are therefore named as apo-carotenoids. For example, cleavage of β-carotene at the double bond adjacent to the central double bond results in formation of two possible apo-carotenoid products, 14'-apo-β-caroten-14'-al (**21**; Fig. 1.4) and 13-apo-β-caroten-13-one (**22**; Fig. 1.4), more commonly referred to as β-apo-14'-carotenal and β-apo-13-carotenone.

While the agreed upon nomenclature used to define carotenoids has been useful to facilitate communication and has been reasonably stable since the mid-1970's, new examples of carotenoid natural products are discovered regularly. Thus, it is possible that new carotenoid chemotypes will be found that will necessitate extension or modification of the currently accepted nomenclature rules. Otherwise, it appears likely that interest will continue in determining the occurrence and biological relevance of the apocarotenoids. For example, surprisingly, the β-apo-13-carotenone (**22**; Fig. 1.4) mentioned above has been found to be a potent antagonist of the actions of RA at both the nuclear retinoid X receptors (RXR) and the RA receptors (RAR) [13, 14] and is present in foods and human plasma. Similarly, there is interest in determining whether apocarotenoids derived from lycopene (apolycopenoids) are also biologically active species derived from the parent carotenoid [28].

Retinoid/Carotenoid Chemistry

Synthesis

Development of the Field

The natural retinoids, as well as their carotenoid precursors, all contain relatively long chain conjugated polyene units. Because of the presence of this type of unit, the molecules are fairly sensitive to allylic oxidation and epoxidation, as well as isomerization which can be effected primarily by acid, light, or heat. Given the presence of these polyene units, olefin forming reactions feature prominently in the syntheses of retinoids. Classically, the Wittig reaction has been so extensively employed in retinoid and carotenoid syntheses that an important early review of the Wittig reaction shows many applications in this specific retinoid/carotenoid field [30]. For example, as shown in Scheme 1.1, the base-catalyzed Wittig reaction of phosphonium salts such as **23** (Scheme 1.1) with carbonyl-containing compounds can be used to prepare the ROL precursor retinyl acetate (**4**; Fig. 1.3) as well as the ethyl ester of RA (**24**; Scheme 1.1).

In most instances it is *trans* olefin linkages that are present in natural retinoids, such as **1–3** (Fig. 1.3), and *trans*-selective olefin forming reactions are desired. One difficulty with employing the Wittig reaction in these retinoid and related syntheses is a tendency to form isomer mixtures at the newly created olefinic linkage, often with the *cis* isomer predominating. Thus, a number of modifications and alternatives to the Wittig olefination have been developed. For example, the replacement of the phosphonium salts such as **23** (Scheme 1.1) with phosphonate esters (the Horner-Wadsworth-Emmons [HWE] modification), results in more nucleophilic phosphonate stabilized carbanions which in many retinoid syntheses can strongly favor formation of the *trans* or *E*-olefin. For example, the HWE modification can be used to prepare the ethyl ester of RA **24** (compare Schemes 1.2 and 1.1) via the base-

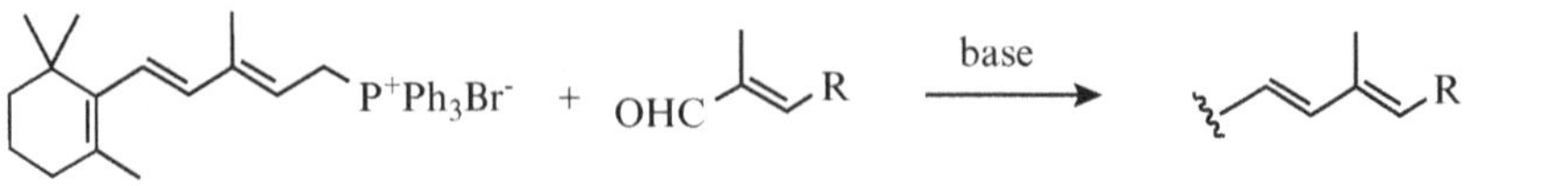

23

R = CH_2OOCCH_3
R = $COOCH_2CH_3$

4 R = CH_2OOCCH_3
24 R = $COOCH_2CH_3$

Scheme 1.1 Synthesis of retinoids via Wittig reaction

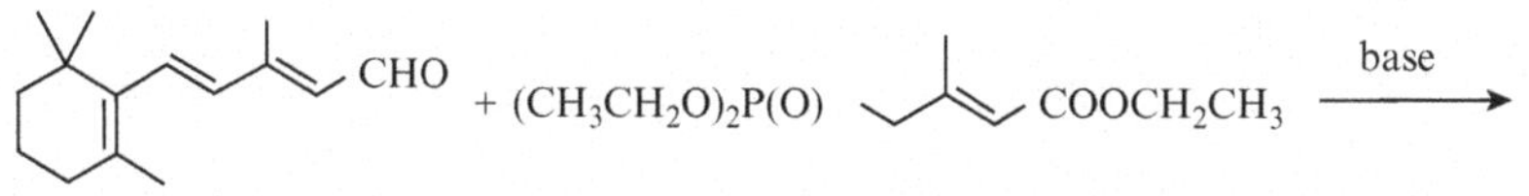

24

25 26

Scheme 1.2 Synthesis of retinoids via Horner-Wadsworth-Emmons modification

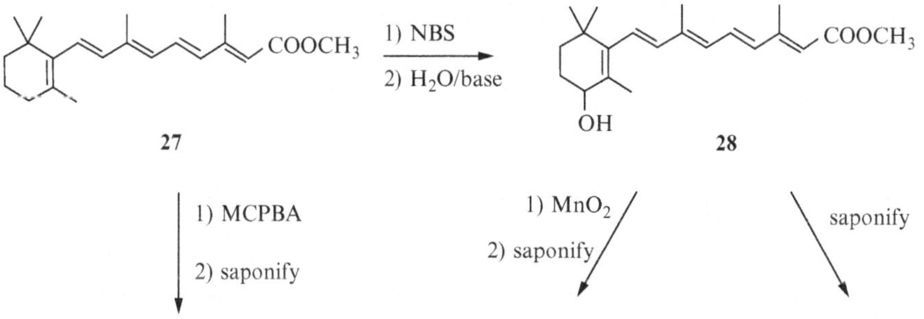

27 28

13 12 11

Scheme 1.3 Synthesis of retinoic acid metabolites; NBS = *N*-bromosuccinimide, MnO_2 = manganese dioxide, MCPBA = *meta*-chloroperbenzoic acid

catalyzed coupling of β-ionylidene acetaldehyde (**25**; Scheme 1.2) and phosphonate **26** (Scheme 1.2) with high *trans*-selectivity. Other conceptually similar olefin-forming reactions have also been used in retinoid syntheses such as the Julia [20] and Peterson [25] olefinations.

With the parent retinoids in hand, preparation of many of the oxygenated metabolites of these molecules is relatively straightforward. As others have similarly shown, we demonstrated that the 4-oxygenated RA metabolites such as 4-hydroxyretinoic acid (**11**; Fig. 1.3) and 4-oxoretinoic acid (**12**; Fig. 1.3) can be prepared (see Scheme 1.3) by allylic bromination of methyl retinoate (**27**; Scheme 1.3) with *N*-bromosuccinimide (NBS) followed by solvolysis of the unstable bromide with aqueous base, giving **28** (Scheme 1.3), and ester hydrolysis to give 4-hydroxyretinoic acid (**11**; see Scheme 1.3 and Fig. 1.3). Allylic oxidation of **28** (Scheme 1.3) with activated manganese dioxide (MnO_2) and saponification smoothly provides 4-oxoretinoic acid (**12**; see Scheme 1.3 and Fig. 1.3) [9]. The tetrasubstituted ring double bond of RA can also be directly epoxidized with peracid to provide 5,6-epoxyretinoic acid (**13**; see Scheme 1.3 and Fig. 1.3) [19].

There have been many approaches to the synthesis of the carotenoids. However, for the symmetrical carotenoids such as β-carotene (**15**; Fig. 1.4), the most effective method for synthesis has been the base-catalyzed double Wittig condensation of, for example, the C_{10} dialdehyde **29** (Scheme 1.4) with a C_{15} phosphonium salt such

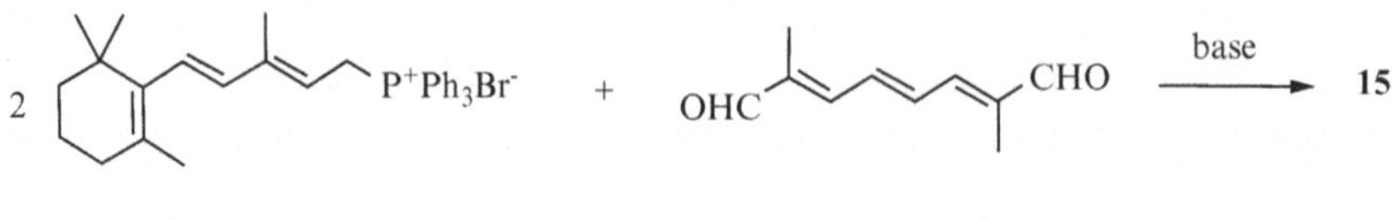

Scheme 1.4 Synthesis of β-Carotene via Wittig reaction

as **23** as shown in Scheme 1.4. At least 34 examples of this approach using **29** (Scheme 1.4) and various cyclic end group phosphonium salts were described in a review a number of years ago [39].

Current State of the Field

As chemists and biologists began to recognize that structures without the classic extended polyene chain could be effective retinoids, the range of structures that were synthesized for study expanded dramatically. The forerunner of this development is generally agreed to be the polycyclic structure already mentioned above: TTNPB (**14**; Fig. 1.3). While it was synthesized in 1980, TTNB was eventually found to be a highly potent ROL/RA mimic [35] due ultimately to its metabolism resistance and high affinity for the later discovered nuclear RARs.

A broadened array of structures and chemistry used to make analogs now exists for making molecules called retinoids. The outlines of the range of chemistry used to make these molecules has been extensively reviewed in the classic text "The Retinoids: Biology, Chemistry, and Medicine" [10]. Using TTNB to illustrate the much different chemistry that can be used to prepare these synthetic retinoids, we [1] and others [29] have used strategies such as those shown in Scheme 1.5 making use of both the Friedel-Crafts alkylation (reaction of **31** with **32**) and acylation (joining of **33** and **34**) and our old friend the HWE modification (coupling of **35** and **36**). More recently, very comprehensive updates of the chemistry used to synthesize retinoids and precursor carotenoids have been published [11, 37]. Particularly useful for readers here may be the sections in the Dominguez reference [11] that describe the use of various modern metal-mediated coupling reactions being used in C-C bond forming processes in retinoid syntheses.

Relevance and Future Directions

Robust syntheses of the natural retinoids and carotenoids have been necessary to confirm their structure and explore their biological activity. Significant progress has been made in applying older and more modern techniques to the preparation of these molecules. However, in the future, it is probable that conceptually new C-C single bond, and in particular C-C double bond forming methods that show good control of stereochemistry, will be developed for application to this field.

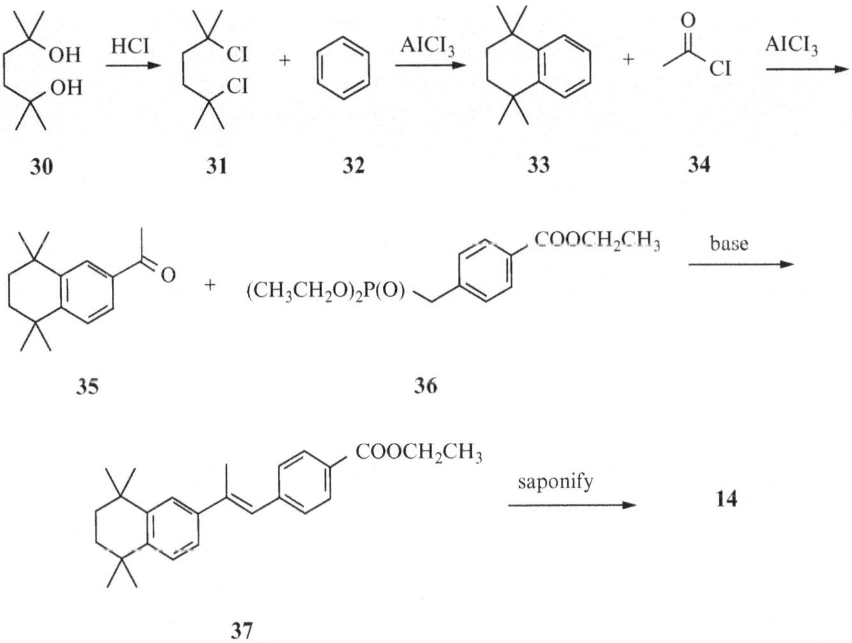

Scheme 1.5 Synthesis of TTNPB

Once the nuclear RARs/RXRs were discovered, there was a tremendous burst in efforts to prepare synthetic retinoids, particularly with receptor subtype selectivity (not only for the RARs vs. RXRs but also for each of the α, β, or γ isoforms of each receptor subclass). In the 1990s and into the early 2000s a large array of these types of molecules were synthesized, many of which had the desired subtype and isoform selectivity. Because of their role in epithelial differentiation, there was expectation that retinoids would find use in cancer therapy or prevention. However, with only a few exceptions, clinical trials of retinoids and rexinoids (a term analogous to "retinoid" now used for RXR selective retinoids) showed limited, if any, therapeutic value for these new single agent drugs.

Developments in the recent past have revived interest in retinoid/rexinoid analog studies. These new efforts were brought about by a recognition that there might be a role for combination therapies that include retinoids or rexinoids. For example, the modestly RXR-selective agonist bexarotene (**38**; Fig. 1.3), which has only been approved for treatment of the relatively rare cutaneous T-cell lymphoma (CTCL), has shown good activity when combined with the epidermal growth factor tyrosine kinase inhibitor drug, erlotinib, in lung cancer [12]. There have also been some surprisingly successful, but controversial studies of this rexinoid in a mouse model of Alzheimer's disease which, if confirmed, may be an important advance [8]. The modest selectivity and residual toxicity of bexarotene suggests that reexamination of other RXR-selective agonists may be warranted.

Some recent efforts are also being made to fine tune the structure of retinoid receptor binders in order to exploit subtle differences in the interactions of receptors with, for example, heterodimeric receptor protein partners as well as accessory protein coregulators in order to more selectively and successfully target disease processes [27]. The recognition that retinoids affect stem cell and cancer cell differentiation in part by effects on epigenetic modifications suggests the possibility that a role for the many previously or yet to be discovered receptor subtype selective retinoids may emerge as a consequence of new knowledge regarding molecular processes that influence aberrant cell proliferation and development [17].

Analysis

Development of the Field

The analysis and quantification of individual retinoids and carotenoids begins with the separation of the individual analytes from synthetic reaction mixtures or from extracts of complex biological samples. For both retinoids and carotenoids this has almost always been done by some form of liquid chromatography. Indeed, the development of chromatography itself is often credited to the Russian botanist Mikhail Tswett, who in 1903 achieved a colorful separation of plant pigments by passing extracts of green leaves through a column of calcium carbonate [15]. He found two green pigments (chlorophylls) and several yellow ones (carotenoids).

Liquid-adsorption chromatography methods using insoluble substances, including alumina, charcoal, and silica gel, were critical to the early development of both the carotenoid and vitamin A fields. These methods were crucial in work conducted by both Karrer in Switzerland and Kuhn in Germany, who were awarded the Nobel Prize in Chemistry in 1937 and 1938, respectively, for their independent work in carotenoids and vitamin A. Later, liquid-liquid partition chromatography (paper chromatography, thin-layer chromatography, gas-liquid chromatography) methods were developed and used to separate vitamin A and carotenoids. Quantitative analysis was achieved primarily by either determining the UV absorbance of the purified analyte or by specific colorimetric measurements of chemically-derivatized analytes.

Current State of the Field

Today, almost all quantitative analyses of retinoids, carotenoids, their metabolites and derivatives use high-performance liquid chromatography (HPLC) coupled with various modes of detection [16, 33]. The HPLC columns most often used are so-called "reverse-phase" columns packed with hydrophobic (C18) matrices. Compounds elute in order of polarity with more water-soluble compounds eluting first and more non-polar compounds eluting later. Separations are optimized by using gradient elution with mixtures of various organic solvents. In the case of carotenoids, a special "C30" column has been developed to facilitate resolution of the cis-isomers of carotenoids. For more polar compounds (viz., RA and its derivatives and apocarotenoic acids) "normal-phase" silica-based matrices are often used, and compounds elute in reverse order of polarity.

Quantitation of compounds present in sufficient amounts is commonly achieved by using UV absorption detection, often with a diode array detector that allows recording of the complete UV absorption spectrum of each peak. Thus, peak purity, identity and amount of analyte can all be assessed, provided that appropriate purified standards are available.

Because the sensitivity of UV detection precludes unambiguous quantitation of analytes, including RA, apocarotenoids, and minor isomers that are present in only very small amounts [21], more sensitive methods of analysis and detection have been developed. Today, the most powerful analytical techniques available for the unambiguous quantitative analysis of retinoids and carotenoids and their metabolites are those involving HPLC coupled with various modes of mass spectral detection, broadly called LC-MS. A number of mass spectrometer configurations can be used with differences in mode of ionization and configurations of single or multiple mass analyzers. See Kane [21] for a review of this topic as it specifically applies to RA analysis, which is beyond the scope of the current chapter.

In general, the application of LC/MS-MS techniques with Multiple Reaction Monitoring (MRM) gives the greatest reliability in terms of sensitivity and specificity of analyses. By selecting a major initial "parent" ion of the molecule of choice and then simultaneously detecting one or more specifically produced "daughter ions", the interference by other isobaric molecules in the mixture is greatly minimized. These methods still depend critically on the availability of purified standards. Indeed, the most rigorous quantitation is best achieved by having standards of the compound of interest that are labelled with stable isotopes such as ^{13}C or ^{2}H. Figure 1.5 shows an example of the quantitative analysis of the endogenous β-apo-13-carotenone (**22**; Fig. 1.4) in human plasma, a β-carotene derivative present at nanomolar concentrations [14]. The biggest drawback to LC/MS-MS analyses is that the instrumentation is very expensive and complex and requires a high degree of skill for its maintenance and operation. Also, it requires rigorous and time-consuming assay validation.

Because mass spectrometry is the only analytical technique that can detect stable isotopes in trace amounts, it is the only practical way to study the intestinal absorption, transport and metabolism of dietary carotenoids and retinoids in living humans. Indeed, most of what we know about the metabolism of dietary provitamin A carotenoids and vitamin A has come from such studies [36]. An example of using deuterated β-carotene to study its intestinal absorption and conversion to retinyl esters is shown in Fig. 1.6. When plasma kinetics of isotopically labelled retinol are combined with compartmental analysis, we can begin to understand whole body retinoid metabolism in both experimental animals and humans in quantitative terms [4].

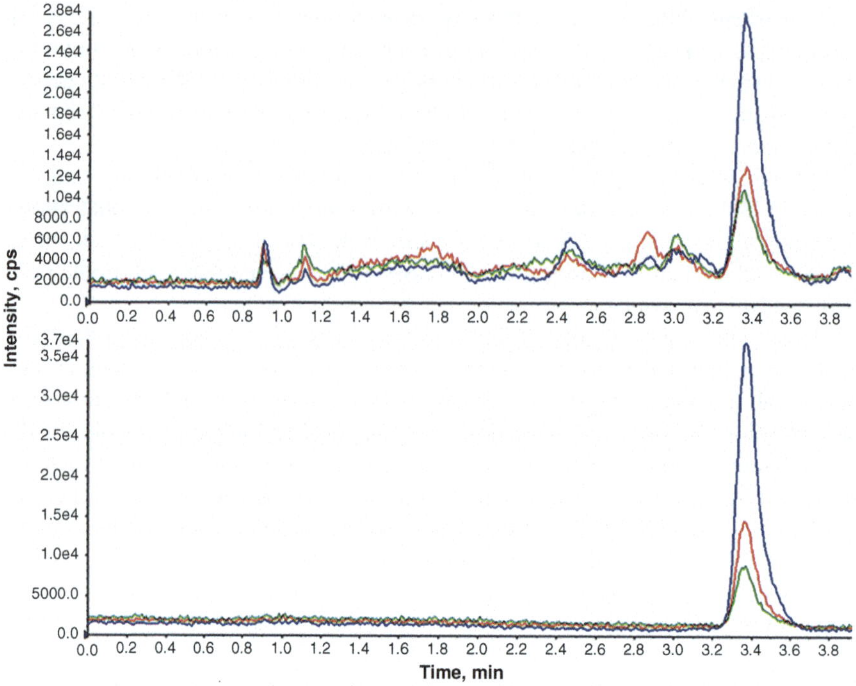

Fig. 1.5 Analysis of β-apo-13-carotenone in human plasma by HPLC/MS. Multiple reaction monitoring (MRM) chromatogram of β-apo-13-carotenone in blood plasma (*top*) and a standard (*bottom*) as analyzed by atmospheric pressure chemical ionization in positive mode after C30 HPLC. The MRM was composed of three transitions – m/z 259.2 > 175.1 (*blue*), 119.1 (*red*) and 69.0 (*green*) and the matching elution time and relative intensities of the transitions confirm the peak identity (from [14])

Relevance and Future Directions

It is clear that advances in chemical separations and analyses have been critical at all stages of the development of the retinoid field throughout the twentieth century and now into the twenty-first. We can expect that rapid advances in mass spectrometry and chromatography will continue and development of this technology will have profound effects on progress in retinoid and carotenoid research. We can anticipate that LC/MS-MS will be less costly and generally easier to carry out and hence, more laboratories will be able to adopt the technology. Increases in speed and lowering of costs will make large scale population studies of retinoid and carotenoid metabolism more feasible. Increases in sensitivity and specificity will likely allow us to discover "new" metabolites and to better quantify those that we already know. Increased sensitivity will also allow the analysis of smaller samples from experimental animals or isolated cells. Ultimately, we may see unanticipated advances that will someday make it possible to study the metabolism and action of retinoids and carotenoids at the level of the single cell.

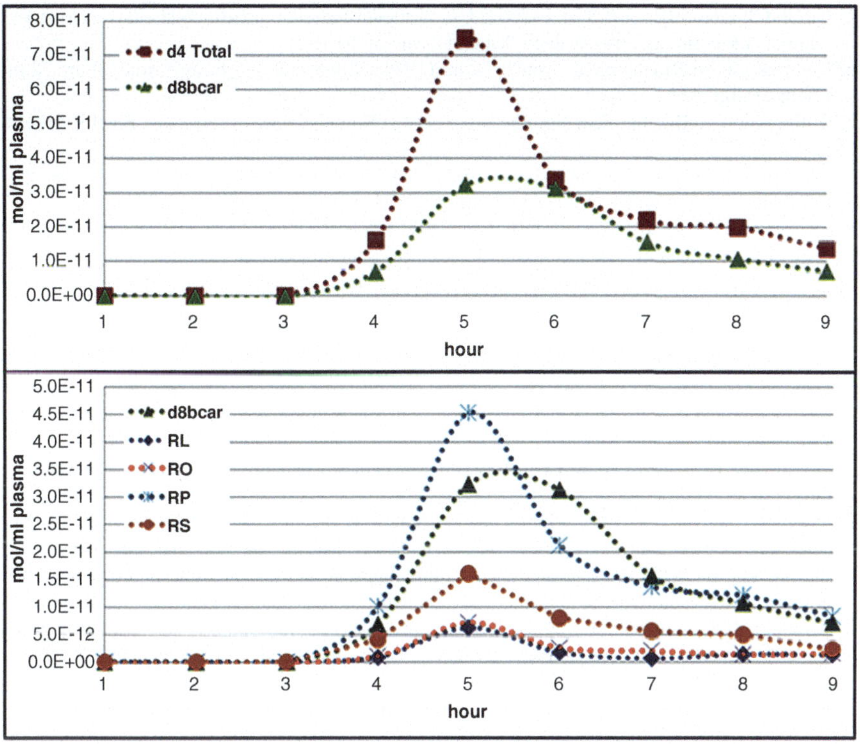

Fig. 1.6 Quantitative HPLC-MS of β-carotene and retinoids in a chylomicron fraction from the plasma of an individual given an oral dose of d8-β-carotene which is converted to d4-retinyl esters in the intestine. X-Axis shows time after dose. The top panel shows the total d4-retinyl esters (d4 Total) and d8-βC (d8bcar) while the bottom panel contains the individual d4-retinyl esters (retinyl linoleate, RL; retinyl oleate, RO; retinyl palmitate, RP; retinyl stearate, RS) and d8-βC for the same subject. Individual retinyl esters followed the same trend as the total retinyl esters and β-carotene with the peak concentrations of both d4-RE and d8-βC at 5–6 h post-dose (from Fleshman et al., J Lipid Res 53:820–827, 2012)

References

1. Anding AL, Nieves NJ, Abzianidze VV, Collins MD, Curley RW Jr, Clagett-Dame M (2011) 4-Hydroxybenzyl modification of the highly teratogenic retinoid, 4-[(1E)-2-(5,5,8,8-tetramethyl-5,6,7,8-tetrahydro 2 naphthalenyl)-1-propen-1-yl]benzoic acid (TTNPB), yields a compound that induces apoptosis in breast cancer cells and shows reduced teratogenicity. Chem Res Toxicol 24:1853–1861
2. Blount JD, McGraw KJ (2008) Signal functions of carotenoid colouration. In: Britton G, Liaaen-Jensen S, Pfander H (eds) Carotenoids, Volume 4: natural functions. Birkhauser Verlag, Basel, pp 213–236
3. Britton G (2008) Functions of intact carotenoids. In: Britton G, Liaaen-Jensen S, Pfander H (eds) Carotenoids, Volume 4: natural functions. Birkhauser Verlag, Basel, pp 189–212

4. Cifelli CJ, Green JB, Green MH (2007) Use of model-based compartmental analysis to study vitamin A kinetics and metabolism. Vitam Horm 75:161–195
5. Commission on Biochemical Nomenclature (1974) Nomenclature of carotenoids. Pure Appl Chem 41:405–431
6. Commission on Biochemical Nomenclature (1982) Nomenclature of retinoids. Eur J Biochem 129:1–5
7. Commission on the Nomenclature of Biological Chemistry (1960) Definitive rules for the nomenclature of vitamins. J Am Chem Soc 82:5575–5581
8. Cramer PE, Cirrito JR, Wesson DW, Lee CYD, Karlo JC, Zinn AE, Casali BT, Restivo JL, Goebel WD, James MJ, Brunden KR, Wilson DA, Landreth GE (2012) ApoE-directed therapeutics rapidly clear β-amyloid and reverse deficits in AD mouse models. Science 335:1503–1506
9. Curley RW Jr, Carson DL (1987) Synthesis of the 4-oxygenated retinoid metabolites. Drug Des Deliv 1:219–224
10. Dawson MI, Hobbs PD (1994) The synthetic chemistry of retinoids. In: Sporn MB, Roberts AB, Goodman DS (eds) The retinoids: biology, chemistry, and medicine, 2nd edn. Raven Press Ltd, New York, pp 5–178
11. Dominguez B, Alvarez A, de Lera AR (2003) Recent advances in the synthesis of retinoids. Org Prep Int 35:239–306
12. Dragnev KH, Ma T, Cyrus J, Galimberti F, Memoli V, Busch AM, Tsongalis GJ, Seltzer M, Johnstone D, Erkmen CP, Nugent W, Rigas JR, Liu X, Freemantle SJ, Kurie JM, Waxman S, Dmitrovsky E (2011) Bexarotene plus erlotinib suppress lung carcinogenesis independent of KRAS mutations in two clinical trials and transgenic models. Cancer Prev Res 4:818–828
13. Eroglu A, Hruszkewycz DP, Curley RW Jr, Harrison EH (2010) The eccentric cleavage product of β-carotene, β-apo-13-carotenone, functions as an antagonist of RXRα. Arch Biochem Biophys 504:11–16
14. Eroglu A, Hruszkewycz DP, dela Sena C, Narayanasamy S, Riedl KM, Kopec RE, Schwartz SJ, Curley RW Jr, Harrison EH (2012) Naturally occurring eccentric cleavage products of provitamin A β-carotene function as antagonists of retinoic acid receptors. J Biol Chem 287:15886–15895
15. Ettre LS (1980) Evolution of liquid chromatography: a historical overview. In: Horvath C (ed) High-performance liquid chromatography: advances and perspectives. Academic Press Inc, New York, pp 2–74
16. Furr HC (2004) Analysis of retinoids and carotenoids: problems resolved and unsolved. J Nutr 134:281S–285S
17. Gudas LJ (2013) Retinoids induce stem cell differentiation via epigenetic changes. Semin Cell Dev Biol 24:701–705
18. Harrison EH (2012) Mechanisms involved in the intestinal absorption of dietary vitamin A and provitamin A carotenoids. Biochim Biophys Acta 1821:70–77
19. Hartman DA, Basil JB, Robertson LW, Curley RW Jr (1990) Microbial transformation of retinoic acid by *Cunninghamella echinulata* and *Cunninghamella blakesleeana*. Pharm Res 7:270–273
20. Julia M, Arnould D (1973) Use of sulfones in synthesis. III. Synthesis of vitamin A. Bull Soc Chim Fr 2:746–750
21. Kane MA (2012) Analysis, occurrence, and function of 9-cis-retinoic acid. Biochim Biophys Acta 1821:10–20
22. Karrer P, Helfenstein A, Wehrli H, Wettstein A (1930) Pflanzenfarbstoffe XXV. Uber die constitution des lycopins und carotins. Helv Chim Acta 13:1084–1099
23. Karrer P, Morf R, Schopp K (1931) Zur kenntnis des vitamins-A aus fischtranen. II. Helv Chim Acta 14:1431–1436
24. Kirby J, Keasling JD (2009) Biosynthesis of plant isoprenoids: perspectives for microbial engineering. Annu Rev Plant Biol 60:335–355

25. Koch D, Gartner W (1997) Steric hindrance between chromophore substituents as the driving force of rhodopsin isomerization: 10-methyl-13-demethyl retinal containing rhodopsin. Photochem Photobiol 65:181–186
26. Landrum JT (2013) Reactive oxygen and nitrogen species in biological systems: reactions and regulation by carotenoids. In: Tanumihardjo SA (ed) Carotenoids and human health. Springer, New York, pp 57–101
27. le Maire A, Alvarez S, Shankaranarayanan P, de Lera AR, Bourguet W, Gronemeyer H (2012) Retinoid receptors and therapeutic applications of RAR/RXR modulators. Curr Top Med Chem 12:505–527
28. Lindshield BL, Canene-Adams K, Erdman JW (2007) Lycopenoids: are lycopene metabolites bioactive? Arch Biochem Biophys 458:136–140
29. Loeliger P, Bollag W, Mayer H (1980) Arotinoids, a new class of highly active retinoids. Eur J Med Chem 15:9–15
30. Maercker A (1965) The Wittig reaction. Org React 14:270–490
31. McCollum EV, Davis M (1913) Necessity of certain lipins in the diet during growth. J Biol Chem 15:167–175
32. Morton RA (1944) Chemical aspects of the visual process. Nature 153:69–71
33. Rivera SM, Canela-Garayoa R (2012) Analytical tools for the analysis of carotenoids in diverse materials. J Chromatogr A 1224:1–10
34. Sporn MB, Dunlop NM, Newton DL, Smith JS (1976) Prevention of chemical carcinogenesis by a vitamin A and its synthetic analogs (retinoids). Fed Proc Fed Am Soc Exp Biol 35:1332–1338
35. Stephens-Jarnagin A, Miller DA, DeLuca HF (1985) The growth supporting activity of a retinoidal benzoic acid derivative and 4,4-difluororetinoic acid. Arch Biochem Biophys 237:11–16
36. Tang G (2012) Techniques for measuring vitamin A activity from β-carotene. Am J Clin Nutr 96(suppl):1185S–1188S
37. Valla AR, Cartier DL, Labia R (2004) Chemistry of natural retinoids and carotenoids: challenges for the future. Curr Org Synth 1:167–209
38. Wald G (1934) Carotenoids and the vitamin A cycle in vision. Nature 134:65
39. Widmer E (1985) Synthetic advances in the carotenoid field. Pure Appl Chem 57:741–752
40. Yeum K-J, Aldini G, Russell RM, Krinsky NI (2009) Antioxidant/pro-oxidant actions of carotenoids. In: Britton G, Liaaen-Jensen S, Pfander H (eds) Carotenoids, Volume 5: nutrition and health. Birkhauser Verlag, Basel, pp 235–268

Chapter 2
Functions of Intracellular Retinoid Binding-Proteins

Joseph L. Napoli

Abstract Multiple binding and transport proteins facilitate many aspects of retinoid biology through effects on retinoid transport, cellular uptake, metabolism, and nuclear delivery. These include the serum retinol binding protein sRBP (aka *Rbp4*), the plasma membrane sRBP receptor Stra6, and the intracellular retinoid binding-proteins such as cellular retinol-binding proteins (CRBP) and cellular retinoic acid binding-proteins (CRABP). sRBP transports the highly lipophilic retinol through an aqueous medium. The major intracellular retinol-binding protein, CRBP1, likely enhances efficient retinoid use by providing a sink to facilitate retinol uptake from sRBP through the plasma membrane or via Stra6, delivering retinol or retinal to select enzymes that generate retinyl esters or retinoic acid, and protecting retinol/retinal from excess catabolism or opportunistic metabolism. Intracellular retinoic acid binding-proteins (CRABP1 and 2, and FABP5) seem to have more diverse functions distinctive to each, such as directing retinoic acid to catabolism, delivering retinoic acid to specific nuclear receptors, and generating non-canonical actions. Gene ablation of intracellular retinoid binding-proteins does not cause embryonic lethality or gross morphological defects. Metabolic and functional defects manifested in knockouts of CRBP1, CRBP2 and CRBP3, however, illustrate their essentiality to health, and in the case of CRBP2, to survival during limited dietary vitamin A. Future studies should continue to address the specific molecular interactions that occur between retinoid binding-proteins and their targets and their precise physiologic contributions to retinoid homeostasis and function.

Keywords Acyl-CoA:retinol acyltransferase • Cellular retinol binding-protein • Cellular retinoic acid binding-protein • Cytochrome P-450 • Acyl-CoA:diacylglycerol acyltransferase • Acyl-CoA:monoacylglycerol acyltransferase • Lecithin:retinol acyltransferase • Peroxisomal proliferator activated receptor δ/β • Retinal dehydrogenase • Retinol dehydrogenase • Retinal • Retinol • Retinoic acid • Retinoic acid receptor • Serum retinol binding-protein • Retinyl ester hydrolase

---

J.L. Napoli (✉)
Graduate Program in Metabolic Biology, Department of Nutritional Sciences and Toxicology, University of California, 119 Morgan Hall, 94720 Berkeley, CA, USA
e-mail: jna@berkeley.edu

M.A. Asson-Batres, C. Rochette-Egly (eds.), *The Biochemistry of Retinoid Signaling II*, Subcellular Biochemistry 81, DOI 10.1007/978-94-024-0945-1_2

Abbreviations

ADH	alcohol dehydrogenase(s) (medium chain ADH gene family)
ARAT	acyl-CoA: retinol acyltransferase
bsiREH	bile-salt independent REH
CRBP1	cellular retinol binding-protein type 1
CRBP2	cellular retinol binding-protein type 2
CRABP1	cellular retinoic acid binding-protein type 1
CRABP2	cellular retinoic acid binding-protein type 2
CYP	cytochrome P-450
DGAT1	acyl-CoA:diacylglycerol acyltransferase 1
FABP5	fatty acid binding-protein type 5
k_{cat}	catalytic constant (defines the efficiency of an enzyme for product formation)
k_d	equilibrium dissociation constant
LRAT	lecithin:retinol acyltransferase
MGAT	acyl-CoA:monoacylglycerol acyltransferase
PPARδ/β	peroxisomal proliferator activated receptor δ/β
RA	all-*trans*-retinoic acid
RALDH	retinal dehydrogenase encoded by *Aldh* gene family 1A
RAR	RA receptor
Rbp1	gene that encodes CRBP1
Rbp2	gene that encodes CRBP2
Rbp4	gene that encodes serum retinol binding-protein (sRBP)
RDH	retinol dehydrogenase(s) (SDR gene family)
RE	retinyl ester(s)
REH	retinyl ester hydrolase(s)
SDR	short-chain dehydrogenase/reductase
sRBP	serum retinol binding-protein
TTR	transthyretin
WT	wild-type

Introduction

Multiple binding-proteins display high specificity for retinoids. These binding-proteins apparently are involved in most facets of retinoid biology through mediating retinoid transport, metabolism and function. They belong to at least three gene families. The serum retinol binding-protein, sRBP (encoded by *Rbp4*), the main transporter of retinol through serum, belongs to the lipocalin gene family [3, 75]. Cellular retinaldehyde-binding protein, CRALBP (encoded by *Rlbp1*), which transports 11-*cis*-retinal in the eye, belongs to the CRAL_Trio gene family [240]. The intracellular retinoid-binding proteins, which transport and mediate metabolism and function of retinol, retinal and retinoic acid, belong to the fatty acid binding-protein gene family [263]. Data increasingly support occurrence of retinoid metabolons consisting of physiological interactions among intracellular retinoid binding-proteins and retinoid metabolizing enzymes that chaperone retinoid uptake, metabolism and action to enhance efficiency of retinoid use. This chapter focuses on intracellular retinoid binding-proteins of the fatty acid binding-protein gene family that have the most understood functions in retinoid transport and metabolism: cellular retinol binding-protein, type 1 (CRBP1), cellular retinol binding-protein, type 2 (CRBP2), cellular retinoic acid binding-protein, type 1 (CRABP1), cellular retinoic acid binding-protein, type 2 (CRABP2) and fatty acid binding-protein, type 5 (FABP5), and their interactions with enzymes that regulate retinoid homeostasis and nuclear receptors that mediate RA action.

History

Discovery of the Intracellular Retinoid Binding-Proteins

The existence of intracellular retinoid binding-proteins of the FABP gene family was established by the lab of Frank Chytil reporting what is currently known as CRBP1 (aka CRBP) [6]. Prompted by knowledge that steroids interact with specific intracellular receptors, Bashor et al. demonstrated the occurrence of a protein in cytosol of multiple rat tissues that binds retinol with high affinity and specificity. They noted that the 16 kDa molecular weight distinguished it from the steroid hormone receptors known at the time. CRBP1 expression was extended to additional tissues and to fetuses [5]. This was followed quickly by demonstration of two distinct retinoid binding-proteins in human uterus—one that binds retinol specifically (CRBP1) and another that binds RA specifically, currently known as CRABP1 [49, 185, 232, 241], and an RA binding-protein in skin that was likely CRABP2 [245]. Reviews of this early work with these two proteins have been published [46, 47, 232].

The next phase of work focused on purifying CRBP1 and CRABP1 [186, 187, 233, 234]. This was followed by determining their areas of expression in vertebrate tissues [67, 115, 188, 208]. Early work suggested that CRBP1 mediated uptake of retinol into cells and protected it from catabolism, whereas CRABP1 was postulated to serve as a potential mediator of RA action [48]. These deductions have been verified by work from multiple labs.

Approximately 10 years after the discovery of CRBP1 and CRABP1, a second retinol binding-protein, CRBP2, was discovered and purified [183]. Several tissues of the neonatal rat express CRBP2, with neonatal liver and intestine having ~100-fold higher expression than other tissues, and adult rat intestine having ~500-fold higher expression than other tissues. Expression of CRBP2 in adult tissues is more limited than that of CRBP1, both in location and in intensity, with the exception of intestinal mucosa, where CRBP2 levels are ~1000-fold higher than those of CRBP1. The earliest hypothesis for CRBP2 function suggested that it contributes to retinol absorption, which it does.

A second RA binding-protein was identified and cloned in 1990 and named CRABP2 [78], which likely is the same CRABP originally identified in skin [245]. Embryos and skin express CRABP2 most intensely, but low expression is widespread. RA induces the mRNA of CRABP2, suggesting that CRABP2 is involved in the mechanism of RA action.

In 1990, a third CRBP, CRBP3, was isolated from fish eyes [176]. In 2001, two groups reported additional retinol binding-proteins, also referred to as CRBP3 [71, 267]. The relationship of the CRBP3 reported by Nishiwaki et al. to the other two is not clear. It seems, however, that the proteins described by Vogel et al. and Foli et al. are not identical.

Mouse heart, muscle and epididymal white adipose express the CRBP3 reported by Vogel et al. most intensely. This protein binds retinol with a k_d value of ~109 nM. It has similar affinities for 13-*cis* and 9-*cis*-retinol. This CRBP3 functions as a substrate for lecithin:retinol acyltransferase (LRAT), and seems necessary for optimal incorporation of RE into milk, as demonstrated by knocking out its gene [205]. Knocking out CRBP3 also results in reduced food intake, decreased adiposity and increased lean body mass. The knockout mice are more cold tolerant than WT when fed a high-fat diet, owing to increased mitochondrial fatty acid oxidation in brown adipose tissue. In contrast, kidney and liver express the human CRBP3 reported by Folli et al. most intensely: this CRBP3 has a k_d value of ~60 nM for retinol. It is not the same protein as the CRBP3 reported by Vogel et al., and has no mouse ortholog. Its precise function remains unknown, as does the function and relationship of the fish CRBP3 to other retinoid binding-proteins.

Multiple reviews summarize the molecular characteristics, expression locations, and amino acid conservation of these binding proteins in detail. Readers are referred to these reviews for detailed information [134, 164, 174, 178, 184, 231].

Development of the Field

Affinity of CRBP for Retinoids

Even though initial purification of CRBP1 was slow, it still relied on monitoring fluorescence of bound retinol, despite opportunity for dissociation and distribution of retinol throughout membranes or other matrices. This established a high-affinity interaction with a slow off-rate between the ligand and the binding-protein (low dissociation constant) [182, 187, 233, 242]. Originally, a k_d value of 16 nM was calculated using a standard fluorescence titration approach [187]. High affinity, however, presents a formidable problem for establishing a k_d value, because a system must be at equilibrium to calculate an accurate equilibrium constant. Older fluorimeters

Table 2.1 Binding affinities (k_d values, nM) of retinoid isomers for CRBP1 and 2. RA and its isomers do not bind with CRBP1 or 2

Retinoid	CRBP1	CRBP2
all-*trans*-retinol	3±2	10±3
13-*cis*-retinol	3±1	30±12
9-*cis*-retinol	11±2	68±7
all-*trans*-retinal	9±4	11±4
13-*cis*-retinal	–	–
9-*cis*-retinal	8±4	5±3

Data are from Kane et al. [112]

were insufficiently sensitive to monitor CRBP1 titration with retinol using a protein concentration in the range of the anticipated k_d value. Therefore, a CRBP1 concentration much higher than the k_d value (~μM) had to be used. This determined the number of binding sites accurately, but allowed only an upper estimate of the k_d. Initially, there was also disagreement whether CRBP1 and 2 could bind retinal, but ultimately, various studies demonstrated that both binding-proteins recognize retinal [133, 135, 145]. As fluorimeters improved, lower estimates of k_d values were generated. In the 90s, CRBP1 k_d values were reported to be 2 nM for retinol and 30 nM for retinal [149]. With further improvements in equipment and application of non-linear regression analyses to the raw data, the latest k_d values for a variety of retinoids have been reported in the low nM range for CRBP1 and somewhat higher for CRBP2 (Table 2.1) [112]. These values also likely represent upper limits, because the fluorescence titrations were done with 150 nM CRBP1 for most retinoids, still well above the calculated k_d values, owing to improved, yet not sufficiently sensitive fluorimeters. Reasonably, the true values may be an as much as an order of magnitude lower. Estimates of ~0.1 nM have been published [135].

Affinities of CRABP1 and 2 for RA and Its Metabolites

CRABP1, purified to homogeneity from tissues, binds all-*trans*-RA (RA) with a k_d value of 4 nM (rat testes) and 16 nM (rat liver) [186, 187]. Expression of bovine CRABP1 in *E. coli* provided an opportunity to extend binding analysis studies [68, 70]. Purified CRABP1 isolated from *E. coli* has ~threefold greater affinity for RA than *E. coli* –expressed CRABP2. CRABPs also bind other forms of RA. A retinoid active in skin, 3,4-didehyro-RA [215], binds to both CRABP1 and 2 with an affinity similar to RA. The RA metabolites, 18-OH-RA, 4-OH-RA, 4-oxo-RA and 16-OH-4-oxo-RA, also have affinity constants similar to RA. These values were established with CRABP 1 and 2 concentrations in the high nanomolar range. Similar to CRBP1, fluorescent titration was used to estimate k_d values for CRABP 1 and 2, which provides only an upper limit k_d value for high-affinity ligands when the protein concentration is much higher than the k_d value, because the ligand saturates the protein, i.e. does not establish equilibrium. For lower affinity ligands, equilibrium can be established, allowing a more accurate value. Thus, the affinities of RA and its phase I *trans* metabolites could be much less, revealing greater differences relative to the *cis* isomers. This expectation was verified by re-visiting k_d values of CRABP1 and 2 for RA, 9-*cis*-RA and 13-*cis*-RA, also generated by fluorescence titration, but with CRABP1 and 2 concentrations from 5 to 16 nM. This approach, along with non-linear regression analyses of the data, generated k_d values of 0.4 and 2 nM RA for CRABP1 and CRABP2, respectively, and affirmed the order of binding: RA>9-*cis*-RA>13-*cis*-RA [177]. The k_d values of 9-*cis*-RA for both RA binding proteins were ~200 nM, whereas 13-*cis*-RA showed little to no affinity.

Retinoids Have Low Solubility in Aqueous Media

Retinoids have a limited aqueous solubility. Aqueous media accommodate a maximum retinol concentration of ~60 nM [260]. Therefore, μM intracellular concentrations of retinol in solution in the aqueous phase are not possible unless retinol is esterified or "chaperoned" by binding proteins. The high-affinity association of retinol with binding-proteins provides background for appreciating a chaperone model.

In rat liver, 95 % of retinoids are esterified and most occur within lipid droplets. The remaining unesterified retinol [15] occurs predominantly in the microsomal or cytosolic fractions [91], where the concentrations of retinoid binding-proteins (sRBP in microsomes and CRBP1 in cytosol) match or exceed the amount of retinol, supporting the conclusion that retinol is binding-protein-bound. In retinal pigment epithelia cells, which harbor large amounts of retinoids for use by the visual cycle, 92 % of total retinol occurs as retinyl esters (RE) and 8 % as CRBP1-retinol. Thus, intracellular retinol is either esterified or associated with a binding protein.

Insights from Knocking Out Rbp1, the Gene that Encodes CRBP1

The CRBP1-null mouse, created by disrupting the *Rbp1* gene though homologous recombination, illustrates the systemic physiological functions of CRBP1. Although *Rbp1*-null mice show no obvious morphological impairments, retinoid homeostasis is abnormal in multiple tissues [113].

Rbp1-null mice display severe vitamin A loss, but not deficiency, even when fed standard rodent chow, which contains copious vitamin A [77]. At the time of the studies, rodent chow supplied at least 25 IU vitamin A/g diet in the form of retinyl esters. Chow likely had a higher vitamin A equivalent value because of carotenoid content, as chow diets include plant material. Thus, the minimum vitamin A content was sixfold greater than the amount of vitamin A recommended for rodents [219, 220].

Despite copious vitamin A in the dam's diet, livers of *Rbp1*-null mice had threefold decreased retinyl palmitate (the major RE) in day 16.5 and 18.5 fetuses relative to WT mice. Starting at 4 weeks of age, both retinyl palmitate and retinol were ~50 % lower. Hepatic stellate cells, which store most of the vitamin A in liver as RE and express CRBP1 and retinoid metabolizing enzymes, contained fewer and smaller cytoplasmic lipid droplets in 8-week-old knockouts relative to WT. The elimination half-life ($t_{½}$) of total [^{3}H]retinoids in liver was more rapid (10 days in *Rbp1*-null mice compared to 60 days in WT). Six hours after dosing [^{3}H]retinol, livers of *Rbp1*-null mice retained 5 % of the dose and a retinyl palmitate/retinol ratio of 2, compared to WT, which retained 10 % of the dose with a retinyl palmitate/retinol ratio of 5. Differences in uptake, esterification, and mobilization were not determined. *Rbp1* ablation resulted in 50–60 % lower liver RA concentrations, and allowed increased susceptibility to hepatic retinoid depletion upon dioxin treatment [97].

Retinoid homeostasis also is abnormal in mammary tissue of *Rbp1*-null mice. Relative to controls, RA is reduced 40 % in mammary tissue of *Rbp1-null* mice, a decrease attributed to reduced retinol dehydrogenase (RDH) activity [65, 124, 157, 207]. This aberration in retinoid metabolism is accompanied by morphologic abnormalities, such as epithelial hyperplasia and stromal hypercellularity, which promote tumor progression. These data are significant because ~ 25 % of human breast cancers silence *Rbp1* epigenetically [123].

The pancreas of the *Rbp1*-null mouse has increased retinol, intense ectopic expression of *Rbp2* mRNA (encodes CRBP2), defective islet expression of glucose sensing and insulin secretion genes, α-cell infiltration into the β-cell interior of islets, diminished glucose-stimulated insulin secretion, high glucagon secretion, abnormally high gluconeogenesis, and hyperglycemia. Conversely, CRBP1 attenuates the negative impact of copious dietary retinol on glucose tolerance. Thus, glucose homeostasis and energy metabolism rely on CRBP1 moderating retinoid homeostasis.

Consistent with the foregoing observations, overexpression of *Rbp1* in 3T3L1 pre-adipocytes resulted in a threefold increase in ability of RA to induce the expression of target genes [122]. This observation suggests irregular adipogenesis and lipid metabolism in the *Rbp1*-null mouse, and explains the more efficient adipocyte differentiation observed with embryonic fibroblasts from the *Rbp1*-null mouse [290]. The phenotype of the *Rbp1*-null mouse, the conservation of the CRBP1 primary amino acid sequence in vertebrates, and the widespread tissue distribution and expression of CRBP1 in multiple cell types [26, 115, 189, 190], indicate the survival value of CRBP1 and suggest its function as a chaperone that protects retinol from rapid degradation and *efficient* use of retinol.

Current State of the Field

CRBP1 Properties and Structure

The ability of select enzymes to recognize the CRBP1-retinol or CRBP1-retinal "cassette" and tease the retinoid from the binding protein into their own active sites provides a solution for regulating retinol flux into RE vs. RA and sparing bound retinol from metabolism by enzymes that do not recognize the holo-binding proteins. Enzyme interactions with holo-CRBP may occur through direct protein-protein contact or in a microenvironment of membrane lipids (cholesterol, phospholipids, ceramides, sphingosines, e.g.) that contribute to transfer of retinol by influencing protein conformations and/or interactions.

The structures of the fatty acid binding-protein gene family members, including CRBP1 and CRBP2, are similar but not identical. They fold similarly, but have different residues in the internal binding pockets that determine ligand specificity. This discussion will focus on CRBP1, but the major points apply to CRBP2.

The CRBP1 structure was solved first by X-ray crystallography [53]. Differences in the structures of apo- and holo-CRBP1 were apparent immediately. Apo-CRBP1 has a more flexible structure, relative to the more rigid holo-CRBP1. This difference was confirmed by limited proteolysis, which demonstrated resistance of holo-CRBP1 to multiple proteases, whereas apo-CRBP1 was digested by the endopeptidase Arg-C at R30 in α-helix II [100, 175]. Since the determination of the structure by X-ray, several NMR studies and a mass spectrometry-based study have confirmed and expanded insight into the flexibility and the ligand entry portal [38, 72, 73, 134, 135, 144, 158]. CRBP1 has a flattened β-barrel (aka β-clam) shape formed by two orthogonal β-sheets (Fig. 2.1). Each β-sheet consists of five anti-parallel β-strands. The N-terminus blocks one end of the barrel and a cap consisting of two short α-helices (helix-turn-helix) between βA and βB blocks the ligand entrance portal. The binding pocket exists as a closed cavity completely isolated from the external solvent. Retinol assumes a flattened (planar) conformation inside the binding pocket, as indicated by a 25 nm red shift in absorbance of bound retinol relative to retinol in solution [187]. The hydroxyl group points into the interior, hydrogen bonding with N108, which contributes to ligand specificity and affinity. Surprisingly, much of the CRBP1 binding cavity presents a hydrophilic environment with structured water molecules surrounding the isoprenoid side-chain and hydroxyl group. In contrast, the β-ionone ring exists in a hydrophobic region created by L29, I32, L36, F57 and I77. The helix-turn-helix "cap" has few interactions with the rest of the binding protein, supporting the likelihood that reduced flexibility of holo-CRBP1 stems from the β-ionone ring of retinol engaging with hydrophobic residues in αII (L29, I32, A33), the link between αII and βB (L36), the βC-βD hairpin turn (F57), and the βE-βF hairpin turn (I77). These areas have been identified by NMR as more disordered in apo-CRBP1, relative to the rest of the molecule. Thus, the β-ionone ring holds αII and the other disordered regions in place, imparting greater rigidity to holo- relative to apo-CRBP1. Retinol accesses apo-CRBP1 because of its

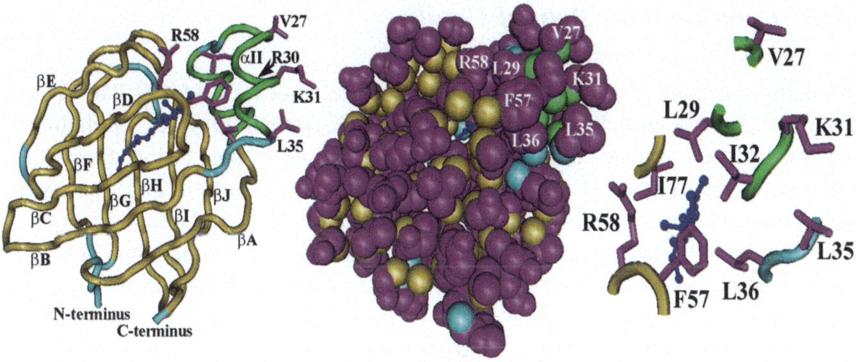

Fig. 2.1 CRBP1 structure. The "worm" diagram at the left shows the two α-helices (*green*) and the ten anti-parallel β-strands (*tan*) arranged in two orthogonal sets of five strands each. Light blue denotes links between the strands or strands and helices. Dark blue indicates retinol. Magenta shows selected exterior residues V27, K31 and L35 in αII. Some interior residues also are shown. Between interior residue R58 and the exterior residues are shown (*unmarked*) from *top* to *bottom* L29, F57 and L36. The space-filling diagram in the middle presents a similar perspective as the "worm" diagram, illustrating seclusion of retinol from the cellular milieu. The depiction of partial residues at the right shows a different perspective from that at the far left to reveal more clearly association or relativity close proximity of L29, I32, L36, F57, R58, and I77 with the β-ionone ring of retinol as they point into the interior of CRBP1, and the outward projections of V27, K31 and L35. Structures were generated with the program Cn3D (http://www.ncbi.nlm.nih.gov/Structure/CN3D/cn3d.shtml) with data downloaded from http://www.ncbi.nlm.nih.gov/Structure/mmdb/mmdbsrv.cgi?uid=24299

relatively open conformation resulting from flexibility of αII and these hairpin turns.

Only small sections of retinol are visible within CRBP1, when viewing a space-filling molecular model. Encapsulation explains observations made during purification of CRBP1, namely, after saturating crude preparations of CRBP1 from animal tissues with retinol, purification of holo-CRBP I by column chromatography could be monitored by following fluorescence of bound retinol in the eluates [182, 187, 233]. One purification even monitored fluorescence of endogenously bound retinol [242].

What would prompt CRBP1 to surrender its ligand? The hydrophilic environment deep in the binding pocket would exert repealing force on the isoprenoid side chain, which would "push" the β-ionone ring against CRBP1 "cap" residues, to force partial opening of the cap, and partial egress of retinol. The three larger residues on or near αII pointing out from CRBP1 (V27, K31, L35) also would contribute to retinol egress. Two of these are hydrophobic, posing the question of why hydrophobic residues projecting into the hydrophilic milieu occur on the surface of a soluble protein. Complementary surfaces on select enzymes might interact with these residues to enhance dissociation of the "disorderly" domains that close around retinol, to prompt retinoid release.

CRBP1 and Cellular Retinol Uptake

Ottonello et al. studied cellular retinol uptake in a cell-free model [195]. They chose plasma membrane-enriched fractions from RPE (retinal pigment epithelium), because RPE accumulates large quantities of retinol to support vision. Incubation with sRBP-retinol resulted in the membrane fractions absorbing and esterifying retinol in a concerted process. Inhibiting RE formation decreased retinol incorporation ~80 %, confirming coupling of retinol uptake and RE formation. They concluded that the overall process occurred without free retinol ever having to leave the binding-protein. They also showed that apo-CRBP1 promoted hydrolysis of membrane-bound RE, generating CRBP1-retinol. The latter point was developed further by the demonstration that apo-CRBP1 activates a bile salt-independent REH (bsiREH) that catalyzes hydrolysis of endogenous RE in liver microsomes, forming holo-CRBP1 [28].

The Ottonello data complemented conclusions of previous reports of a high-affinity (k_d=5 pM) membrane receptor in RPE and intestine that recognizes sRBP [94, 95, 217]. Supporting this conclusion were the observations that retinol uptake from sRBP is saturable and temperature dependent, and apo-sRBP and holo-sRBP compete with each other. This engendered the proposal that relative concentrations of apo- and holo-sRBP in blood control cellular retinol uptake. sRBP-mediated retinol uptake into Sertoli cells is also saturable, and labeled sRBP competes with unlabeled sRBP, again suggesting the presence of a membrane receptor [252]. The rate of uptake from sRBP by Sertoli cells was linear until the concentration of retinol reached the intracellular CRBP1 concentration, and then it decreased. Ultimately, cells contained both free retinol and RE, with RE biosynthesis presumably reflecting the second rate phase, which required discharge of CRBP1 to allow further retinol uptake.

Although efforts were made to isolate and characterize a membrane receptor for sRBP, many years passed before success was achieved. Discovery of an RA-induced gene, *Stra6* (stimulated by RA gene 6), in an unbiased screen of RA-stimulated murine P19 embryonal carcinoma cells [34, 41] provided foundation for identification of an sRBP membrane receptor. The deduced amino acid sequence of STRA6 showed that it was not a known protein. STRA6 is expressed widely during mouse embryogenesis and has strong expression in adult brain, kidney, spleen, the female genital tract, and testis; weak expression in heart and lung; and little to no expression in liver. The research team concluded that *Stra6* encoded a "new type" of membrane protein, expressed intensely in blood-organ barriers, which serves as a "component" of "transport machinery". It wasn't until 2007, however, that Stra6 was verified as an sRBP receptor [116]. Consistent with previous reports of retinol uptake, Kawaguchi et al. showed that LRAT-catalyzed formation of RE and enhanced retinol uptake via Stra6. This conclusion was reinforced by studies in an *LRAT*-null mouse, which also showed that retinol uptake via Stra6 was coupled with LRAT activity [4]. Wolf has published a detailed history of sRBP receptor studies and cellular retinol uptake [276].

Functions of Stra6 are complex and controversial [16, 258]. Although an original report of the *Stra6*-knockout concluded that it was embryonic lethal [99], subsequent reports showed this was not the case [22, 239]. It must be noted that the first *Stra6*-knockout retained a neo^R cassette in the gene and that Neo^R insertion into target genes can influence expression of neighboring genes, producing phenotypes related to nearby genes, rather than the target gene [204]. A second Stra6 knockout mouse was created that did not retain the neo^R cassette. These later produced *Stra6*-null mice were born in Mendelian frequency and did show differences in the amounts of retinol and RE in multiple tissues relative to WT controls. When these *Stra6*-null mice that had been fed a diet copious in vitamin A for 12 weeks were switched to a vitamin A-deficient diet for another 4 weeks, total retinol plus RE levels decreased ~50 % in white adipose tissue and ~10–30 % in heart, kidney and testis. Absence of *Stra6* marginally slowed the rate of retinol uptake into multiple, but not all, tissues. The only physiological dysfunction noted in these *Stra6*-null mice was in the eye, which experienced a hyperplastic primary vitreous body, distorted rod photoreceptor segments, and reduced numbers of cone cells. The newer work demonstrated that Stra6 does not protect cells from retinol toxicity by fostering egress of retinol as had been postulated by the Isken report [99]. Instead, the results were consistent with an essential function of Stra6 in retinoid uptake by the eye, but not for maintaining retinoid homeostasis in other tissues, at least in mice fed a diet copious in vitamin A. These conclusions were confirmed by the group that produced the original *Stra6* knockout with a redeveloped knockout [4]. Together, available data indicate Stra6 is not the sole mechanism for retinol uptake by extraocular tissues in mice fed a diet with copious vitamin A.

Stra6 allows retinol to exit cells when presented with apo-sRBP, but cells that express CRBP1 and/or LRAT presented with holo-sRBP experience net uptake of retinol via STRA6. Inward flow is driven by CRBP1 sequestering retinol and LRAT accessing CRBP1-bound retinol for RE synthesis [21, 117, 118]. sRBP circulates bound with one of the thyroid hormone carriers transthyretin (TTR). This association prevents sRBP from rapid clearance by the kidney. But TTR blocks binding of sRBP with Stra6, indicating that sRBP functions only when its concentration exceeds that of TTR and/or with the free sRBP pool in equilibrium with TTR [19]. This suggests TTR regulates the impact of sRBP on Stra6.

Insulin-resistant mice and humans with obesity and type 2 diabetes have increased circulating sRBP. Overexpression of sRBP causes insulin resistance and ablation of *Rbp4* (encodes sRBP) enhances insulin sensitivity [280]. The mechanism of these effects was unknown until Berry et al. reported that Stra6 functions as a signaling receptor activated by sRBP [18]. Binding of sRBP-retinol with Stra6 initiates a signaling cascade that ultimately induces STAT5 regulated genes. One of these, suppressor of cytokine signaling 3 (SOCS3), inhibits insulin signaling, providing a mechanistic connection between sRBP and insulin resistance. This work, along with previous studies, coupled retinol uptake with sRBP-stimulated cellular signaling. Notably, Stra6 signaling requires retinol flux through it to stimulate SOCS3, as revealed by protection from sRBP-induced insulin resistance in *Lrat*-null mice [152]. The cumulative evidence unites retinol transport by sRBP through blood, cellular retinol uptake, metabolism umpired by CRBP1, LRAT, and Stra6-mediated signaling functions [19, 21, 152].

CRBP and Retinoid-Metabolizing Enzymes

Intracellular CRBP1 Chaperoning of Retinol

Retinol and retinal in many cell types occur bound with CRBP, which therefore provides the most abundant substrates for retinoid metabolism, because rapid dissociation from CRBP does not occur through aqueous media [179]. Indeed, rapid dissociation of retinoid from CRBP would be self-defeating, because it would quickly deplete holo-CRBP of retinoids and expose the "free" retinoids to non-enzymatic degradation and to metabolism catalyzed by any enzyme that recognizes free retinoid as substrate, such as xenobiotic-metabolizing enzymes, which have evolved with low substrate specificity. As stated "…most enzymes are not perfectly specific for a single substrate…" [52]. Also, increased concentrations of "free" retinoids would increase RA biosynthesis by mass action, and a modest increase in RA underlies retinoid toxicity [51]. Enzymes that recognize holo-CRBP have access to relatively high substrate concentrations, while permitting CRBP to ensure efficient retinoid use.

K_m values for RDH, retinal dehydrogenases, and LRAT with CRBP-bound retinoids are lower than their values with unbound retinoids [170, 172, 184]. An apparent exception is RDH10, for which a K_m value was reported as 0.035 μM for unbound retinol, obtained with the lowest substrate concentration assayed greater than the extrapolated K_m [12], which cannot generated a reliable value, because the substrate range did not include the curve's inflection point. In contrast to this preparation of modified RDH10 (RDH10-His$_6$) expressed as 10 % of total insect (Sf9) cell membrane protein, native human RDH10 assayed in membranes of COS-1 cells had an K_m value of ~4 μM for unbound all-*trans*-retinol, indicating the importance of context concerning RDH activity [261].

RE Biosynthesis in the Presence and Absence of CRBP

Early efforts to characterize the enzymology of RE formation identified activity in liver and mammary gland microsomes that relied on acyl-CoA as co-substrate, which was referred to as acyl-CoA:retinol acyltransferase (ARAT) [229, 230]. Liver ARAT activity has a K_m value of ~25 μM for retinol and a Vm ranging from 450 to 675 pmol/min/mg of rat liver microsomal protein. The reaction catalyzed by microsomes from the lactating rat mammary has a K_m value of ~39 μM for retinol and a Vm ~270 pmol/min/mg protein. ARAT does not recognize CRBP1 as substrate. CRBP1 levels are very low in the mammary gland, consistent with the conclusion that ARAT would contribute to RE formation for milk [216]. Knockout of LRAT, which recognizes CRBP1 and 2 as substrates, however, resulted in a transition from mostly RE to mostly unesterified retinol in milk, supporting the conclusion that LRAT generates RE for milk.

ARAT activity also was reported in microsomes of rat small intestine with a K_m value ~5 μM and a Vm of ~300 pmol/mg microsomal protein/mg [93]. Subsequently, ARAT activity was confirmed in intact rat liver cells, which had a combined retinol uptake and esterification maximum rate ~35 pmol/min/mg protein, and a retinol concentration that supported a half-maximal combined rate of uptake and esterification that occurred at ~50 μM [59]. Based on rat liver retinol concentrations, the K_m values of liver ARAT suggest that ARAT works most effectively at higher retinol concentrations. One measurement reported retinol in rat liver as ~6 μM [247]. But other measurements place retinol in the range of 13–30 μM in rat liver [63]. Measurements in mouse liver yielded values from 5 to 50 μM [108, 141, 181, 247]. The reasons for these differences are unclear, as all measurements were done on rodents fed chow diets. The point remains, the K_m values seem reasonable in view of total unesterified retinol, but not in terms of the physiological form of unesterified retinol, which occurs predominantly as CRBP1 or 2-bound. This suggested that other retinol esterifying enzymes function as a major source of RE in CRBP-expressing cells--a prediction that was validated by the discovery of LRAT.

Although LRAT catalyzes most RE biosynthesis in most tissues, a physiological contribution of ARAT activity to RE formation has been revealed by the presence of RE in tissues of *Lrat*-null mice fed a vitamin A-deficient diet [141]. RE levels were normal in kidney and brain, increased in adipose, but <1 to ~5 % in liver, lung and eye. The ARAT activities in these tissues were identified as acyl-CoA:diacylglycerol acyltransferase 1 (DGAT1) and acyl-CoA:monoacylglycerol acyltransferase 1 (MGAT1), with the former having the higher specific enzymatic activity in the human intestine CACO2 cell line and in mouse liver, testis, kidney, and intestine, and in substrate studies with recombinant DGAT1 [194, 277, 281]. It is now accepted that ARAT activity is produced by these two enzymes, especially DGAT1.

The importance of DGAT1 to RE formation in skin was demonstrated using the *Dgat1*-null mouse, which showed that DGAT1 accounts for most ARAT activity [251]. In skin of *Dgat1*-null mice fed a vitamin A-sufficient diet (4 IU vitamin A/g), unesterified retinol increased ~22 % and RA increased ~40 %, but RE levels remained similar to WT, which likely reflects LRAT activity. In contrast, liver retinol, RE and RA levels were similar to WT, consistent with local events driving the changes in skin retinoid homeostasis. When mice were fed a vitamin A-deficient diet, skin retinoids did not differ between null and WT. When mice were fed a diet copious in vitamin A (i.e. standard lab chow, which currently has ~15 IU vitamin A/g), *Dgat1*-deficiency induced a multifold increase in RA-regulated gene expression in skin, consistent with increased RA levels. Therefore, DGAT1 functions as an ARAT in skin to maintain retinoid homeostasis and prevent toxicity, but makes a minimal contribution in liver. These data generated with the *Lrat* and *Dgat1*-null mice demonstrate that LRAT and DGAT1 produce tissue-specific effects on retinoid homeostasis that are influenced by the amount of dietary retinol.

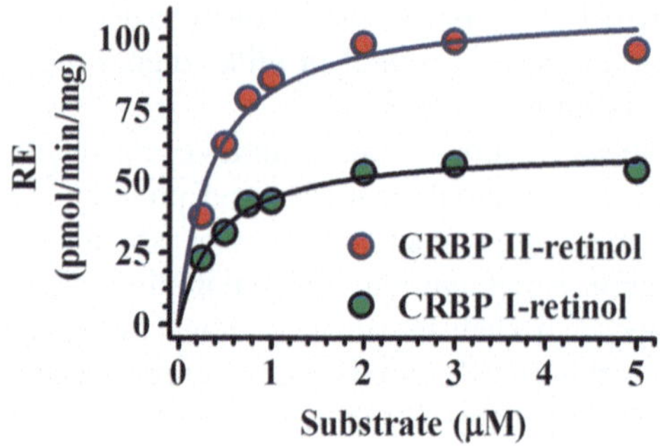

Fig. 2.2 Kinetics of RE formation by LRAT catalyzed by rat liver microsomes with holo-CRBP substrates. Each reaction was run with CRBP1 in a 1:1 ratio with retinol [192]. The K_m values were ~1 and 0.7 μM for holo-CRBP1 and holo-CRBP2, respectively, with adult rat liver microsomes. Use of holo-CRBP did not support RE formation from ARAT activity

Table 2.2 Kinetic values of LRAT for retinol vs. CRBP-bound retinol (holo-CRBP). K_m values in μM; V_m values in pmol/min/mg protein

Substrate	Intestinal microsomes K_m; V_m	Liver microsomes K_m; V_m
Retinol	0.6±0.3; 54±13	0.4±0.2; 80±15
CRBP1-retinol	0.2±0.1; 33±12	0.8±0.1; 76±6
CRBP2-retinol	0.2±0.04; 47±3	0.3±0.02; 61±5

CRBP1 and CRBP2 Contribute to RE Biosynthesis via LRAT

Work on retinol esterification in the mid 1980's established the existence of an activity in intestine and liver that is independent of ARAT (acyl-CoA) activity [74, 148]. Subsequently, CRBP1 and 2 were evaluated as substrates to test the hypothesis that the two binding-proteins serve as the actual substrates for RE biosynthesis and to determine the relative importance of acyl-CoA dependent and independent activities [96, 146, 147, 192, 253, 282]. With holo-CRBP as substrate, retinol esterification by rat and human liver, rat Sertoli cells, and rat intestinal microsomes proceeded in an acyl-CoA independent reaction that relied on the SN1 position of phosphotidyl choline (lecithin) as fatty acid donor: i.e. a lecithin:retinol acyltransferase or LRAT. The acyl-compositions of RE generated by LRAT were similar to the acyl-composition of RE that occurs in tissues. Specifically inhibiting LRAT reduced RE formation in cultured cells (Caco-2, Sertoli) by as much as 90 %, even though acyl-CoA supported activity was not affected [214, 253].

Reaction kinetics with CRBP1 and 2 as substrates using microsomes and a 1:1 ratio of CRBP/retinol produced K_m values lower than ARAT activity. The K_m values were below the concentrations of holo-CRBP1 in liver and holo-CRBP2 in intestine (Fig. 2.2, Table 2.2) and with both CRBP1 and 2, were close to those of free retinol, indicating that dissociation of retinol from CRBP is not required to supply LRAT with retinol. This conclusion ensues because the extreme reduction in unbound retinol in the presence of CRBP would profoundly increase the apparent K_m and lower the Vm values if only free retinol served as substrate.

LRAT and CRBP1 Associate with Lipid Droplets

When cells are not undergoing lipid droplet formation, LRAT embeds in the endoplasmic reticulum (microsomal fraction) via a C-terminal transmembrane domain, with most of the enzyme, including its active site, projecting into the cytoplasm [159, 192, 282]. LRAT surrounds CRBP1, which co-localizes with mitotracker, suggesting association with the outer mitochondrial membrane or mitochondria associated membranes [102]. During lipid droplet formation stimulated by oleate or retinol, LRAT surrounds growing lipid droplets, colocalizing with the lipid droplet surface proteins, desnutrin/adipose triglyceride lipase and perilipin 2, and CRBP1 co-localizes with LRAT. Association of LRAT with lipid droplets is specific as shown by (1) a requirement for two amino acids residues (K36 and R38) in its N-terminus for lipid droplet association and (2) the enhancement of LRAT specific activity upon lipid droplet association [102]. These data place both LRAT and CRBP1 at the surfaces of lipid droplets during acyl ester formation, which is consistent with the kinetic data that reveals a substrate-enzyme relationship between the two proteins. The data also illustrate the mobility of the proteins, which do not remain fixed but migrate during metabolism.

Genetic Ablation of LRAT

Knockout of *Lrat* revealed that LRAT serves as the major, but not sole retinol esterifying enzyme in several mouse tissues [8, 121, 141, 143, 180, 238]. *Lrat*-null mice fed chow diets (copious vitamin A) have severely depleted RE levels in liver, lung, eye and testes, but not in kidney, adipose, pancreas or brain. Despite reduced RE, chow-fed *Lrat*--null mice develop normally, but have degraded rod and cone functions of the neural retina because of a need for profuse vitamin A flux into the eye to support vision. The *Lrat*-null mouse also responds to excess dietary retinol by increasing retinol excretion, RE deposition in adipose, and induction of *Cyp26A1*, which accelerates RA catabolism [143]. Thus, LRAT contributes to regulating vitamin A homeostasis in multiple tissues during copious vitamin A intake.

When *Lrat*-null mice are fed a vitamin A-deficient diet for 6 weeks, retinol is undetectable <0.007 pmol/mg) in most tissues. Liver and kidney have detectable retinol, albeit <6 % of WT. This observation should be evaluated in context of the well-established phenomenon that mice bred from dams fed a chow diet do not become vitamin A-deficient during 36 weeks of feeding a vitamin A deficient diet from weaning (Napoli, unpublished data). These data demonstrate that LRAT is essential for maintaining retinoid levels in select tissues during restricted dietary vitamin A—the normal situation throughout evolution.

Recent work has confirmed the primacy of LRAT as the RE forming enzyme in liver and its function in preventing retinol loss [278]. Interestingly, although low, RE was 50-fold higher in livers of the *Lrat/Rbp1*double knock-out mice compared to the single *Lrat* knockout (Table 1 of [278]), whereas unesterified retinol was ~90 % lower (Fig. 3 of [278]), indicating CRBP1 restricts access of acyl transferases other than LRAT to retinol in vivo. Because a chow diet was fed, the impact of this partnership during low amounts of retinol, a situation of potentially greater CRBP1 impact, remains unclear. Regardless, these data indicate that the CRBP1/LRAT partnership promotes retinol sequestration by creating a gradient driven by RE biosynthesis. Although this channeling would protect retinol from ARAT activity, RE formation in the absence of LRAT and CRBP1 is quantitatively minor relative to WT, reinforcing the model that the channeling partnership promotes *efficient* intracellular retinol sequestration.

Retinyl Ester Hydrolysis

Initial studies of retinyl ester hydrolysis (REH) in vitro focused on activity that required cholate [86]. This bile salt-dependent activity has a specific requirement for ≥ 10 mM cholate or taurocholate, hydrolyzes triacylglycerol and cholesteryl oleate at higher rates than RE, and is remarkably similar to pancreatic carboxyester hydrolases [85]. The specific activity varies radically from animal to animal, suggesting confounding influences, such as erratic contamination by secreted and/or plasma hydrolases. Because most tissues that store and mobilize RE and generate RA are unlikely to contain millimolar bile salts, attention turned to alternative enzymes. In 1989 two labs independently demonstrated occurrence of a bile-salt *in*dependent microsomal REH (bsiREH) activity in liver and in other vitamin A target tissues, including lung, intestine, kidney and testes [89, 90, 169]. The bsiREH did not hydrolyze cholesterol esters and had higher activity with RE than triacylglycerol.

Several intracellular REH activities have been isolated from the intestinal brush border [226–228], liver and other tissues [140, 259]. They belong to the ES or CES family (i.e. ES-2, and ES-10). In adipocytes hormone-sensitive lipase has higher specific activity with RE compared to diacylglycerol, and its ablation reduces RA signaling, indicating its importance to generation of adipocyte RA [257]. Likely, multiple membrane-associated enzymes function as intracellular REH with tissue-specific expression [87].

The Ratio of Holo-CRBP1 to Apo-CRBP1 Influences Retinol Metabolism

Apo-CRBP1 activates bsiREH-catalyzed hydrolysis of resident RE in liver microsomes and displays Michaelis-Menten kinetics [28]. RE hydrolysis occurs with 2.6 μM apo-CRBP1 stimulating a half-maximum rate (Fig. 2.3). The effect is specific because other proteins capable of sequestering retinol are ineffective, such as

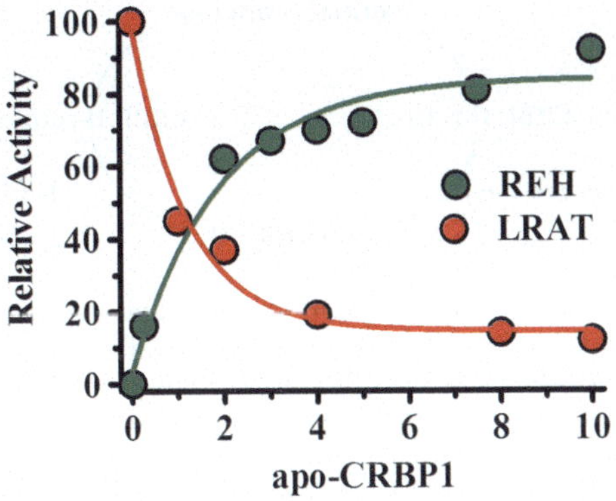

Fig. 2.3 Apo-CRBP1 effects on retinol esterification and RE mobilization. Increasing concentrations of apo-CRBP1 increase the hydrolysis of resident RE in liver microsomes, whereas titrating apo-CRBP1 into an LRAT reaction using holo-CRBP1 as substrate inhibits RE formation. These data indicate that the ratio holo-CRBP1/apo-CRBP1 influence flux between retinol and RE. Fully-charged CRBP1 would favor RE formation, while still allowing retinal formation for RA biosynthesis; whereas appreciable apo-CRBP1 would stimulate RE hydrolysis to maintain holo-CRBP1 as substrate for RA biosynthesis

(bovine serum albumin and β-lactoglobulin). apo-CRBP1 functions as a competitive inhibitor of LRAT, with a K_i value ~0.2 μM [96]. These actions suggest that the ratio of apo-CRBP1 to holo-CRBP1 senses intracellular retinol status and modifies retinol channeling. Increasing apo-CRBP1 during declining retinol concentrations would stimulate RE hydrolysis to prevent total retinol sequestration as RE and maintain sufficient holo-CRBP1 to support RA biosynthesis.

CRBP2 Favors Retinal Reduction and Retinol Esterification in the Intestine

The first demonstration that a retinoid binding-protein was involved in retinol metabolism was reported with CRBP2 [191]. The high concentration of CRBP2 (~1 % of total soluble protein in the intestinal enterocyte) predicts a vital contribution for managing the efficiency of dietary retinoid absorption. This prediction was verified by demonstrating that LRAT in rat intestinal microsomes converts retinol bound to CRBP2 into RE with a fatty-acyl composition that reflects the composition of RE in chylomicrons. The apparent K_m for the CRBP2-retinol complex (~0.2 μM) is likely well below the concentration of the complex generated from retinol absorption or β-carotene metabolism.

Carotenoids provide most dietary vitamin A. A dioxygenase centrally cleaves β-carotene, the prototypical dietary provitamin A carotenoid, into two molecules of retinal [55, 88]. Retinal can undergo reduction into retinol for esterification and distribution as RE in chylomicrons, or conversion into RA. What prevents generation of large amounts of RA rather than predominantly RE? The answer was

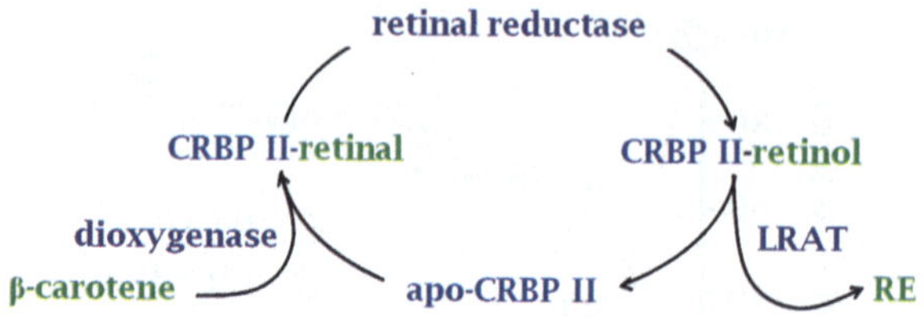

Fig. 2.4 Channeling of intestinal retinal metabolism to direct pro-retinoid and retinoid flux into RE. β-Carotene and pro-vitamin A carotenoids provide the major sources of dietary vitamin A. A dioxygenase (BCO1) generates retinal from β-carotene. Retinal can then be reduced by retinal reductase or dehydrogenated into RA. The dehydrogenation rate of retinal bound to CRBP2 is <300-fold the rate of reduction into retinol. β-Carotene is not toxic, whereas RA is toxic. Moreover, large scale conversion of retinal in the intestine into RA likely would not serve the vitamin A needs of the animal. CRBP2-directed metabolism of retinal through retinol into RE for incorporation into chylomicrons provides a mechanism for efficient conservation of vitamin A as its ester, systemic delivery of RE via chylomicrons, and limiting RA biosynthesis

provided by the demonstration that CRBP2-bound retinal serves as substrate for retinal reduction, catalyzed by rat intestinal microsomes with a K_m value ~0.5 μM [106]. Retinol generated by the reductase remains bound to CRBP2, available for esterification without diffusion into the aqueous medium. Significantly, CRBP2-bound retinol is not oxidized efficiently (<0.3 pmol/min/mg protein). compared to the reduction of CRBP2-bound retinal (~300 pmol/min/mg protein). These data reveal the mechanism that directs most intestinal carotenoid metabolites into RE, while avoiding generating large amounts of RA (Fig. 2.4). The phenotype of the *Rbp2*-null mouse (*RBP2* encodes CRBP2) supports this conclusion. *Rbp2*-null mice fed a chow diet (copious vitamin A) have RE levels that are 40 % of the levels in WT mice, but show no symptoms of vitamin A-deficiency. When fed marginal dietary vitamin A (0.6 IU/g), *Rbp2*-null pups suffer from 100 % mortality within 24 h after birth vs 0 % mortality for WT. These results illustrate two points: (1) cellular retinoid-binding proteins are essential for efficient use of vitamin A; (2) contributions of cellular retinoid-binding proteins to retinoid homeostasis and function can be overwhelmed by diets copious in vitamin A. Table 2.3 provides a summary of rationale supporting interactions of CRBP with retinoid-metabolizing enzymes.

RA Biosynthesis Independent of CRBP: ADH

The earliest studies demonstrating conversion of retinol into retinal outside of the visual cycle were done before RA was appreciated as essential for the systemic functions of vitamin A. Lacking this insight, early studies did not focus on investigating de novo pathway(s) that generate RA from physiological substrates, but rather, focused on catalysis by members of the medium-chain alcohol dehydrogenase gene family (ADH) [27, 223]. Later work, which concluded that ethanol inhibits generation of "active retinal", relied on 60 μM retinol (exceeds the testis concentration of retinol by 750-fold) in a crude cytosolic fraction [265]. The latter

Table 2.3 Rationale supporting substrate-product relationships between CRBP and retinoid-metabolizing enzymes. These observations have been developed mostly with CRBP1, but several pertain to CRBP2, CRABP1, and CRABP2

High affinity	CRBP binds retinoids with affinities near those of nuclear hormone receptors and engulfs retinoids in an interior pocket protected from the cellular milieu.
Abundance	Holo-CRBP embodies the major form of non-esterified retinol in many tissues and therefore the substrate of highest concentration.
Regulation	By allowing only specific enzymes access to its tightly-bound retinoids, CRBP would control retinoid metabolism (see *CRBP-null* below).
Security	CRBP sheltering enhances retinol stability and protects retinal from forming adventitious Schiff's bases with amine residues.
Kinetics	Prototypical enzyme-substrate kinetic relationships occur between holo-CRBP (retinol or retinal) concentrations and rates of retinoid metabolism.
Crosslinking	Protein crosslinking reagents form covalent bonds between holo-CRBP1 and RDH and LRAT in microsomes (many possible targets, only two interact).
CRBP mutants	A single exterior amino acid mutation (L35A) of holo-CRBP1 decreases the rate of retinal formation without affecting retinol binding affinity.
Gene knockouts	*Rbp1* and Rbp2-null mice experience depleted retinol uptake and/or enhanced depletion and metabolic disease, related to changes in retinoid metabolism.

report hypothesized that competitive inhibition by ethanol reduced retinal formation and caused the testicular atrophy and aspermatogenesis experienced by alcoholics. These studies were done before: (1) quantification of retinol concentrations in tissues; (2) identification of specific high-affinity retinoid binding-proteins; (3) there was an appreciation for RA's occurrence and importance. Therefore, there was no background to question whether xenobiotic clearing enzymes would represent credible candidates for maintaining RA homeostasis. Also, the short-chain dehydrogenase/reductase (SDR) gene family was not known. If it were, earlier investigators might have inquired about its involvement in RA biosynthesis, because SDR metabolize hormones and autocoids, such as steroids and prostanoids [193].

The presumption that ADH catalyze retinol metabolism spawned a hypothesis that ethanol interferes with retinoid metabolism by competitive inhibition. This hypothesis has been long-lived, despite not having been tested by analytically rigorous assays of ethanol effects on RA levels in vivo, and despite much evidence to the contrary [165]. For example, a spontaneous mutation in ADH1 in *peromyscus* (aka deermouse) was reported in 1978 [37]. *Adh1*-null deermice have normal testes histology and reproduction, which indicate the animals have normal vitamin A-supported processes [131]. Cytosol of the *Adh1*-null deermouse catalyzes RA biosynthesis from physiological concentrations of retinol with rates from 0.3 to 0.7 nmol/h/mg of protein, more than sufficient to fulfill cellular needs [209, 210]. Finally, *ADH1* expression in vivo does not correlate with sites of RA biosynthesis [268].

Further, a concentration of ethanol (1000 mM) that far exceeds the blood alcohol concentration of chronic alcoholics (~25 mM) did not inhibit RA synthesis in kidney cytosol from the *Adh1*-null deermouse. Analytically robust assays also demonstrated that ethanol does not decrease RA concentrations in vivo in multiple tissues of WT mice [111, 155, 165], findings that support earlier observations showing that a 10,000–30,000-fold excess of ethanol (100–300 mM) relative to retinol only partially inhibits (54–69 %) RA biosynthesis from 10 μM retinol in rat liver cytosol [168]. Nor is RA biosynthesis from retinol inhibited by ethanol in intact cells in culture [111, 155, 162, 165, 254]. In fact, ethanol causes massive loss of hepatic RE, a phenomenon established earlier, but for which a mechanism has not been determined [127, 246]. One potential mechanism seems to be RA catabolism by Cyp2e1, a xenobiotic clearing enzyme with over 85 substrates [136, 142, 270]. Ethanol-induced RA catabolism causes a compensatory mechanism to replenish RA by drawing on RE stores [111]. Ethanol and its metabolite acetaldehyde, however, affect multiple binding proteins and receptors that participate in generating RA and mediating vitamin A function, and undoubtedly many related processes. Further discussions of ethanol effects on retinoid metabolism are available [165, 270, 284].

ADH-null house mice have not been informative concerning whether they contribute to RA biosynthesis under physiological conditions. Studies of *ADH* knockout mice have not reported essential experiments, such as RA levels during normal vitamin A nutriture, pathology resulting from reduced RA, complete rescue with reasonable amounts of RA, or gene expression changes indicating changes in RA signaling and/or compensation for decreased RA [56]. Thus, claims of ADH involvement in the *physiological* metabolism of retinol remain unsubstantiated, despite the long-term availability of ADH knockout mice.

A revised hypothesis proposed that ADH1 and ADH3 catabolize "excess" retinol to alleviate retinol toxicity. This hypothesis overlooks a large body of data and contradicts the original hypothesis by the same group that ADH produce RA to serve physiological functions [160]. Throughout evolution, vitamin A has been a scarce but essential nutrient—retinol insufficiency remains the third most prevalent nutrient deficiency worldwide. Mammals evolved to sequester retinol efficiently, as shown by the *Crbp2*-null mouse, not to eliminate an "excess". There was no excess and therefore no evolutionary pressure to catabolize an excess. Abundant dietary retinol is stored as RE in liver, which can achieve concentrations exceeding 1 mM. Other tissues, such as adipose, also can accumulate relatively high RE concentrations. It is unlikely that ADH evolved to "detoxify" an essential nutrient that can be stored in substantial amounts, and do so by sending it down a path that generates a metabolite (RA) with greater toxicity.

A single 3 mg retinol/kg dose induces 71 % incidence of cleft palate and a 39 mg/kg dose induces 76 % incidence of neural tube defects [23]. The data used to support the contention that ADH detoxifies retinol were generated by measuring RA in *Adh1*-null or *Adh3*-null mice after administering a retinol dose of 50 mg/kg or 100 mg/kg. The 50 mg/kg dose delivers ~300-fold more vitamin A than the recommended daily intake for mice, and was given to mice fed a diet copious in vitamin A. This dose produces serum RA concentrations ~1600–2330-fold greater than the steady-state value (~2 nM) in WT mice, ~100–200-fold greater than normal in the *Adh1*-null mouse, and ~650-fold greater than normal in the *Adh3*-null mouse. These

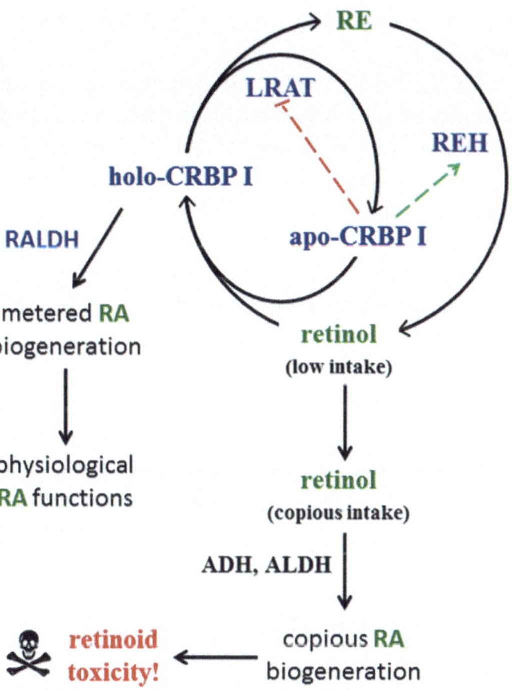

Fig. 2.5 Disposition of low amounts of dietary retinol vs. copious or toxic dietary retinol. This figure presents a model for retinol toxicity consistent with saturation of retinoid binding-proteins by copious dietary retinol resulting in unmetered generation of RA by xenobiotic clearing enzymes. This path is in contrast to channeling of dietary sufficient amounts of vitamin A by retinoid binding-proteins, such as CRBP1. This path complements the path imposed upon the intestine by CRBP2, which directs dietary retinol into RE, rather than RA

data do not support ADH1 or 3 as detoxifying retinol because prolonged exposure to even modestly increased concentrations of RA causes teratology [50, 173, 264]. Generation of RA from toxic amounts of retinol likely contributes to retinoid toxicity. Figure 2.5 summarizes the distinction between metabolism of low amounts of dietary retinol involving binding-proteins vs. copious or toxic amounts of retinol that overwhelm binding-proteins and become accessible to xenobiotic-clearing enzymes.

A De Novo Approach to RA Biosynthesis

Study of retinol metabolism, using analytically validated retinoid quantification by HPLC and GC/MS, established that conversion of retinol into retinal is rate-limiting in intact cells (pig kidney cell line LLC-PK_1) [162]. Reduction of retinal into retinol and dehydrogenation of retinal into RA occur at rates four–eightfold and 30–60-fold faster, respectively, than retinol dehydrogenation. Identification of the rate-limiting step was reinforced in studies of RA biosynthesis by primary hippocampus astrocytes [273].

The work with LLC-PK_1 cells also showed, using the ADH inhibitor 4-methylpyrazole, that ethanol metabolizing enzymes do not catalyze RA formation in intact cells in the physiological range of retinol concentrations. Cultured human keratinocytes provided the same insights: conversion of retinol into retinal is rate-limiting and ADH are not involved in the RA biosynthesis in intact cells [254].

CRBP1 as Bait to Identify Physiologically Relevant Retinol Dehydrogenases *(RDH)*

Many cell types, subcellular fractions, and enzyme types convert free retinol into retinal [163, 168]. The issue was identifying the enzyme(s) that catalyze the rate-limiting step *in vivo* under physiological conditions. Based on awareness that holo-CRBP1 provides the major potential substrate for retinol metabolism, holo-CRBP1 was used as bait to identify physiologically significant RDH [29, 211].

Technical Issues with CRBP and Retinol as Substrates

Experiments that evaluate free vs. bound retinol as substrate require careful attention to preparing holo-CRBP. It is important to prepare holo-CRBP1 frequently and to pass it through a Sephadex column to insure a 1:1 ratio of CRBP1:retinol. Only this assures a true 1:1 ratio. Apo-CRBP1 is then added to give the most accurate ratio of apo/holo. Storing holo-CRBP1 for longer times, even at −80 °C, allows conversion of some all-*trans*-retinol into 9-*cis*-retinol (Napoli, unpublished data). This became obvious because reactions with "old" holo-CRBP1 produced some 9-*cis*-RA, which our HPLC system distinguishes from RA. Freshly prepared holo-CRBP1 does not produce 9-*cis*-RA. Experiments done in my lab used CRBP1 prepared as just described.

Other issues arise from the lability of retinol in aqueous media and impurities that contaminate commercially available retinol, including retinal. In the absence of CRBP1, 60 % of the total retinal generated in vitro was artifactual (did not rely on presence of cofactor or enzyme), but the retinal generated from CRBP1-retinol exceeded background by >tenfold. Thus, it is imperative to purify retinol through HPLC or by binding it to CRBP and purifying the holo-CRBP.

CRBP1 and Retinal Biosynthesis

Multiple experimental approaches indicate that RDH recognize holo-CRBP1 as substrate. Typical Michaelis-Menten enzyme kinetics are observed between concentrations of holo-CRBP1 and rates of retinal formation catalyzed by liver microsomes [211]. (Liver microsomes do not produce RA). Rates of retinal formation

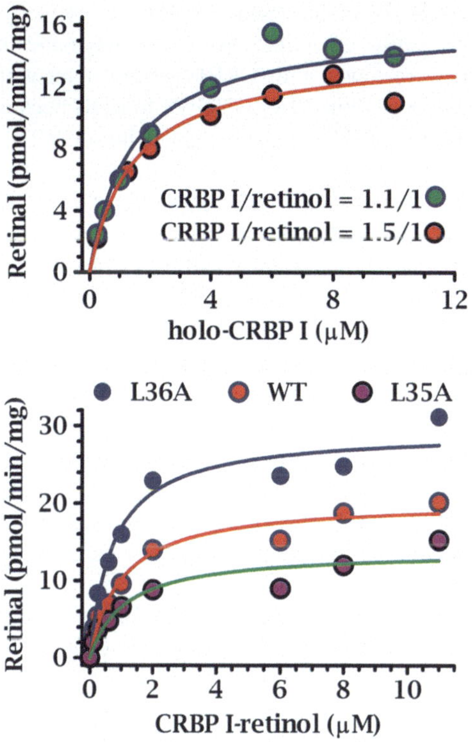

Fig. 2.6 Kinetic relationships between RDH and CRBP1-retinol. The top panel shows the Michaelis-Menten relationship (substrate-concentration dependent, saturable kinetics, initial velocity conditions) between holo-CRBP1 and the rate of retinal formation catalyzed by microsomes. The kinetic constants generated were independent of the ratio total CRBP1/retinol, indicating ability of CRBP1 to deliver retinol without diffusion. The bottom panel shows the impact of mutating external residues of CRBP1 on kinetics of retinal formation using microsomes as source of RDH. The sensitivity of the kinetics to changes in one external CRBP amino acid residue, without affecting affinity for retinol, corroborates an interaction between RDH and CRBP1

tested with two different ratios of total CRBP1 to retinol, 1:1 and 1.5:1, each produced the same K_m and Vm values, which could not occur if only free retinol were substrate (Fig. 2.6). The Vm would have decreased markedly with the nearly five-fold decrease in free retinol after adding the additional apo-CRBP1. Furthermore, the nanomolar concentrations of free retinol present with the micromolar amounts of holo-CRBP1 were not sufficient to drive the observed rates of retinal biosynthesis. N-ethylmaleamide preincubated with RDH inhibited retinal generation from holo-CRBP1, but did not prevent RDH from converting unbound retinol into retinal, consistent with obstructing access of CRBP1 to RDH without affecting RDH catalytic activity [30]. In contrast, N-ethylmaleamide preincubated with holo-CRBP1 did not prevent retinal biosynthesis. With free retinol in the absence of cofactor, retinal was produced at ~75 % of the cofactor-supported rate owing to non-enzymatic oxidation. With holo-CRBP1 in the absence of cofactor, retinal generation was negligible, indicating protection of retinol from non-enzymatic oxidation.oxidation. Velocities, however, were lower with CRBP1-retinol than with unbound retinol, possibly because transfer of retinol from CRBP1 to RDH became rate limiting [30, 171]. These data indicate that RDH must recognize holo-CRBP1, in addition to free retinol.

A similar approach, also using two different ratios of total CRBP1 to retinol (1:1 and 2:1) demonstrated a Michaelis-Menten relationship with a K_m value of ~0.8 μM between holo-CRBP1 and RDH activity in calf liver cytosol [196]. Cytosol contains RALDH isoforms, so cytosol can convert retinal generated from retinol into RA. In this preparation, increasing the ratio of CRBP1/retinol to 2 decreased the Vm ~50 %. Because increasing the ratio decreased the free retinol concentration from 120 to 3 nM (40-fold) at a nominal CRBP1-retinol concentration of 5 μM, this result confirmed that holo-CRBP1 acts as substrate and that apo-CRBP1 competes with holo-CRBP1. The K_m value for CRBP1-bound retinol was lower than that of free retinol, but was greater than the k_d value for retinol dissociation from CRBP1. These phenomena cannot occur if retinol must dissociate from CRBP1 prior to association with RDH. This work also showed that apo-CRBP1 captured the retinal generated from CRBP1-retinol.

To resolve the relative contributions of microsomal vs cytosolic RDH, both subcellular fractions from liver, kidney, testes, and lung were assayed for activity with 5 μM holo-CRBP1 and with 5 μM holo-CRBP1 plus 2 μM apo-CRBP1 [31]. The latter was done to model RA biosynthesis during states of moderate vitamin A-status in which CRBP1 would not be saturated with retinol, because apo-CRBP1 is a potent inhibitor of cytosolic RDH (IC_{50} ~ 1 μM). In the absence of apo-CRBP1, microsomes accounted for 60–83 % of the RDH activity (total enzyme units). In the presence of apo-CRBP1, microsomes accounted for 80–94 % of the combined microsomal plus cytosolic activity. Specific activities of microsomal RDH in presence of 5 μM holo-CRBP1 plus 2 μM apo-CRBP1 ranged from 13 to 48-fold higher than those in cytosol, depending on the tissue. This showed that microsomal RDH generates the quantitatively major share of retinal for RA biogenesis from the predominant physiological substrate, holo-CRBP1. The SDR inhibitor carbenoxolone and apo-CRBP1 confirmed the cytosolic activity as resulting from an SDR. Possibly, cytosolic RDH that recognize holo-CRBP1 function as reductases (see below). If so, inhibition by apo-CRBP1 would allow continued RA biosynthesis during restricted cellular retinol. Thus, work reported in 1986 and continued through 1996 established that RDHs serve as physiological generators of retinal for RA biosynthesis in multiple species and tissues, and excluded ADH participation in RA biosynthesis under normal conditions.

External Residue Mutation Affects CRBP1 as Substrate Without Altering Affinity for Retinol

Specific CRBP1 residues were mutated to assess ligand binding interactions, conformation flexibility, and the kinetics of retinal formation catalyzed by microsomes [198]. Data generated from these mutants show that residues L29, I32, and R58 do not affect substrate efficiency, but contribute to ligand binding and protein rigidity. In contrast, mutation of L36 (L36A) increased substrate efficiency, but the mutant also was more flexible and had a lower ligand binding affinity (Fig. 2.1).

Surface residue L35 projects from the body of CRBP1. Converting it into an alanine residue (L35A) did not change affinity for retinol, but decreased the Vm of retinal formation by 50 % (Fig. 2.6). This suggests RDH interacts with the L35 side-chain to counteract interactions of interior residues with the β-ionone ring of retinol. Enzyme binding to L35 apparently alters the position of the cap to allow transfer of retinol from CRBP1. Residues V27 and K31 in αII may also contribute to interactions with RDH, because they are bulky and project from the CRBP1 surface. Unfortunately, these residues were not mutated. Individual mutations of V27 and K31, as well as triple mutation of V27, K31, and L35 would test this hypothesis further. Mutant L35A was somewhat less rigid than WT, as judged by Arg-C digestion of holo-CRBP1, which cleaves at R30 in α-helix II (αII). Charged mutations L35E and L35R had the same efficiency for retinal formation as WT, consistent with a van der Waals interaction between L35 and RDH. Each charged L35 mutation has affinity for retinol equivalent to WT, but each had much more flexibility than L35A, presumably because of charge repulsion from nearby residues. These are remarkable kinetic results for mutation of a single exterior residue that does not determine affinity of ligand binding.

In contrast to L35A, CRBP1 mutants (L29A, I32A, L36A, F57A, F58A, R58E) each had ~two–fivefold reduced affinity for retinol and were proteolyzed as much as ~fivefold more rapidly than WT. These data connect affinity of retinol binding with protein flexibility and are consistent with the greater flexibility of apo- vis-à-vis holo-CRBP1. None of these mutations, however, had a Vm for retinal formation lower than WT. In fact, mutants L36A and R58A had 50 % increases in Vm values and >tenfold increases in efficiency relative to WT, likely reflecting increased ease of portal opening and retinol release. Taken together, these data suggest that the β-ionone ring grasps the "handle" provided by the interior residues discussed, offset by the hydrophilic cavity exerting repulsive forces on the isoprene side chain. Enzyme interaction with the αII exterior residues would contribute sufficient force to promote retinol egress. Thus, pressure for the conservation of the *external* residues of CRBP seems to reflect a need for interaction with enzymes that rely on CRBP1 to provide retinol, not merely for protein conformation and/or ligand binding.

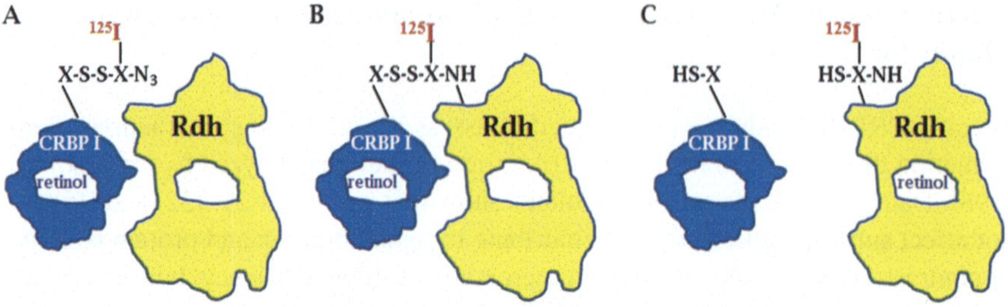

Fig. 2.7 Holo-CRBP1 crosslinks with RDH. A. CRBP1 was covalently modified with a heterobifunctional, cleavable crosslinking reagent. The crosslinker was radioiodinated. B. UV irradiation activated the azide of the crosslinker to a nitrene residue that formed a covalent bond between CRBP1 and closely associating RDH (and LRAT). C) Cleavage of the crosslinking reagent left the radioiodine on the target protein, effectively resulting in a transfer of radioiodine from holo-CRBP1 to RDH and LRAT. CRBP1 cross-linked only with two proteins in microsomes, RDH and LRAT, indicating specificity of the interaction. Crosslinking with RDH required presence of $NAD(P)^+$, consistent with the ordered bisubstrate reaction mechanism of a dehydrogenase

CRBPI Identified *SDR as RDH*

Microsomal RDH that recognize holo-CRBP1 occur in at least two forms: integral and peripheral membrane [29]. RDH1 has been identified as an integral microsomal enzyme that anchors in the membrane facing the cytoplasm [272, 288]. The more stable peripheral form was purified to two major bands on SDS-PGE, one ~35 kDa and the other ~54 kDa [29]. To determine which might represent RDH, either microsomes or a glycerol extract of microsomes was presented with CRBP1 covalently bound with a radioiodinated, heterobifunctional, cleavable crosslinking arm (Fig. 2.7). This allowed transfer of radioiodine from the donor (CRBP1) to specific acceptors. Only two molecules cross-linked. One was ~35 k Da and the other was ~25 kDa. Crosslinking the ~35 kDa protein required pyridine nucleotide cofactor, consistent with the ordered bisubstrate reaction mechanism of a $NAD(P)^+$-supported dehydrogenase, which must bind cofactor before accepting substrate. The ~25 kDa protein did not require cofactor to crosslink, and likely represented LRAT, as purification of LRAT showed that its molecular weight was ~25 kDa [237]. Crosslinking was specific, because only these two proteins in microsomes cross-linked, and crosslinking occurred with each protein based on their individual cofactor requirements.

The 35 kDa protein had a molecular weight and other attributes of an SDR, including the highly conserved sequence WXLVNNAG, Zn^{2+} independence, inhibition by carbenoxolone ($IC_{50}=55$ pM), and insensitivity to inhibition by ethanol and the ADH inhibitor 4-methylpyrazole. Multiple different cDNA were identified. One of these was cloned first and referred to as RODHI. The cDNA validated it as an SDR that recognized holo-CRBP1 (1.4:1 ratio total CRBP1/retinol, to ensure virtually no free retinol) with a Michaelis-Menten relationship with a K_m value <1 μM [39]. Soon, other RDH/SDR were cloned based on the data generated from RODHI [39, 40, 43, 256, 287]. Subsequently, human orthologs were cloned [25, 80, 105,

Table 2.4 Dehydrogenases and reductases likely to contribute physiologically to RA homeostasis[1]

Mouse	Rat	Human	Function
Rdh1 (9C17)	Rdh2, Rdh7 (9C28, 9C29)	RDH16 (9C8) (RODH4, RDH-E)	Dehydrogenase
	(RodhII, RoDHIII)		
	Rdh10 (16C15)	RDH10 (16C4)	Dehydrogenase
Rdh10 (16C10)	Dhrs9 (9C26)	DHRS9 (9C4)	Dehydrogenase
	(eRolDH, Rdhl)	(retSDR8, RDHL, Rdh TBE, RoDH-E_2, 3α-HSD)	
	SDR16C5 (16C114)	RDHE2 (16C5)	Dehydrogenase
Dhrs9 (9C12) (Rdh15)	SDR16c6 (16C113)	[a]SDR (16C6P)	Dehydrogenase
Rdhe2 (16C11)	–	RDH14 (7C4)	Dehydrogenase
(Scdr9, Sdr16c5)			
Rdhe2S (16C12)		(PAN2)	Reductase
(SDR16c6)			
Rdh14 (7C12)	Dhrs3 (16C16)	DHRS3 (16C1) (RDH17, SDR1, SDR16C1, retSDR1, Rsdr1)	
Dhrs3 (16C9)	Dhrs4 (25C7) (PSCD)	DHRS4 (25C2)	Reductase
		(SCAD-SRL, SDR-SRL)	
		Rdh11 (7C1)	
(retSDR1, Rsdr1)	Rdh11 (7C14)	(PSDR1, RalR1)	Reductase
Dhrs4 (25C5) (RRD)	–	RDH13 (7C3)	Reductase
Rdh11 (7C9)			
(SCALD, Psdr1)			
Rdh13 (7C11)			

[1]All belong to the SDR gene family. Orthologs are aligned in rows. "Family designations" are given in parentheses next to each human gene, each is preceded by SDR, e.g. SDR9C8 [202, 203]. Alternative names are given in parentheses below. Several genes were cloned multiple times and named differently. Various nomenclature initiatives introduced further confusion by assigning different common names to orthologs (e.g. mRDH1, rRDH2, hRDH16)

[a]This is a pseudo gene in human. –, no entry for rat

References: Rdh1, Rodh [29, 39, 40, 287]; RDH16 [80, 105]; Rdh10 [12, 279]; Rdhe2 [129, 153]; Rdhe2S [14]; Dhrs3 [84]; Dhrs4 [130]; Dhrs9 [151, 222]; Rdh11 [114, 138]; Rdh14 [9]; RDH13 [13]

150] (Table 2.4). Of these, RDH1, RDH10 and DHRS9 have been associated with vitamin A-dependent processes as indicated by studies of gene ablation in the mouse or in zebrafish, or overexpression in cells.

RDH10 is one of multiple members of the SDR gene family that catalyze interconversion of retinol and retinal in the visual cycle, but also contributes to RA biosynthesis in multiple tissues [197, 206]. RDH10 was first cloned from the RPE and identified as important to the visual cycle [279], and recognizes two other proteins that mediate retinoid metabolism in the eye, RPE65 and CRALBP [66]. RDH10 immunoprecipitates with both proteins and catalyzes the conversion of retinol bound to a twofold molar excess of CRALBP into retinal at the same rate as with free retinol. Outside the eye, RDH10 is broadly distributed in the cell in microsomes and mitochondria and mitochondrial associated membranes, but during acyl ester and lipid droplet biosynthesis, it translocates to lipid droplets and co-localizes with LRAT and CRBP1 [103]. This suggests that lipid droplets, which store RE, also serve as a source of RA, and again illustrates the intracellular mobility of retinoid binding-proteins and enzymes.

Retinal Reductases Contribute to Retinol and RA Homeostasis Outside of Intestine and Eye

Reductases of the SDR gene family have been identified that convert retinal into retinol [9, 10, 84, 114, 120, 130] (Table 2.4). Most have not been tested for recognition of CRBP1-retinal as substrate, with the exception of DHRS4 (originally RRD), a peroxisomal reductase that leaks into cytosol. DHRS4 recognizes CRBP1-bound retinal as evidenced by a Michaelis-Menten relationship with CRBP1-retinal (twofold molar excess of total CRBP1 to retinal), but with a Vm ~25 % of the rate of unbound retinal [130]. To provide insight into whether competitive inhibition between apo- and holo-CRBP1 produced these results, apo-CRBP1 was titrated into a reaction with unbound retinal as substrate. The rate of reduction did not decrease markedly as long as the total CRBP1 concentration was less than the retinal concentration, i.e. as long as no apo-CRBP1 was present. Adding apo-CRBP1 concentrations greater than the retinal concentration decreased reduction rates with an IC_{50} ~ 0.6 μM CRBP1, indicating competitive inhibition between CRBP1-retinal and apo-CRBP1.

Of the reductases, only DHRS3 has been knocked out in mice [24]. Ablation caused a 40 % increase in RA and 55-60 % decreases in retinol and RE during embryogenesis. These results show that regulation of RA homeostasis requires both retinol dehydrogenation and retinal reduction, achieved through the actions of SDRs.

Retinoid Metabolizing Enzymes Have Interactions Beyond Binding-Proteins

Knockdown of *Dhrs9* in primary astrocytes increased RA biosynthesis ~ 40 % by increasing RALDH1 expression [273]. It may seem odd that increasing RALDH1 would increase RA biosynthesis, because RDH catalyze the rate-limiting reaction. But the Vm of the slowest enzyme in a branched pathway is not necessarily rate *determining* for any one final product, and the pathway from retinol to RA is non-linear. Retinal can be fed into the path from retinol dehydrogenation or carotene cleavage, and can be removed by reduction or dehydrogenation. Therefore, at least four reactions determine the concentration of the retinal pool and the rate of RA biosynthesis. Increasing the concentration of RALDH1 would allow more efficient competition for the retinal pool (reaction rate ∝ [enzyme] x k_{cat} x [substrate]), especially because the reaction is irreversible.

Repression of RALDH1 expression complements the observation that DHRS9 "moonlights" as a transcriptional repressor [151]. Another form of interaction may be through direct binding to other proteins. During purification, rat RDH activity required binding to a 54 kDa protein, identified as CYP2D1 [29, 98]. In another example, recombinant tagged RDH10 and DHRS3 expressed in HEK293 cells form a heterodimer that enhances activities of both [1]. These data indicate the complexity of regulating RA biosynthesis and maintaining RDH structure to retain enzymatic activity and interactions of enzymes with retinoid binding-proteins.

Features of RDH Knockouts

The *Rdh1*-knockout produced the unexpected results of no apparent developmental or morphological phenotype, and no decrease in liver RA, even though *Rdh1* expression correlates with RA targets. The *Rdh1*-null mouse, however, has altered retinoid homeostasis [285, 286, 289]. Decreases in enzymes that catabolize RA, i.e. liver *Cyp26A1* and brown adipose *Cyp26B1,* and increases in liver RDH10 and brown adipose DHRS9 and RALDH1 modify the phenotype of the *Rdh1*-null mouse. When fed low-fat diets with modest amounts of vitamin A, the *Rdh1*-null mouse gains up to 37 % more weight than WT, due to increased adiposity. Rescue of the phenotype with diets high in vitamin A, and the ameliorating effect of pharmacological RA doses on weight gain in mice challenged with a high-fat diet corroborate the phenotype of the *Rdh1*-null mice as RA related [20]. In another model, RA was reduced 50 % in the kidney cortex of Tg26 mice (HIV-1 transgenic) and 70 % in their glomeruli, associated with a 70 % reduction in *Rdh1* mRNA and ~twofold increase in *Cyp26A1* mRNA [218].

Although *Rdh10* ablation is embryonic lethal, it does not produce complete vitamin A-deficiency, indicating essentiality of additional RDH during development [224]. This activity may be supported by either RDH1 or DHRS9.

Both free- and CRBP1-bound retinol serve as substrates for DHRS9 [150]. Transfection of *Dhrs9* into COS cells increases their ability to biosynthesize RA [221, 273]. *Dhrs9* expression seems to be estrogen regulated, as the rat uterine lining epithelium expresses *Dhrs9* only during estrus and co-expresses it with *CRBP1* and *CRABP2* [64, 221]. *Dhrs9* has not been knocked out in mice, but zebra fish require *Dhsr9* for normal gut development and differentiation, both RA dependent processes [161]. Introduction of the tumor suppressor adenomatous polyposis coli into the colon carcinoma cell line HT29 induces *Dhrs9* mRNA fivefold and increases RA biosynthesis ~twofold [101].

Controversies and Confounding Factors in Evaluating CRBP1 Function and Retinol Metabolism

One concern is the tendency of endogenous RDH in their native mammalian membrane environments generating retinal from holo-CRBP1 at lower rates than from unbound retinol, possibly because release of retinol from holo-CRBP1 imposes a rate-limiting event slower than the release of cofactor from RDH, the normal rate-limiting event with dehydrogenases [150, 196, 211]. Nevertheless, holo-CRBP1 supports retinal formation in vitro at rates sufficient to provide the 1–50 nmol/g RA observed in serum and tissues [107].

Because the Vm values in vitro supported by holo-CRBP1 are lower than the values supported by unbound retinol, an alternative perspective proposes that RDH recognize only unbound (free) retinol [119]. This contention does not address the rate of RDH catalysis possible in vivo from the low intracellular concentrations of free retinol established by CRBP1 and other modes of sequestering retinol (Tables 2.1 and 2.2). A practical assessment requires comparing the rate that could be supported by the fraction of intracellular free retinol *actually* available to the rate that holo-CRBP1 would support. In addition, arguments that RDH recognize only unbound retinol should address all data supporting participation of retinoid binding-proteins in the RDH reaction. It also would be prudent to address the data that support participation of retinoid binding-proteins in the *entire pathway* of retinol/RA metabolism; otherwise the inference seems that the rate-limiting step is the only one that does not involve a binding-protein.

Some rate evidence used to challenge participation of holo-CRBP1 actually may support participation, or leave the question unresolved. Reduction of 2 μM retinal into retinol by Sf9-generated RDH12 with a C-terminal His_6-tag was monitored with increasing concentrations of apo-CRBP1 from 0 to 40 μM [11]. Increasing apo-CRBP1 up to 10 μM decreased the rate of retinol formation to ~25 % of the rate observed in the absence of CRBP1. Based on a 50 mM k_d of retinal with CRBP1 (one of the values in the literature at the time, the other being 30 nM), unbound retinal in the solution of 2 μM retinal and 10 μM CRBP1 was calculated as sufficient to account for product generation. The recently revised k_d value of ~9 nM (Table 2.1, undoubtedly still an underestimate of affinity) indicates that the rate observed actually was ~fivefold greater than could be supported by the free retinal in this experiment. The decrease in rate between 0 and 10 μM CRBP1 might result from two substrates contributing to retinol formation—bound and free retinal, each supporting a different Vm. A Michaelis-Menten experiment with varying ratios of CRBP1 at each concentration of retinal would have addressed this possibility (assuming that an SDR over expressed in an insect cell would behave identically to one in its native mammalian cell environment). Significantly, the rate did not decrease further when total CRBP1 was increased from 10 to 20 to 40 μM (5–20-fold excess over retinal). This fourfold increase in the ratio CRBP1 to retinal would have decreased unbound retinal by fivefold, which would have decreased the reaction rate by a little over fourfold—based on the K_m of RDH12 with free retinal—if free retinal were the only substrate. These experimental conditions and data generated do not support the paper's conclusion that only free retinal functioned as substrate.

Lower rates of retinal production catalyzed by RDH16 (human ortholog of RDH1) and RDH10 from holo-CRBP1 relative to unbound retinol also have prompted arguments that these enzymes recognize only free retinol. In both cases possible Michaelis-Menten relationships between holo-CRBP1 and product were not reported. In fact, the RDH16 assayed had a k_{cat} of 1.1 min^{-1} with free reinol, i.e. generated one mole of retinal per mole of enzyme in 50 s, indicating an enzyme with severely impaired activity. RDH with impaired activity may not *appear* to interact with holo-CRBP1 because retinal generation falls below limits of detection, or may not interact because the changes that impaired activity may have impaired interaction.

Another confounding factor was overexpression of RDH16 and RDH10 in Sf9 cells with RDH10 representing 10 % of total insect cell protein. Over expression of DHRS9 in Sf9 cells also resulted in low enzymatic activity for retinol relative to steroids, and relative to activity of endogenous enzyme in its natural membrane environment [221]. Membrane-associated proteins depend on membrane composition and their stoichiometry with membrane lipids and proteins for activity and function [2, 35, 92, 126]. Rat RODH activity during purification was maintained only with phosphatidylcholine (no other phospholipids were as effective), and required binding to the membrane protein CYP2D1, illustrating the importance of the membrane environment to activity [29, 30, 98]. This suggests that RDH expressed in insect cells might have inherently low activity because of the unnatural environment and/or stoichiometry.

Modification of RDH and/or CRBP1 introduced further confounding factors. RDH16 was reacted with an N-terminal fusion of CRBP1 with glutathione S-transferase or with a C-terminal fusion of CRBP1 with the chitin binding domain. Mutating single exterior residues of CRBP1 altered ability to serve as substrate (see above). A His_6 construct was used with RDH10. Modifying microsomal RDH activity with N-ethylmaleimide inhibited holo-CRBP1-supported retinal biosynthesis by 90 %, but did not inhibit RDH activity with unbound retinol, consistent with obstructing access of CRBP1 to RDH without affecting RDH catalytic activity [30]. This latter observation provides further support for RDH recognizing holo-CRBP, and indicates caution when testing the premise with modified RDH or CRBP1.

Lower Vm values are difficult to interpret without a context of supporting data, including assaying for a Michaelis-Menten relationship, determining whether apo-binding proteins inhibit the reaction, allowing for the possibility of rate-limiting transfer of retinoid from binding protein to enzyme, and attention to the environment of the enzyme.

Table 2.5 RA biosynthesis catalyzed by microsomes (RDH activity) and a cytosolic fraction (RALDH activity) using CRBP1-retinol as substrate[1]

	RA generated (pmol ± SD)											
Addition	0	0	0	0	0	0	0	~2	5±4	9±5	20±2	49±4
µsomes	+	+	+	−	−	−	+	−	+	+	+	+
P1	−	−	−	+	+	+	+	+	−	+	+	+
NAD+	−	+	−	−	+	−	−	+	+	+	−	+
NADP+	−	−	+	−	−	+	−	+	+	−	+	+

[1]A total of 7.5 µM CRBP1 was used with 5 µM retinol to insure minimal free retinol [212]. With a k_d of 3 nM, the unbound retinol concentration was 6 nM, whereas the holo-CRBP1 concentration was 4994 nM. The liver cytosol fraction P1 with RALDH activity was isolated by anion-exchange chromatography

CRBP1 Participates in Irreversible RA Biosynthesis Catalyzed by RALDH

Three cytosolic RALDH catalyze irreversible conversion of retinal into RA: RALDH1 (*Aldh1A1*); RALDH2 (*Aldh1A2*); RALDH3 (*Aldh1A3*) [166]. RALDH1 and 2 recognize CRBP1-retinal as substrate, whereas RALDH3 has not been tested with CRBP1 [199, 200, 212, 271]. A fourth, RALDH4 (*ALDH8A1*) recognizes only 9-*cis*-retinal [137, 139]. Though RALDH4 is widely expressed and has especially intense expression in liver, its function is uncertain because 9-*cis*-RA has not been detected outside pancreas [110].

Multiple RALDH candidates occur in cytosol using free retinal as substrate [125, 128, 154, 212]. To identify physiologically relevant candidates RALDH, CRBP1-retinol was incubated with liver microsomes (as a source of RDH) to generate retinal in situ. Liver cytosolic fractions, resolved via anion-exchange chromatography, were added and the combination was assayed for RALDH activity [212]. RA biosynthesis from CRBP1-retinol required both microsomes and the cytosolic fraction P1 (Table 2.5). Adding pyridine nucleotide cofactors generated maximum RA. This experiment achieved the dual purposes of verifying that the CRBP1-retinol/microsomal RDH partnership generates retinal, *and* showing that retinal generated by microsomes supports RA biosynthesis by cytosolic RALDH. These data suggested that CRBP1 acts as conduit of retinal from microsomes to cytosol to overcome low solubility of retinal in aqueous media. These data were complemented by the observation that retinal generated from CRBP1-retinol occurs bound with CRBP1 [196]. The ability of microsomes and CRBP1-retinol to provide retinal for RA biosynthesis was revisited with purified recombinant RALDH1 [199, 200]. Data very similar to those in Table 2.5 were generated, reinforcing the notion of an RA-generating metabolon consisting of CRBP1, microsomal RDH and cytosolic RALDH1.

Kinetics of CRBP1-delivered retinal with RALDH1 and 2 are consistent with the binding protein "cassette" serving as substrate (Fig. 2.8). CRBP1-retinal produces a Michaelis-Menten relationship with RALDH1-catalyzed RA formation, with a $K_{0.5}$ value less than half that obtained with unbound retinal (Table 2.6). The Vm using CRBP1-retinal was much lower than with unbound retinal under the conditions used, which consisted of a twofold molar excess of CRBP1 at each retinal

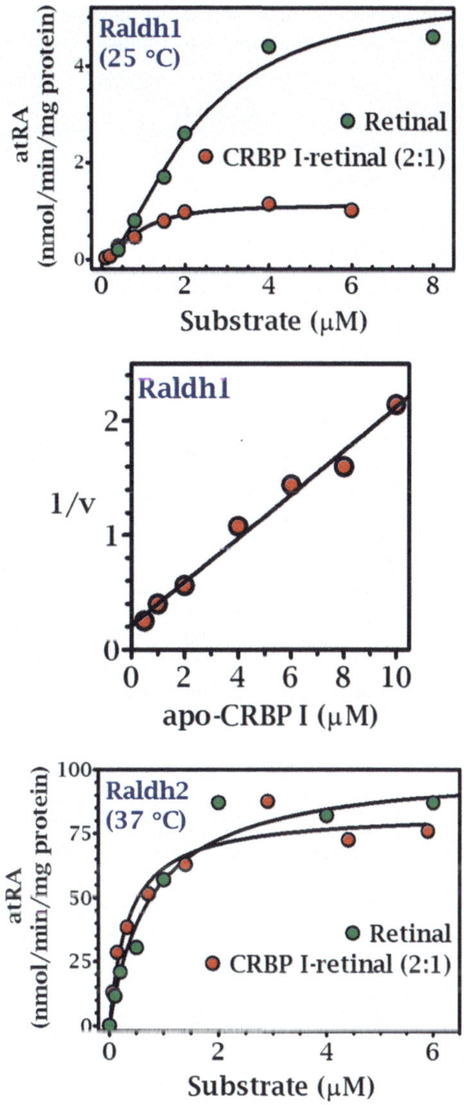

Fig. 2.8 Kinetic relationships between CRBP1 and recombinant RALDH1 or RALDH2. Top: conversion of retinal into RA by RALDH1 using CRBP1-retinal (2:1 ratio) or unbound retinal. Middle: Dixon plot of the effect of titrating apo-CRBP1 into 2 μM retinal on generation of RA by RALDH1 illustrating inhibition by apo-CRBP1. These data allow calculating the uninhibited Vm of RALDH1 with CRBP1-retinal, which was ~90 % of the Vm with unbound retinal. The reaction with RAlDH1 was run at 25 °C to reduce its rate to preserve initial velocity conditions. The IC_{50} of apo-CRBP1 for inhibiting RALDH1 was 1.4 μM. Bottom: conversion of retinal into RA by RALDH2. A twofold excess of binding protein to retinal did not affect the reaction rate, indicating that RALDH2 interacts with CRBP1 [199]

concentration [199]. This ratio practically eliminates any free retinal. For example, at 4 μM retinal, the 8 μM total CRBP1 would reduce free retinal >130-fold, but resulted in only a fourfold decrease in the rate of RA production relative to the reaction with 4000 nM free retinal, which indicates that CRBP1 acts as substrate. The decrease in Vm was caused by the ~4000 nM apo-CRBP1 inhibiting RALDH1, as revealed by a Dixon plot (Fig. 2.8). A Dixon plot allows calculating the rate that occurs at a precise 1:1 ratio of CRBP1 to retinal, and can be more reliable than trying to achieve a precise 1:1 ratio experimentally. The Dixon plot revealed that the

Table 2.6 Kinetic values of RALDH substrates. The Vm/Km values for RALDH1 and 2 are each twofold more efficient with CRBP1-retinal as substrate compared to free retinal

	Raldh1[a]	Raldh2[b]	Raldh3[c]	Raldh4[d]
Substrate	$K_{0.5}$ (μM)			
CRBP1-retinal	0.8	0.2	not tested	not tested
all-*trans*-retinal	1.7	0.7	0.2–0.3	<2 % of 9-*cis*-retinal
9-*cis*-retinal	5.2	0.5	not active	2.3

[a][62, 199, 200, 212]
[b][271]
[c][81, 82, 255]
[d][137, 139, 255]

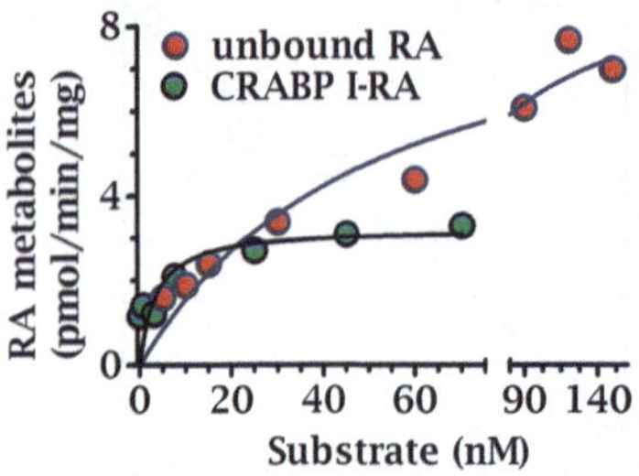

	K_m (nM)	Vm (pmol/min/mg)	V_m/K
$RA_{unbound}$	50	10	0.2
holo-CRABP I	1.8	2.6	1.4

Elimination half-life (min)		
Retinoid	Unbound	CRABP I-bound
RA	35	40
4-OH-RA	40	180
4-oxo-RA	9	630

Fig. 2.9 Kinetic relationship between CRABP1 and RA metabolism. Testis microsomes were used as source of RA catabolic activity, because testis has one of the highest intracellular RA concentrations [181]. Kinetics with CRABP1 were generated with a threefold molar excess of CRABP1 to RA to reduce free RA to negligible amounts [68]. Note that the reaction with CRABP1-RA is more efficient than with unbound RA (higher Vm/Km). Note also that the elimination half-life of RA does not change between unbound and CRABP1-bound catabolism, again suggesting interaction between enzyme and binding protein. The rates of RA oxidized at C4 do change, however, suggesting that these bound retinoids would inhibit RA catabolism if in sufficiently high concentrations. With physiological amounts of RA, the 4-oxidized derivatives have very low concentrations, but with toxic amounts of RA, the concentrations of 4-oxidized derivatives increase and contribute to RA toxicity

Vm value for 1:1 ratio CRBP1-retinal (i.e. no apo-CRBP1) was 89 % of the value with unbound retinal. Thus, delivering retinal bound to CRBP1 allowed RALDH1 to function ~twofold more efficiently than with unbound retinal (by comparing Vm/K_m). This outcome illustrates the problem encountered when interpreting lower rates produced by titrating CRBP1 into a fixed concentration of retinoid.

In contrast to the impact on RALDH1 activity, adding a twofold molar ratio of CRBP1 to each retinal concentration produced RADLH2 kinetics similar to those obtained with unbound retinal. The Vm/K_m for RALDH2 was twofold higher with CRBP1-retinal, again indicating a more efficient reaction. The ability of apo-CRBP1 to inhibit RALDH1, but not RALDH2, suggests each operates at different levels of vitamin A nutriture, and reveals the complexity of apo- and holo-CRBP1nteractions with retinoid metabolizing enzymes.

CRABPs Offer RA for Catabolism

Since the demonstrations of CRABP1 and 2 expression in multiple species and tissues, much effort has gone into understanding their functions [49, 185, 241, 245]. One purpose seems to be regulating RA concentrations.

Both unbound RA and CRABP1-bound RA have a Michaelis-Menten relationship with the rate of RA metabolism catalyzed by testes microsomes (Fig. 2.9). Kinetic constants for total metabolite formation are K_m ~50 nM and Vm ~10 pmol/min/mg protein for unbound RA, compared to 1.8 nM and 2.6 pmol/min/mg protein, respectively, for CRABP1-RA. Efficiency of the reaction with CRABP1-RA (3/1 ratio at all substrate concentrations) is ~sevenfold higher than with unbound RA, based on values of $Vm/K_m = 1.4$ vs. 0.2 for bound vs. free, respectively. Titrating apo-CRABP1 into a solution of 120 nM CRABP1-RA (1:1) reduces the rate of reaction by 35 % at a fivefold ratio CRABP1/RA (total CRABP1 = 600 nM). Using a K_d value of 0.4 nM, the decrease in unbound RA would have reduced the rate 60-fold if only unbound RA were substrate (calculating from a K_m value for unbound RA of 50 nM and a concentration of unbound RA decreasing from 7 nM (1:1 ratio CRABP:RA) to 0.1 nM (5:1 ratio CRABP1:RA). Overall, three criteria indicate that holo-CRABP1 serves as substrate: (1) the rate of RA metabolism in the presence of CRABP1 is faster than the rate supported by any unbound RA present; (2) increasing the concentration of apo-CRABP1 does not decrease the rate of metabolism to the extent predicted if only free RA were substrate; (3) the CRABP1 reaction is more efficient (Vm/K_m) than the reaction with unbound RA. These results were also observed with CRABP2 (Fiorella and Napoli, unpublished).

CRABP1 also serves as substrate for RA metabolism with microsomes from other tissues that express the binding-protein intensely, such as kidney and lung [69]. The composition of RA metabolites reflected the tissue source of microsomes, consistent with a previous report of tissue specificity in RA metabolism [167]. 4-OH-RA had a much higher concentration in the RA metabolite pool generated by testes when RA was presented bound to CRABP1 than when RA was presented unbound. This occurred because RA metabolites, such as 4-OH-RA and 4-oxo-RA, bind with CRABP1 upon formation in the incubation medium, and are not accessible to further metabolism when bound. The nearly complete arrest of 4-OH-RA and 4-oxo-RA degradation by CRABP1 contrasts with CRABP's lack of interference in the overall rate of RA catabolism, indicating that pathways other than those initiated by C4 oxidation occur, such as 18-hydroxylation and epoxidation.

Similar results have been demonstrated in the murine embryonic stem cell line, AB1, and a mutant AB1 cell line with *CRABP1*-ablated [44]. Deletion of CRABP1 caused a 50 % decrease in intracellular RA when medium RA concentrations were in the range 1–10 nM (serum concentrations are ~1-4 nM) indicating in this experiment CRABP1 served as a reservoir to draw RA into cells. Deletion of CRABP1 did not change the elimination $t_{1/2}$ of RA, however. Loss of CRABP1 increased the polar metabolites in cells ~30 % (1 nM RA treatment) relative to RA, but with no prior RA treatment or prior treatment with 1 nM RA, RA remained the quantitatively major intracellular retinoid. These data should be considered in context of the demonstration that F9 cells incubated with 50 nM RA has 99 % of the total retinoids in the medium, but only 44 % of the 99 % occurs as unmetabolized RA, whereas unmetabolized RA accounts for 76 % of the retinoids in cells, and 94 % of the retinoids in nuclei [274]. In other words, despite metabolism, RA accounts for most of the retinoids in the nucleus. Thus, the importance of increased 4-OH-RA and 4-oxo-RA is not certain. Their k_d values for RAR rival that of RA, but their low concentrations under physiological conditions (as opposed to during toxicity) do not suggest a major role in nuclear retinoid signaling. These data suggest that CRABP1 modulates the intracellular concentrations of RA and increases levels of some polar metabolites, but the physiological impact is uncertain. CRABP1-binding could prevent cell membranes from absorbing RA and diminishing its functional availability.

The *CRABP1*-null mouse lacks an obvious morphological phenotype, as tested in mice fed a diet containing copious vitamin A [79], which rescues phenotypes of the *Rbp4*-null mouse (*Rbp4* encodes sRBP), the *Rbp2* null mouse (*Rbp2* encodes CRBP2), and the *Rdh1*-null mouse [60, 213, 289]. Moreover, no other phenotypes related to retinoid action were studied, such as immune dysfunction, cancer, or short-term memory.

13-*cis*-RA (isotretinoin) exhibits about tenfold lower teratogenicity than RA (aka, tretinoin) [236]. To examine mechanisms, teratogenic amounts of RA and 13-*cis*-RA were dosed to mice during gestation, and retinoid distribution was monitored in embryos. More 13-*cis*-RA localized to the nucleus than RA, and nuclear 13-*cis*-RA gave rise to nuclear RA. Two conclusions were reached from these data: (1) the low affinity of 13-*cis*-RA for CRABP1 allowed greater nuclear localization; (2) conversion of 13-*cis*-RA into RA allowed RAR binding, thereby inducing teratogenic effects. These data are consistent with CRABP1 regulating access of RA to the nucleus.

Although 4-OH-RA and 4-oxo-RA have similar affinity for RAR as RA, they do not seem to affect RAR-gene regulation under normal circumstances because their concentrations are very low relative to RA. But with high doses of retinol or RA, the concentrations of 4-oxidized RA metabolites increase to the point where they likely contribute to retinoid toxicity [54, 61]. This suggests that during toxic vitamin A exposure, the ability of CRABP1 to repress metabolism of 4-OH-RA and 4-oxo-RA would contribute to toxicity, which could be augmented by the ability of CRABP1 to deliver retinoids to the nucleus.

Ketoconazole, an imidazole antimycotic and a potent inhibitor of CYP, inhibits metabolism of unbound and CRABP1-bound RA catalyzed by testes microsomes with IC_{50} values of 2 and 0.7 μM, respectively [162, 274, 275]. When this work was reported, the specific CYPs responsible for catalyzing RA metabolism were not known. Inhibition of RA metabolism by imidazole antimycotics, however, established the principle that arresting RA metabolism increased its potency. This principle was the foundation for using RA metabolism blocking agents (RAMBA) to treat RA-responsive diseases, as an alternative to dosing with RA or synthetic retinoids [156].

CRABPs Direct RA to Nuclear Receptors and/or Mediate Non-canonical Actions

Additional functions of RA binding-proteins involve delivering RA to the nucleus and mediating non-genomic mechanisms of RA action. Several labs have verified that CRABP1 localizes in nuclei as well in extra-nuclear structures [76, 83, 235, 243, 244, 262]. Notably, the precise subcellular locus of CRABP1 depends to a large extent on the fixation procedure used [76, 235]. The significance of CRABP1's nuclear localization (in terms of activating receptors) was cast into doubt with the demonstration that CRABP1 did not enhance interaction of RARα with RAR response-elements, nor affect transcriptional activation by RA. these data supported a conclusion that CRABP1 did not directly deliver RA to RAR [266]. This observation was taken a step further with the demonstration that RA was transferred from CRABP1 to RARα only through diffusion in a cell-free system—a report that also confirmed the inability of CRABP1 to stimulate RA-induced transcription [58]. It is likely there is more to this story.

RA does not function exclusively through transcriptional activation, however. In hippocampus neurons, RA stimulates dendritic growth within minutes by activating translation through MAPK, ERK1/2, mTOR and their targets 4E-BP, p70S6K, and ribosome protein S6 [42, 45]. RARα, within dendritic RNA transport granules, binds the RNA-binding protein PURα. RA binding to RARα frees PURα to allow translation. Another mechanism of non-canonical (non-transcriptional) regulation by RA involves the CRABP1-RA complex activating ERK1/2 in embryonic stem cells [201].

In conclusion, non-nuclear CRABP1 serves to control RA concentrations, deliver RA for catabolism, affect the intracellular steady-state composition of RA metabolites, mediate rapid non-canonical actions of RA, and may contribute to toxicity during vitamin A excess by impeding clearance of 4-oxidized RA metabolites.

CRABP2 Functions Differently from CRABP1 and Delivers RA to RARs in the Nucleus

Analysis of CRABP2 mRNA expression in 12 breast cell lines revealed a correlation with ability of RA to inhibit growth. Overexpression of CRABP2 enhanced transcriptional activity of RA in these cells [104]. These observations are consistent with nuclear detection of CRABP2 in transfected COS cells, mouse embryos and other cell lines [76]. Further study led to the conclusion that nuclear CRABP2 in its *un*liganded state associates with RAR and confirmed the observation of CRABP2 enhancing transcription potency of RA [7, 57]. Remembering that most CRABP2 resides in cytosol, a plausible model for its dual location is provided by the observation that after binding with RA, the CRABP2-RA complex translocates from cytosol to the nucleus and channels RA to RAR through protein-protein interactions [36, 58]. The CRABP2-RAR interaction was reported to be short-lived, allowing recycling of CRABP2 back to the cytosol. As with previous reports, these studies also demonstrated that overexpression of CRABP2 enhanced cell sensitivity to RA-induced growth inhibition in MCF7 mammary carcinoma cells, whereas reducing CRABP2 expression reduced RA action.

FABP5 and Not CRABP2 Delivers RA to PPARδ/β

RA has pleiotropic actions, including diverse actions in different cell types. One mechanism for these diverse actions could depend on expression of dissimilar nuclear coactivators. Another could include differential activation of nuclear hormone receptors. The demonstration that RA binds and activates PPARδ/β (hereafter referred to as PPARδ) provided insight into mechanisms underlying these diverse cell responses. RA binds to PPARδ with high-affinity ($k_d = 17$ nM; an affinity likely underestimated because fluorescent titration was done with 0.77 μM PPARδ [250]. RA selectively transcriptionally activates PPARδ (vs. PPARα and PPARγ) with an EC_{50} between 0.2 and 0.3 μM in COS-7 cells transfected with PPRE-driven luciferase reporters. The EC_{50} is higher than typical concentrations of RA in tissues, which fall between 1 and 50 nM [107, 109], which may be the result of RA insolubility in aqueous media and its sequestration by medium components and in cell membranes.

Binding of RA to PPARδ prompted a model for two intracellular RA binding-proteins, CRABP2 and FABP5, delivering RA to either RAR or PPARδ, respectively [248]. In cells with high CRABP2/FABP5, RA would function mainly through RAR as a proapoptotic agent, but in cells with relatively low CRABP2/FABP5, RA would function through PPARδ to promote cell survival and proliferation. In HaCaT keratinocytes, 100 nM RA causes fivefold activation of a PPAR response element, but the synthetic RAR-specific ligand TTNPB does not. Both the PPARδ-specific agonist GW0742 and RA induce expression of three PPARδ-regulated genes from 1.5 to ~3-fold: *FIAF*, *ADRP* and *PDK-1*. Notably, one of these, the PPARδ-target *PDK-1*, contributes to the anti-apoptotic activity of PPARδ. This work also demonstrated that RA serves as a ligand for FABP5 (k_d ~35 nM) and endogenous FABP5 is predominantly cytosol, until RA exposure, which causes translocation to the nucleus. Remarkably, increasing the ratio CRABP2/FABP5 converts RA from a survival to a proapoptotic agent in HaCaT and NaF (mouse mammary tumor) cells. These results support the hypothesis that CRABP2 delivers RA to RAR, which causes cell cycle arrest, apoptosis and growth inhibition, whereas FABP5 delivers RA to PPARδ, which stimulates survival paths to enhance proliferation.

This model was extended in the RA-resistant mouse model of breast cancer, MMTV-neu, to show that RA activates PPARδ through an "aberrantly" high ratio of FABP5/CRABP2 and changing this ratio reduced RA-resistance [248, 249]. A switch of RA action mediated by binding proteins and their associated receptors also seems to function during neuronal [283] and preadipocyte differentiation [17]. This was verified in the P19 neuronal stem cell model, where RA induced neurogenesis via CRABP2/RAR, whereas FABP5/PPARδ promotes the development of mature neurons as differentiation proceeds. In contrast, in NIH3T3 L1 preadipocytes, down regulation of CRABP2 impairs ability of RA to *prevent* differentiation into mature adipocytes.

Collectively, these data support a model of CRABP2 and FABP5 influencing RA action by sorting delivery to either RAR or PPARδ, which in turn prompts selective actions. Not everyone embraces the aspect of the model that includes RA binding FABP5 and activating PPARδ, however. One report challenged the notions that RA activates PPARδ and that PPARδ stimulates proliferation in HaCaT keratinocytes [32]. Both RA and the "highly specific" PPARδ ligands (GW0742; GW501516) were observed to inhibit proliferation, but only the PPARδ ligands increased expression of PPARδ-target genes. Another group reported that concentrations of GW501516 from 0.1 to 1 μM induced transcription ~17-fold of an PPARδ-activated luciferase reporter driven by a PPRE in NIH3T3 cells, but the same concentration range of RA had no effect [225]. In HaCaT cells, this report concluded that GW5051516 induced expression of the PPARδ-target genes *ADRP* (adipocyte differentiation-related protein) and *ANGPTL4* (angiopoietin-like protein 4), but RA did not, even though RA induced the RAR-target gene *Cyp26A1*. This report found no binding of RA to PPARδ in vitro by fluorescence assay in concentrations to 100 μM, in contrast to GW501516, which saturated PPARδ at 10 nM. A third report observed enhanced ligand-induced expression of *ADRP* and *ANGPTL4* in HaCaT cells that over-expressed PPARδ [33], but reported that RA did not increase expression of *ANGPTL4,* even though cells expressed a high FABP5/CRABP2 ratio. The reasons for disparate results produced by different labs with the (nominally) same cell lines are not clear.

The lab that originally reported RA binding to FABP5 and delivery by FABP5 to PPARδ revisited the model [132]. They again showed that RA binds recombinant FABP5 with a k_d ~42 nM by fluorescence assay. They also crossbred *FABP5*-null mice with MMTV-ErbB2/HER2 oncomice. The latter strain develops mammary tumors spontaneously. The double genetically modified model developed tumors in only 6 of 14 mice, whereas 100 % of the MMTV-ErbB2/HER2 mice developed tumors. Survival time of the latter did not exceed 39 weeks, but mice with the double mutation lived until at least 64 weeks, and had tumor volumes ~3.5-fold less than the control group. These data attest to the oncogenic (proliferative) actions of FABP5 in vivo. This work did not test the impact of RA on tumor growth in either strain of mouse, but showed that RA and FABP5 promote oncogenic metrics in NIH3T3 cells. Resolving the role of FABP5 and PPARδ in RA action is crucial to understanding mechanisms of RA action, and will have essential impact on therapeutic use of retinoids and PPARδ agonists.

CRABP2 Stabilizes mRNA

Apo-CRABP2 also binds with the RNA binding protein HuR and increases its affinity for the 3'-untranslated regions of mRNA, which delays degradation. This suggests that RA binding would cause CRABP2 release from HuR, allowing translocation to the nucleus [269].

Relevance

Data from multiple labs generated during the last 40 years indicate participation of retinoid binding-proteins during each step from retinol uptake into cells to its metabolism into RE or RA, with binding proteins also participating in RA delivery to nuclei, non-genomic (non-canonical) retinoid actions, and RA catabolism (Fig. 2.10).

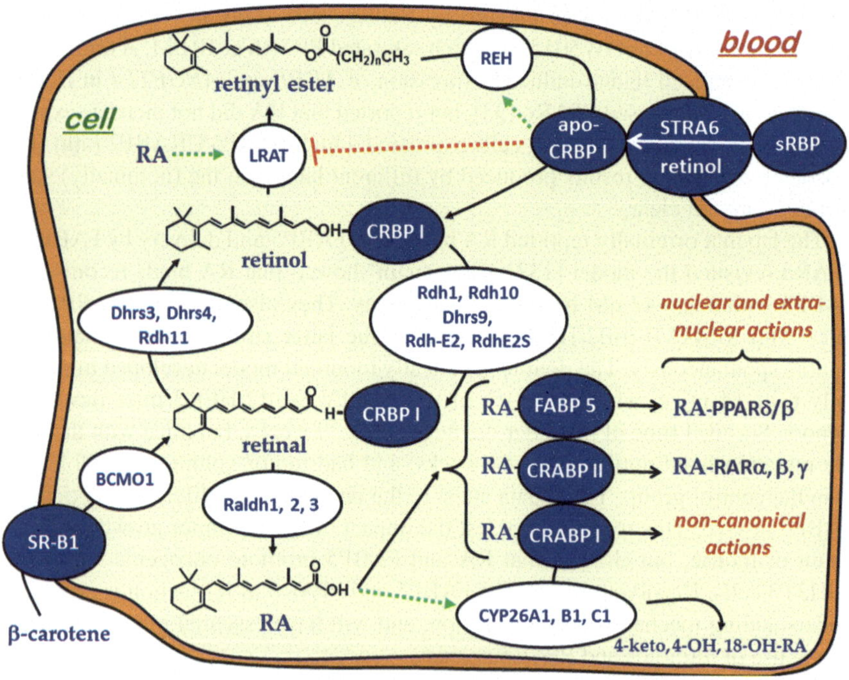

Fig. 2.10 Model of retinoid homeostasis illustrating roles of retinoid binding-proteins. This model illustrates contributions of retinoid binding-proteins to multiple aspects of retinol uptake and metabolism, and RA catabolism and action. The complexity of storing and activating retinol undoubtedly contributes to the multiplicity of retinoid actions. This chaperoning increases the efficiency of retinoid metabolism, but is not obligatory for life. Note, however, that the *CRBP1*-null mouse suffers from disrupted retinol metabolism, retinol wasting, and dysfunctions in intermediary metabolism. The dotted lines represent regulatory points

A number of aspects of this model and the data that support it are important to recognize. Neither the model nor the data indicate obligatory participation of binding-proteins in retinoid metabolism (as shown by knockouts). Rather, binding proteins enhance efficiency of metabolism to spare essential and rare (during evolution) retinoids from excess catabolism and to chaperone lipophilic retinoids through aqueous media. The model proposed does not address vitamin A metabolism during intake of copious or toxic amounts of vitamin A. With copious and/or toxic amounts of dietary or dosed vitamin A, enzymes that did not evolve to participate in vitamin A metabolism are presented with substrate as a result of saturated retinoid binding-proteins and high intracellular retinoid concentrations. This conclusion is supported by rescue of at least three knockouts (*Rbp4, Rdh1, Rbp2*) by chow diets (copious in vitamin A). It would seem important, therefore, when testing the contribution of a binding protein or enzyme, to avoid copious dietary vitamin A to reveal their functions (unless necessary for breeding, but returning to restricted amounts afterwards). Also, in light of the complexity of RA generation and homeostasis, and potential "moonlighting" by retinoid-metabolizing enzymes, it would seem important to assess multiple components of the retinoid metabolon before postulating sites that generate RA.

Future Directions

Much work remains to be done concerning specific interactions among retinoid binding-proteins and the enzymes they engage. In reality, functions of the binding-proteins and the enzymes RDH and RALDH are known only in "outline". Not entirely clear are the purposes of multiple CRBP, CRABP, RDH, RALDH and CYP isoforms. Finally, the different phenotypes of the multiple orthologs of binding-proteins, RDH, and RALDH knockouts suggest specialized metabolons generating RA for specific functions. This hypothesis and systemic as well as mechanistic regulation of retinoid homeostasis remain to be explored.

Acknowledgement The author is grateful for thought-provoking suggestions from David Ong, especially concerning but not limited to, the structures and binding characteristics of retinoid binding-proteins.

References

1. Adams MK, Belyaeva OV, Wu L, Kedishvili NY (2014) The retinaldehyde reductase activity of DHRS3 is reciprocally activated by retinol dehydrogenase 10 to control retinoid homeostasis. J Biol Chem 289:14868–14880
2. Ahn T, Kim M, Yun C-H, Chae H-J (2007) Functional regulation of hepatic cytochrome p450 enzymes by physicochemical properties of phospholipids in biological membranes. Curr Protein Pept Sci 8:496–505

3. Akerstrom B, Flower DR, Salier JP (2000) Lipocalins: unity in diversity. Biochim Biophys Acta 1482:1–8
4. Amengual J, Zhang N, Kemerer M, Maeda T, Palczewski K, Von Lintig J (2014) STRA6 is critical for cellular vitamin A uptake and homeostasis. Hum Mol Genet. doi:10.1093/hmg/ddu258
5. Bashor MM, Chytil F (1975) Cellular retinol-binding protein. Biochim Biophys Acta 411:87–96
6. Bashor MM, Toft DO, Chytil F (1973) In vitro binding of retinol to rat-tissue components. Proc Natl Acad Sci U S A 70:3483–3487
7. Bastie JN, Despouy G, Balitrand N, Rochette-Egly C, Chomienne C, Delva L (2001) The novel co-activator CRABPII binds to RARalpha and RXRalpha via two nuclear receptor interacting domains and does not require the AF-2 "core.". FEBS Lett 507:67–73
8. Batten ML, Imanishi Y, Maeda T, Tu DC, Moise AR, Bronson D et al (2004) Lecithin-retinol acyltransferase is essential for accumulation of all-trans-retinyl esters in the eye and in the liver. J Biol Chem 279:10422–10432
9. Belyaeva OV, Kedishvili NY (2002) Human pancreas protein 2 (PAN2) has a retinal reductase activity and is ubiquitously expressed in human tissues. FEBS Lett 531:489–493
10. Belyaeva OV, Stetsenko AV, Nelson P, Kedishvili NY (2003) Properties of short-chain dehydrogenase/reductase RalR1: characterization of purified enzyme, its orientation in the microsomal membrane, and distribution in human tissues and cell lines. Biochemistry (Mosc) 42:14838–14845
11. Belyaeva OV, Korkina OV, Stetsenko AV, Kim T, Nelson PS, Kedishvili NY (2005) Biochemical properties of purified human retinol dehydrogenase 12 (RDH12): catalytic efficiency toward retinoids and C9 aldehydes and effects of cellular retinol-binding protein type I (CRBPI) and cellular retinaldehyde-binding protein (CRALBP) on the oxidation and reduction of retinoids. Biochemistry (Mosc) 44:7035–7047
12. Belyaeva OV, Johnson MP, Kedishvili NY (2008) Kinetic analysis of human enzyme RDH10 defines the characteristics of a physiologically relevant retinol dehydrogenase. J Biol Chem 283:20299–20308
13. Belyaeva OV, Korkina OV, Stetsenko AV, Kedishvili NY (2008) Human retinol dehydrogenase 13 (RDH13) is a mitochondrial short-chain dehydrogenase/reductase with a retinaldehyde reductase activity. FEBS J 275:138–147
14. Belyaeva OV, Lee S-A, Adams MK, Chang C, Kedishvili NY (2012) Short chain dehydrogenase/reductase rdhe2 is a novel retinol dehydrogenase essential for frog embryonic development. J Biol Chem 287:9061–9071
15. Berman ER, Segal N, Feeney L (1979) Subcellular distribution of free and esterified forms of vitamin A in the pigment epithelium of the retina and in liver. Biochim Biophys Acta 572:167–177
16. Berry DC, Noy N (2012) Signaling by vitamin A and retinol-binding protein in regulation of insulin responses and lipid homeostasis. Biochim Biophys Acta 1821:168–176
17. Berry DC, Soltanian H, Noy N (2010) Repression of cellular retinoic acid-binding protein II during adipocyte differentiation. J Biol Chem 285:15324–15332
18. Berry DC, Jin H, Majumdar A, Noy N (2011) Signaling by vitamin A and retinol-binding protein regulates gene expression to inhibit insulin responses. Proc Natl Acad Sci U S A 108:4340–4345
19. Berry DC, Croniger CM, Ghyselinck NB, Noy N (2012) Transthyretin blocks retinol uptake and cell signaling by the holo-retinol-binding protein receptor STRA6. Mol Cell Biol 32:3851–3859
20. Berry DC, DeSantis D, Soltanian H, Croniger CM, Noy N (2012) Retinoic acid upregulates preadipocyte genes to block adipogenesis and suppress diet-induced obesity. Diabetes 61:1112–1121

21. Berry DC, O'Byrne SM, Vreeland AC, Blaner WS, Noy N (2012) Cross talk between signaling and vitamin A transport by the retinol-binding protein receptor STRA6. Mol Cell Biol 32:3164–3175
22. Berry DC, Jacobs H, Marwarha G, Gely-Pernot A, O'Byrne SM, DeSantis D et al (2013) The STRA6 receptor is essential for retinol-binding protein-induced insulin resistance but not for maintaining vitamin A homeostasis in tissues other than the eye. J Biol Chem 288:24528–24539
23. Biesalski HK (1989) Comparative assessment of the toxicology of vitamin A and retinoids in man. Toxicology 57:117–161
24. Billings SE, Pierzchalski K, Butler Tjaden NE, Pang X-Y, Trainor PA, Kane MA et al (2013) The retinaldehyde reductase DHRS3 is essential for preventing the formation of excess retinoic acid during embryonic development. FASEB J Off Publ Fed Am Soc Exp Biol 27:4877–4889
25. Biswas MG, Russell DW (1997) Expression cloning and characterization of oxidative 17beta- and 3alpha-hydroxysteroid dehydrogenases from rat and human prostate. J Biol Chem 272:15959–15966
26. Blaner WS, Das K, Mertz JR, Das SR, Goodman DS (1986) Effects of dietary retinoic acid on cellular retinol- and retinoic acid-binding protein levels in various rat tissues. J Lipid Res 27:1084–1088
27. Bliss AF (1951) The equilibrium between vitamin A alcohol and aldehyde in the presence of alcohol dehydrogenase. Arch Biochem Biophys 31:197–204
28. Boerman MH, Napoli JL (1991) Cholate-independent retinyl ester hydrolysis. Stimulation by Apo-cellular retinol-binding protein. J Biol Chem 266:22273–22278
29. Boerman MH, Napoli JL (1995) Characterization of a microsomal retinol dehydrogenase: a short-chain alcohol dehydrogenase with integral and peripheral membrane forms that interacts with holo-CRBP (type I). Biochemistry (Mosc) 34:7027–7037
30. Boerman MH, Napoli JL (1995) Effects of sulfhydryl reagents, retinoids, and solubilization on the activity of microsomal retinol dehydrogenase. Arch Biochem Biophys 321:434–441
31. Boerman MH, Napoli JL (1996) Cellular retinol-binding protein-supported retinoic acid synthesis. Relative roles of microsomes and cytosol. J Biol Chem 271:5610–5616
32. Borland MG, Foreman JE, Girroir EE, Zolfaghari R, Sharma AK, Amin S et al (2008) Ligand activation of peroxisome proliferator-activated receptor-beta/delta inhibits cell proliferation in human HaCaT keratinocytes. Mol Pharmacol 74:1429–1442
33. Borland MG, Khozoie C, Albrecht PP, Zhu B, Lee C, Lahoti TS et al (2011) Stable overexpression of PPARβ/δ and PPARγ to examine receptor signaling in human HaCaT keratinocytes. Cell Signal 23:2039–2050
34. Bouillet P, Sapin V, Chazaud C, Messaddeq N, Décimo D, Dollé P et al (1997) Developmental expression pattern of Stra6, a retinoic acid-responsive gene encoding a new type of membrane protein. Mech Dev 63:173–186
35. Brenner RR (1984) Effect of unsaturated acids on membrane structure and enzyme kinetics. Prog Lipid Res 23:69–96
36. Budhu AS, Noy N (2002) Direct channeling of retinoic acid between cellular retinoic acid-binding protein II and retinoic acid receptor sensitizes mammary carcinoma cells to retinoic acid-induced growth arrest. Mol Cell Biol 22:2632–2641
37. Burnett KG, Felder MR (1978) Peromyscus alcohol dehydrogenase: lack of cross-reacting material in enzyme-negative animals. Biochem Genet 16:1093–1105
38. Careri M, Elviri L, Mangia A, Zagnoni I, Torta F, Cavazzini D et al (2006) Mass spectrometry techniques for detection of ligand-dependent changes in the conformational flexibility of cellular retinol-binding protein type I localized by hydrogen/deuterium exchange. Rapid Commun Mass Spectrom RCM 20:1973–1980
39. Chai X, Boerman MH, Zhai Y, Napoli JL (1995) Cloning of a cDNA for liver microsomal retinol dehydrogenase. A tissue-specific, short-chain alcohol dehydrogenase. J Biol Chem 270:3900–3904

40. Chai X, Zhai Y, Popescu G, Napoli JL (1995) Cloning of a cDNA for a second retinol dehydrogenase type II. Expression of its mRNA relative to type I. J Biol Chem 270:28408–28412
41. Chazaud C, Bouillet P, Oulad-Abdelghani M, Dollé P (1996) Restricted expression of a novel retinoic acid responsive gene during limb bud dorsoventral patterning and endochondral ossification. Dev Genet 19:66–73
42. Chen N, Napoli JL (2008) All-trans-retinoic acid stimulates translation and induces spine formation in hippocampal neurons through a membrane-associated RARalpha. FASEB J Off Publ Fed Am Soc Exp Biol 22:236–245
43. Chen W, Song M-S, Napoli JL (2002) SDR-O: an orphan short-chain dehydrogenase/reductase localized at mouse chromosome 10/human chromosome 12. Gene 294:141–146
44. Chen AC, Yu K, Lane MA, Gudas LJ (2003) Homozygous deletion of the CRABPI gene in AB1 embryonic stem cells results in increased CRABPII gene expression and decreased intracellular retinoic acid concentration. Arch Biochem Biophys 411:159–173
45. Chen N, Onisko B, Napoli JL (2008) The nuclear transcription factor RARalpha associates with neuronal RNA granules and suppresses translation. J Biol Chem 283:20841–20847
46. Chytil F, Ong DE (1978) Cellular vitamin A binding proteins. Vitam Horm 36:1–32
47. Chytil F, Ong DE (1979) Cellular retinol- and retinoic acid-binding proteins in vitamin A action. Fed Proc 38:2510–2514
48. Chytil F, Ong DE (1987) Intracellular vitamin A--binding proteins. Annu Rev Nutr 7:321–335
49. Chytil F, Page DL, Ong DE (1975) Presence of cellular retinol and retinoic acid binding proteins in human uterus. Int J Vitam Nutr Res Int Z Für Vitam- Ernährungsforschung J Int Vitaminol Nutr 45:293–298
50. Collins MD, Mao GE (1999) Teratology of retinoids. Annu Rev Pharmacol Toxicol 39:399–430
51. Collins MD, Eckhoff C, Chahoud I, Bochert G, Nau H (1992) 4-Methylpyrazole partially ameliorated the teratogenicity of retinol and reduced the metabolic formation of all-trans-retinoic acid in the mouse. Arch Toxicol 66:652–659
52. Cornish-Bowden A (1995) Fundamentals of enzyme kinetics. Portland Press, London
53. Cowan SW, Newcomer ME, Jones TA (1993) Crystallographic studies on a family of cellular lipophilic transport proteins. Refinement of P2 myelin protein and the structure determination and refinement of cellular retinol-binding protein in complex with all-trans-retinol. J Mol Biol 230:1225–1246
54. Creech Kraft J, Löfberg B, Chahoud I, Bochert G, Nau H (1989) Teratogenicity and placental transfer of all-trans-, 13-cis-, 4-oxo-all-trans-, and 4-oxo-13-cis-retinoic acid after administration of a low oral dose during organogenesis in mice. Toxicol Appl Pharmacol 100:162–176
55. Dela Seña C, Riedl KM, Narayanasamy S, Curley RW, Schwartz SJ, Harrison EH (2014) The human enzyme that converts dietary provitamin A carotenoids to vitamin A is a dioxygenase. J Biol Chem 289:13661–13666
56. Deltour L, Foglio MH, Duester G (1999) Metabolic deficiencies in alcohol dehydrogenase Adh1, Adh3, and Adh4 null mutant mice. Overlapping roles of Adh1 and Adh4 in ethanol clearance and metabolism of retinol to retinoic acid. J Biol Chem 274:16796–16801
57. Delva L, Bastie JN, Rochette-Egly C, Kraïba R, Balitrand N, Despouy G et al (1999) Physical and functional interactions between cellular retinoic acid binding protein II and the retinoic acid-dependent nuclear complex. Mol Cell Biol 19:7158–7167
58. Dong D, Ruuska SE, Levinthal DJ, Noy N (1999) Distinct roles for cellular retinoic acid-binding proteins I and II in regulating signaling by retinoic acid. J Biol Chem 274:23695–23698
59. Drevon CA, Blomhoff R, Rasmussen M, Kindberg GM, Berg T, Norum KR (1985) Retinol esterification in cultured rat liver cells. Biochem J 230:617–623

60. E X, Zhang L, Lu J, Tso P, Blaner WS, Levin MS et al (2002) Increased neonatal mortality in mice lacking cellular retinol-binding protein II. J Biol Chem 277:36617–36623
61. Eckhoff C, Nau H (1990) Vitamin A supplementation increases levels of retinoic acid compounds in human plasma: possible implications for teratogenesis. Arch Toxicol 64:502–503
62. El Akawi Z, Napoli JL (1994) Rat liver cytosolic retinal dehydrogenase: comparison of 13-cis-, 9-cis-, and all-trans-retinal as substrates and effects of cellular retinoid-binding proteins and retinoic acid on activity. Biochemistry (Mosc) 33:1938–1943
63. Esteban J, Elabbas LE, Borg D, Herlin M, Akesson A, Barber X et al (2014) Gestational and lactational exposure to the polychlorinated biphenyl mixture Aroclor 1254 modulates retinoid homeostasis in rat offspring. Toxicol Lett 229:41–51
64. Everts HB, Sundberg JP, Ong DE (2005) Immunolocalization of retinoic acid biosynthesis systems in selected sites in rat. Exp Cell Res 308:309–319
65. Farias EF, Ong DE, Ghyselinck NB, Nakajo S, Kuppumbatti YS, Mira Y, Lopez R (2005) Cellular retinol-binding protein I, a regulator of breast epithelial retinoic acid receptor activity, cell differentiation, and tumorigenicity. J Natl Cancer Inst 97:21–29
66. Farjo KM, Moiseyev G, Takahashi Y, Crouch RK, Ma J (2009) The 11-cis-retinol dehydrogenase activity of RDH10 and its interaction with visual cycle proteins. Invest Ophthalmol Vis Sci 50:5089–5097
67. Fex G, Johannesson G (1984) Radioimmunological determination of cellular retinol-binding protein in human tissue extracts. Cancer Res 44:3029–3032
68. Fiorella PD, Napoli JL (1991) Expression of cellular retinoic acid binding protein (CRABP) in Escherichia coli. Characterization and evidence that holo-CRABP is a substrate in retinoic acid metabolism. J Biol Chem 266:16572–16579
69. Fiorella PD, Napoli JL (1994) Microsomal retinoic acid metabolism. Effects of cellular retinoic acid-binding protein (type I) and C18-hydroxylation as an initial step. J Biol Chem 269:10538–10544
70. Fiorella PD, Giguère V, Napoli JL (1993) Expression of cellular retinoic acid-binding protein (type II) in Escherichia coli. Characterization and comparison to cellular retinoic acid-binding protein (type I). J Biol Chem 268:21545–21552
71. Folli C, Calderone V, Ottonello S, Bolchi A, Zanotti G, Stoppini M et al (2001) Identification, retinoid binding, and x-ray analysis of a human retinol-binding protein. Proc Natl Acad Sci U S A 98:3710–3715
72. Franzoni L, Lücke C, Pérez C, Cavazzini D, Rademacher M, Ludwig C et al (2002) Structure and backbone dynamics of Apo- and holo-cellular retinol-binding protein in solution. J Biol Chem 277:21983–21997
73. Franzoni L, Cavazzini D, Rossi GL, Lücke C (2010) New insights on the protein-ligand interaction differences between the two primary cellular retinol carriers. J Lipid Res 51:1332–1343
74. Futterman S, Andrews JS (1964) The composition of liver Vitamin A Ester and the synthesis of Vitamin A Ester by liver microsomes. J Biol Chem 239:4077–4080
75. Ganfornina MD, Gutiérrez G, Bastiani M, Sánchez D (2000) A phylogenetic analysis of the lipocalin protein family. Mol Biol Evol 17:114–126
76. Gaub MP, Lutz Y, Ghyselinck NB, Scheuer I, Pfister V, Chambon P et al (1998) Nuclear detection of cellular retinoic acid binding proteins I and II with new antibodies. J Histochem Cytochem Off J Histochem Soc 46:1103–1111
77. Ghyselinck NB, Båvik C, Sapin V, Mark M, Bonnier D, Hindelang C et al (1999) Cellular retinol-binding protein I is essential for vitamin A homeostasis. EMBO J 18:4903–4914
78. Giguère V, Lyn S, Yip P, Siu CH, Amin S (1990) Molecular cloning of cDNA encoding a second cellular retinoic acid-binding protein. Proc Natl Acad Sci U S A 87:6233–6237
79. Gorry P, Lufkin T, Dierich A, Rochette-Egly C, Décimo D, Dollé P et al (1994) The cellular retinoic acid binding protein I is dispensable. Proc Natl Acad Sci U S A 91:9032–9036

80. Gough WH, VanOoteghem S, Sint T, Kedishvili NY (1998) cDNA cloning and characterization of a new human microsomal NAD+–dependent dehydrogenase that oxidizes all-trans-retinol and 3alpha-hydroxysteroids. J Biol Chem 273:19778–19785
81. Graham CE, Brocklehurst K, Pickersgill RW, Warren MJ (2006) Characterization of retinaldehyde dehydrogenase 3. Biochem J 394:67–75
82. Grün F, Hirose Y, Kawauchi S, Ogura T, Umesono K (2000) Aldehyde dehydrogenase 6, a cytosolic retinaldehyde dehydrogenase prominently expressed in sensory neuroepithelia during development. J Biol Chem 275:41210–41218
83. Gustafson AL, Donovan M, Annerwall E, Dencker L, Eriksson U (1996) Nuclear import of cellular retinoic acid-binding protein type I in mouse embryonic cells. Mech Dev 58:27–38
84. Haeseleer F, Huang J, Lebioda L, Saari JC, Palczewski K (1998) Molecular characterization of a novel short-chain dehydrogenase/reductase that reduces all-trans-retinal. J Biol Chem 273:21790–21799
85. Harrison EH (1988) Bile salt-dependent, neutral cholesteryl ester hydrolase of rat liver: possible relationship with pancreatic cholesteryl ester hydrolase. Biochim Biophys Acta 963:28–34
86. Harrison EH (1998) Lipases and carboxylesterases: possible roles in the hepatic metabolism of retinol. Annu Rev Nutr 18:259–276
87. Harrison EH (2000) Lipases and carboxylesterases: possible roles in the hepatic utilization of vitamin A. J Nutr 130:340S–344S
88. Harrison EH (2012) Mechanisms involved in the intestinal absorption of dietary vitamin A and provitamin A carotenoids. Biochim Biophys Acta 1821:70–77
89. Harrison EH, Gad MZ (1989) Hydrolysis of retinyl palmitate by enzymes of rat pancreas and liver. Differentiation of bile salt-dependent and bile salt-independent, neutral retinyl ester hydrolases in rat liver. J Biol Chem 264:17142–17147
90. Harrison EH, Napoli JL (1990) Bile salt-independent retinyl ester hydrolase activities associated with membranes of rat tissues. Methods Enzymol 189:459–469
91. Harrison EH, Blaner WS, Goodman DS, Ross AC (1987) Subcellular localization of retinoids, retinoid-binding proteins, and acyl-CoA:retinol acyltransferase in rat liver. J Lipid Res 28:973–981
92. Heinemann FS, Ozols J (1998) Isolation and structural analysis of microsomal membrane proteins. Front Biosci J Virtual Libr 3:d483–d493
93. Helgerud P, Petersen LB, Norum KR (1982) Acyl CoA:retinol acyltransferase in rat small intestine: its activity and some properties of the enzymic reaction. J Lipid Res 23:609–618
94. Heller J (1975) Interactions of plasma retinol-binding protein with its receptor. Specific binding of bovine and human retinol-binding protein to pigment epithelium cells from bovine eyes. J Biol Chem 250:3613–3619
95. Heller M, Bok D (1976) A specific receptor for retinol binding protein as detected by the binding of human and bovine retinol binding protein to pigment epithelial cells. Am J Ophthalmol 81:93–97
96. Herr FM, Ong DE (1992) Differential interaction of lecithin-retinol acyltransferase with cellular retinol binding proteins. Biochemistry (Mosc) 31:6748–6755
97. Hoegberg P, Schmidt CK, Fletcher N, Nilsson CB, Trossvik C, Gerlienke Schuur A et al (2005) Retinoid status and responsiveness to 2,3,7,8-tetrachlorodibenzo-p-dioxin (TCDD) in mice lacking retinoid binding protein or retinoid receptor forms. Chem Biol Interact 156:25–39
98. Imaoka S, Wan J, Chow T, Hiroi T, Eyanagi R, Shigematsu H et al (1998) Cloning and characterization of the CYP2D1-binding protein, retinol dehydrogenase. Arch Biochem Biophys 353:331–336
99. Isken A, Golczak M, Oberhauser V, Hunzelmann S, Driever W, Imanishi Y et al (2008) RBP4 disrupts vitamin A uptake homeostasis in a STRA6-deficient animal model for Matthew-Wood syndrome. Cell Metab 7:258–268

100. Jamison RS, Newcomer ME, Ong DE (1994) Cellular retinoid-binding proteins: limited proteolysis reveals a conformational change upon ligand binding. Biochemistry (Mosc) 33:2873–2879
101. Jette C, Peterson PW, Sandoval IT, Manos EJ, Hadley E, Ireland CM et al (2004) The tumor suppressor adenomatous polyposis coli and caudal related homeodomain protein regulate expression of retinol dehydrogenase L. J Biol Chem 279:34397–34405
102. Jiang W, Napoli JL (2012) Reorganization of cellular retinol-binding protein type 1 and lecithin:retinol acyltransferase during retinyl ester biosynthesis. Biochim Biophys Acta 1820:859–869
103. Jiang W, Napoli JL (2013) The retinol dehydrogenase Rdh10 localizes to lipid droplets during acyl ester biosynthesis. J Biol Chem 288:589–597
104. Jing Y, Waxman S, Mira-y-Lopez R (1997) The cellular retinoic acid binding protein II is a positive regulator of retinoic acid signaling in breast cancer cells. Cancer Res 57:1668–1672
105. Jurukovski V, Markova NG, Karaman-Jurukovska N, Randolph RK, Su J, Napoli JL et al (1999) Cloning and characterization of retinol dehydrogenase transcripts expressed in human epidermal keratinocytes. Mol Genet Metab 67:62–73
106. Kakkad BP, Ong DE (1988) Reduction of retinaldehyde bound to cellular retinol-binding protein (type II) by microsomes from rat small intestine. J Biol Chem 263:12916–12919
107. Kane MA, Chen N, Sparks S, Napoli JL (2005) Quantification of endogenous retinoic acid in limited biological samples by LC/MS/MS. Biochem J 388:363–369
108. Kane MA, Folias AE, Napoli JL (2008) HPLC/UV quantitation of retinal, retinol, and retinyl esters in serum and tissues. Anal Biochem 378:71–79
109. Kane MA, Folias AE, Wang C, Napoli JL (2008) Quantitative profiling of endogenous retinoic acid in vivo and in vitro by tandem mass spectrometry. Anal Chem 80:1702–1708
110. Kane MA, Folias AE, Pingitore A, Perri M, Obrochta KM, Krois CR et al (2010) Identification of 9-cis-retinoic acid as a pancreas-specific autacoid that attenuates glucose-stimulated insulin secretion. Proc Natl Acad Sci U S A 107:21884–21889
111. Kane MA, Folias AE, Wang C, Napoli JL (2010) Ethanol elevates physiological all-trans-retinoic acid levels in select loci through altering retinoid metabolism in multiple loci: a potential mechanism of ethanol toxicity. FASEB J Off Publ Fed Am Soc Exp Biol 24:823–832
112. Kane MA, Bright FV, Napoli JL (2011) Binding affinities of CRBPI and CRBPII for 9-cis-retinoids. Biochim Biophys Acta 1810:514–518
113. Kane MA, Folias AE, Pingitore A, Perri M, Krois CR, Ryu JY et al (2011) CrbpI modulates glucose homeostasis and pancreas 9-cis-retinoic acid concentrations. Mol Cell Biol 31:3277–3285
114. Kasus-Jacobi A, Ou J, Bashmakov YK, Shelton JM, Richardson JA, Goldstein JL et al (2003) Characterization of mouse short-chain aldehyde reductase (SCALD), an enzyme regulated by sterol regulatory element-binding proteins. J Biol Chem 278:32380–32389
115. Kato M, Blaner WS, Mertz JR, Das K, Kato K, Goodman DS (1985) Influence of retinoid nutritional status on cellular retinol- and cellular retinoic acid-binding protein concentrations in various rat tissues. J Biol Chem 260:4832–4838
116. Kawaguchi R, Yu J, Honda J, Hu J, Whitelegge J, Ping P et al (2007) A membrane receptor for retinol binding protein mediates cellular uptake of vitamin A. Science 315:820–825
117. Kawaguchi R, Zhong M, Kassai M, Ter-Stepanian M, Sun H (2012) STRA6-catalyzed vitamin A influx, efflux, and exchange. J Membr Biol 245:731–745
118. Kawaguchi R, Zhong M, Sun H (2013) Real-time analyses of retinol transport by the membrane receptor of plasma retinol binding protein. J Vis Exp JoVE 17:e50169
119. Kedishvili NY (2013) Enzymology of retinoic acid biosynthesis and degradation. J Lipid Res 54:1744–1760

120. Kedishvili NY, Chumakova OV, Chetyrkin SV, Belyaeva OV, Lapshina EA, Lin DW et al (2002) Evidence that the human gene for prostate short-chain dehydrogenase/reductase (PSDR1) encodes a novel retinal reductase (RalR1). J Biol Chem 277:28909–28915
121. Kim Y-K, Wassef L, Hamberger L, Piantedosi R, Palczewski K, Blaner WS et al (2008) Retinyl ester formation by lecithin: retinol acyltransferase is a key regulator of retinoid homeostasis in mouse embryogenesis. J Biol Chem 283:5611–5621
122. Kohlstedt K, Gershome C, Trouvain C, Hofmann W-K, Fichtlscherer S, Fleming I (2009) Angiotensin-converting enzyme (ACE) inhibitors modulate cellular retinol-binding protein 1 and adiponectin expression in adipocytes via the ACE-dependent signaling cascade. Mol Pharmacol 75:685–692
123. Kuppumbatti YS, Bleiweiss IJ, Mandeli JP, Waxman S, Mira-Y-Lopez R (2000) Cellular retinol-binding protein expression and breast cancer. J Natl Cancer Inst 92:475–480
124. Kuppumbatti YS, Rexer B, Nakajo S, Nakaya K, Mira-y-Lopez R (2001) CRBP suppresses breast cancer cell survival and anchorage-independent growth. Oncogene 20:7413–7419
125. Labrecque J, Bhat PV, Lacroix A (1993) Purification and partial characterization of a rat kidney aldehyde dehydrogenase that oxidizes retinal to retinoic acid. Biochem Cell Biol Biochim Biol Cell 71:85–89
126. Lagace TA, Ridgway ND (2013) The role of phospholipids in the biological activity and structure of the endoplasmic reticulum. Biochim Biophys Acta 1833:2499–2510
127. Lee M, Lucia SP (1965) Effect of ethanol on the mobilization of vitamin A in the dog and in the rat. Q J Stud Alcohol 26:1–9
128. Lee MO, Manthey CL, Sladek NE (1991) Identification of mouse liver aldehyde dehydrogenases that catalyze the oxidation of retinaldehyde to retinoic acid. Biochem Pharmacol 42:1279–1285
129. Lee S-A, Belyaeva OV, Kedishvili NY (2009) Biochemical characterization of human epidermal retinol dehydrogenase 2. Chem Biol Interact 178:182–187
130. Lei Z, Chen W, Zhang M, Napoli JL (2003) Reduction of all-trans-retinal in the mouse liver peroxisome fraction by the short-chain dehydrogenase/reductase RRD: induction by the PPAR alpha ligand clofibrate. Biochemistry (Mosc) 42:4190–4196
131. Leo MA, Lieber CS (1984) Normal testicular structure and reproductive function in deermice lacking retinol and alcohol dehydrogenase activity. J Clin Invest 73:593–596
132. Levi L, Lobo G, Doud MK, von Lintig J, Seachrist D, Tochtrop GP et al (2013) Genetic ablation of the fatty acid-binding protein FABP5 suppresses HER2-induced mammary tumorigenesis. Cancer Res 73:4770–4780
133. Levin MS, Locke B, Yang NC, Li E, Gordon JI (1988) Comparison of the ligand binding properties of two homologous rat apocellular retinol-binding proteins expressed in Escherichia coli. J Biol Chem 263:17715–17723
134. Li E, Norris AW (1996) Structure/function of cytoplasmic vitamin A-binding proteins. Annu Rev Nutr 16:205–234
135. Li E, Qian SJ, Winter NS, d' Avignon A, Levin MS (1991) Fluorine nuclear magnetic resonance analysis of the ligand binding properties of two homologous rat cellular retinol-binding proteins expressed in Escherichia coli. J Biol Chem 266:3622–3629
136. Lieber CS (1997) Cytochrome P-4502E1: its physiological and pathological role. Physiol Rev 77:517–544
137. Lin M, Napoli JL (2000) cDNA cloning and expression of a human aldehyde dehydrogenase (ALDH) active with 9-cis-retinal and identification of a rat ortholog, ALDH12. J Biol Chem 275:40106–40112
138. Lin B, White JT, Ferguson C, Wang S, Vessella R, Bumgarner R et al (2001) Prostate short-chain dehydrogenase reductase 1 (PSDR1): a new member of the short-chain steroid dehydrogenase/reductase family highly expressed in normal and neoplastic prostate epithelium. Cancer Res 61:1611–1618

139. Lin M, Zhang M, Abraham M, Smith SM, Napoli JL (2003) Mouse retinal dehydrogenase 4 (RALDH4), molecular cloning, cellular expression, and activity in 9-cis-retinoic acid biosynthesis in intact cells. J Biol Chem 278:9856–9861
140. Linke T, Dawson H, Harrison EH (2005) Isolation and characterization of a microsomal acid retinyl ester hydrolase. J Biol Chem 280:23287–23294
141. Liu L, Gudas LJ (2005) Disruption of the lecithin:retinol acyltransferase gene makes mice more susceptible to vitamin A deficiency. J Biol Chem 280:40226–40234
142. Liu C, Russell RM, Seitz HK, Wang XD (2001) Ethanol enhances retinoic acid metabolism into polar metabolites in rat liver via induction of cytochrome P4502E1. Gastroenterology 120:179–189
143. Liu L, Tang X-H, Gudas LJ (2008) Homeostasis of retinol in lecithin: retinol acyltransferase gene knockout mice fed a high retinol diet. Biochem Pharmacol 75:2316–2324
144. Lu J, Cistola DP, Li E (2003) Two homologous rat cellular retinol-binding proteins differ in local conformational flexibility. J Mol Biol 330:799–812
145. MacDonald PN, Ong DE (1987) Binding specificities of cellular retinol-binding protein and cellular retinol-binding protein, type II. J Biol Chem 262:10550–10556
146. MacDonald PN, Ong DE (1988) A lecithin:retinol acyltransferase activity in human and rat liver. Biochem Biophys Res Commun 156:157–163
147. MacDonald PN, Ong DE (1988) Evidence for a lecithin-retinol acyltransferase activity in the rat small intestine. J Biol Chem 263:12478–12482
148. Mahadevan S, Murthy SK, Krishnamurthy S, Ganguly J (1961) Studies on vitamin A esterase. 4. The hydrolysis and synthesis of vitamin A esters by rat intestinal mucosae. Biochem J 79:416–424
149. Malpeli G, Stoppini M, Zapponi MC, Folli C, Berni R (1995) Interactions with retinol and retinoids of bovine cellular retinol-binding protein. Eur J Biochem FEBS 229:486–493
150. Markova NG, Pinkas-Sarafova A, Karaman-Jurukovska N, Jurukovski V, Simon M (2003) Expression pattern and biochemical characteristics of a major epidermal retinol dehydrogenase. Mol Genet Metab 78:119–135
151. Markova NG, Pinkas-Sarafova A, Simon M (2006) A metabolic enzyme of the short-chain dehydrogenase/reductase superfamily may moonlight in the nucleus as a repressor of promoter activity. J Invest Dermatol 126:2019–2031
152. Marwarha G, Berry DC, Croniger CM, Noy N (2014) The retinol esterifying enzyme LRAT supports cell signaling by retinol-binding protein and its receptor STRA6. FASEB J Off Publ Fed Am Soc Exp Biol 28:26–34
153. Matsuzaka Y, Okamoto K, Tsuji H, Mabuchi T, Ozawa A, Tamiya G et al (2002) Identification of the hRDH-E2 gene, a novel member of the SDR family, and its increased expression in psoriatic lesion. Biochem Biophys Res Commun 297:1171–1180
154. McCaffery P, Lee MO, Wagner MA, Sladek NE, Dräger UC (1992) Asymmetrical retinoic acid synthesis in the dorsoventral axis of the retina. Dev Camb Engl 115:371–382
155. McCaffery P, Koul O, Smith D, Napoli JL, Chen N, Ullman MD (2004) Ethanol increases retinoic acid production in cerebellar astrocytes and in cerebellum. Brain Res Dev Brain Res 153:233–241
156. Miller WH (1998) The emerging role of retinoids and retinoic acid metabolism blocking agents in the treatment of cancer. Cancer 83:1471–1482
157. Mira-Y-Lopez R, Zheng WL, Kuppumbatti YS, Rexer B, Jing Y, Ong DE (2000) Retinol conversion to retinoic acid is impaired in breast cancer cell lines relative to normal cells. J Cell Physiol 185:302–309
158. Mittag T, Franzoni L, Cavazzini D, Schaffhausen B, Rossi GL, Günther UL (2006) Retinol modulates site-specific mobility of apo-cellular retinol-binding protein to promote ligand binding. J Am Chem Soc 128:9844–9848
159. Moise AR, Golczak M, Imanishi Y, Palczewski K (2007) Topology and membrane association of lecithin: retinol acyltransferase. J Biol Chem 282:2081–2090

160. Molotkov A, Fan X, Duester G (2002) Excessive vitamin A toxicity in mice genetically deficient in either alcohol dehydrogenase Adh1 or Adh3. Eur J Biochem FEBS 269:2607–2612
161. Nadauld LD, Shelton DN, Chidester S, Yost HJ, Jones DA (2005) The zebrafish retinol dehydrogenase, rdh1l, is essential for intestinal development and is regulated by the tumor suppressor adenomatous polyposis coli. J Biol Chem 280:30490–30495
162. Napoli JL (1986) Retinol metabolism in LLC-PK1 Cells. Characterization of retinoic acid synthesis by an established mammalian cell line. J Biol Chem 261:13592–13597
163. Napoli, JL (1990) The biogenesis of retinoic acid: a physiologically significant promoter of differentiation. Chem Biol Synth Retin. Boca Raton: CRC Press, p. 229–49
164. Napoli JL (1999) Interactions of retinoid binding proteins and enzymes in retinoid metabolism. Biochim Biophys Acta 1440:139–162
165. Napoli JL (2011) Effects of ethanol on physiological retinoic acid levels. IUBMB Life 63:701–706
166. Napoli JL (2012) Physiological insights into all-trans-retinoic acid biosynthesis. Biochim Biophys Acta 1821:152–167
167. Napoli JL, McCormick AM (1981) Tissue dependence of retinoic acid metabolism in vivo. Biochim Biophys Acta 666:165–175
168. Napoli JL, Race KR (1987) The biosynthesis of retinoic acid from retinol by rat tissues in vitro. Arch Biochem Biophys 255:95–101
169. Napoli JL, Pacia EB, Salerno GJ (1989) Cholate effects on all-trans-retinyl palmitate hydrolysis in tissue homogenates: solubilization of multiple kidney membrane hydrolases. Arch Biochem Biophys 274:192–199
170. Napoli JL, Posch KP, Fiorella PD, Boerman MH (1991) Physiological occurrence, biosynthesis and metabolism of retinoic acid: evidence for roles of cellular retinol-binding protein (CRBP) and cellular retinoic acid-binding protein (CRABP) in the pathway of retinoic acid homeostasis. Biomed Pharmacother Bioméd Pharmacothérapie 45:131–143
171. Napoli JL, Posch KC, Burns RD (1992) Microsomal retinal synthesis: retinol vs. holo-CRBP as substrate and evaluation of NADP, NAD and NADPH as cofactors. Biochim Biophys Acta 1120:183–186
172. Napoli JL, Boerman MH, Chai X, Zhai Y, Fiorella PD (1995) Enzymes and binding proteins affecting retinoic acid concentrations. J Steroid Biochem Mol Biol 53:497–502
173. Nau H (2001) Teratogenicity of isotretinoin revisited: species variation and the role of all-trans-retinoic acid. J Am Acad Dermatol 45:S183–S187
174. Newcomer ME (1995) Retinoid-binding proteins: structural determinants important for function. FASEB J Off Publ Fed Am Soc Exp Biol 9:229–239
175. Newcomer ME, Jamison RS, Ong DE (1998) Structure and function of retinoid-binding proteins. Subcell Biochem 30:53–80
176. Nishiwaki S, Kato M, Okuno M, Kanai M, Muto Y (1990) Purification and partial characterization of a novel cellular retinol-binding protein, type three, from the piscine eyes. Biochim Biophys Acta 1037:192–199
177. Norris AW, Cheng L, Giguère V, Rosenberger M, Li E (1994) Measurement of subnanomolar retinoic acid binding affinities for cellular retinoic acid binding proteins by fluorometric titration. Biochim Biophys Acta 1209:10–18
178. Noy N (2000) Retinoid-binding proteins: mediators of retinoid action. Biochem J 348(Pt 3):481–495
179. Noy N, Blaner WS (1991) Interactions of retinol with binding proteins: studies with rat cellular retinol-binding protein and with rat retinol-binding protein. Biochemistry (Mosc) 30:6380–6386
180. O'Byrne SM, Wongsiriroj N, Libien J, Vogel S, Goldberg IJ, Baehr W et al (2005) Retinoid absorption and storage is impaired in mice lacking lecithin:retinol acyltransferase (LRAT). J Biol Chem 280:35647–35657
181. Obrochta KM, Kane MA, Napoli JL (2014) Effects of diet and strain on mouse serum and tissue retinoid concentrations. PLoS One 9:e99435

182. Ong DE (1982) Purification and partial characterization of cellular retinol-binding protein from human liver. Cancer Res 42:1033–1037
183. Ong DE (1984) A novel retinol-binding protein from rat. Purification and partial characterization. J Biol Chem 259:1476–1482
184. Ong DE (1994) Cellular transport and metabolism of vitamin A: roles of the cellular retinoid-binding proteins. Nutr Rev 52:S24–S31
185. Ong DE, Chytil F (1975) Retinoic acid-binding protein in rat tissue. Partial purification and comparison to rat tissue retinol-binding protein. J Biol Chem 250:6113–6117
186. Ong DE, Chytil F (1978) Cellular retinoic acid-binding protein from rat testis. Purification and characterization. J Biol Chem 253:4551–4554
187. Ong DE, Chytil F (1978) Cellular retinol-binding protein from rat liver. Purification and characterization. J Biol Chem 253:828–832
188. Ong DE, Chytil F (1979) Immunochemical comparison of vitamin A binding proteins of rat. J Biol Chem 254:8733–8735
189. Ong DE, Page DL (1986) Quantitation of cellular retinol-binding protein in human organs. Am J Clin Nutr 44:425–430
190. Ong DE, Crow JA, Chytil F (1982) Radioimmunochemical determination of cellular retinol- and cellular retinoic acid-binding proteins in cytosols of rat tissues. J Biol Chem 257:13385–13389
191. Ong DE, Kakkad B, MacDonald PN (1987) Acyl-CoA-independent esterification of retinol bound to cellular retinol-binding protein (type II) by microsomes from rat small intestine. J Biol Chem 262:2729–2736
192. Ong DE, MacDonald PN, Gubitosi AM (1988) Esterification of retinol in rat liver. Possible participation by cellular retinol-binding protein and cellular retinol-binding protein II. J Biol Chem 263:5789–5796
193. Oppermann U, Filling C, Hult M, Shafqat N, Wu X, Lindh M et al (2003) Short-chain dehydrogenases/reductases (SDR): the 2002 update. Chem Biol Interact 143–144:247–253
194. Orland MD, Anwar K, Cromley D, Chu C-H, Chen L, Billheimer JT et al (2005) Acyl coenzyme A dependent retinol esterification by acyl coenzyme A: diacylglycerol acyltransferase 1. Biochim Biophys Acta 1737:76–82
195. Ottonello S, Petrucco S, Maraini G (1987) Vitamin A uptake from retinol-binding protein in a cell-free system from pigment epithelial cells of bovine retina. Retinol transfer from plasma retinol-binding protein to cytoplasmic retinol-binding protein with retinyl-ester formation as the intermediate step. J Biol Chem 262:3975–3981
196. Ottonello S, Scita G, Mantovani G, Cavazzini D, Rossi GL (1993) Retinol bound to cellular retinol-binding protein is a substrate for cytosolic retinoic acid synthesis. J Biol Chem 268:27133–27142
197. Parker RO, Crouch RK (2010) Retinol dehydrogenases (RDHs) in the visual cycle. Exp Eye Res 91:788–792
198. Penzes P, Napoli JL (1999) Holo-cellular retinol-binding protein: distinction of ligand-binding affinity from efficiency as substrate in retinal biosynthesis. Biochemistry (Mosc) 38:2088–2093
199. Penzes P, Wang X, Napoli JL (1997) Enzymatic characteristics of retinal dehydrogenase type I expressed in Escherichia coli. Biochim Biophys Acta 1342:175–181
200. Penzes P, Wang X, Sperkova Z, Napoli JL (1997) Cloning of a rat cDNA encoding retinal dehydrogenase isozyme type I and its expression in E. coli. Gene 191:167–172
201. Persaud SD, Lin Y-W, Wu C-Y, Kagechika H, Wei L-N (2013) Cellular retinoic acid binding protein I mediates rapid non-canonical activation of ERK1/2 by all-trans retinoic acid. Cell Signal 25:19–25
202. Persson B, Kallberg Y (2013) Classification and nomenclature of the superfamily of short-chain dehydrogenases/reductases (SDRs). Chem Biol Interact 202:111–115

203. Persson B, Kallberg Y, Bray JE, Bruford E, Dellaporta SL, Favia AD et al (2009) The SDR (short-chain dehydrogenase/reductase and related enzymes) nomenclature initiative. Chem Biol Interact 178:94–98
204. Pham CT, MacIvor DM, Hug BA, Heusel JW, Ley TJ (1996) Long-range disruption of gene expression by a selectable marker cassette. Proc Natl Acad Sci U S A 93:13090–13095
205. Piantedosi R, Ghyselinck N, Blaner WS, Vogel S (2005) Cellular retinol-binding protein type III is needed for retinoid incorporation into milk. J Biol Chem 280:24286–24292
206. Picozzi P, Marozzi A, Fornasari D, Benfante R, Barisani D, Meneveri R et al (2003) Genomic organization and transcription of the human retinol dehydrogenase 10 (RDH10) gene. FEBS Lett 554:59–66
207. Pierzchalski K, Yu J, Norman V, Kane MA (2013) CrbpI regulates mammary retinoic acid homeostasis and the mammary microenvironment. FASEB J Off Publ Fed Am Soc Exp Biol 27:1904–1916
208. Porter SB, Fraker LD, Chytil F, Ong DE (1983) Localization of cellular retinol-binding protein in several rat tissues. Proc Natl Acad Sci U S A 80:6586–6590
209. Posch KC, Napoli JL (1992) Multiple retinoid dehydrogenases in testes cytosol from alcohol dehydrogenase negative or positive deermice. Biochem Pharmacol 43:2296–2298
210. Posch KC, Enright WJ, Napoli JL (1989) Retinoic acid synthesis by cytosol from the alcohol dehydrogenase negative deermouse. Arch Biochem Biophys 274:171–178
211. Posch KC, Boerman MH, Burns RD, Napoli JL (1991) Holocellular retinol binding protein as a substrate for microsomal retinal synthesis. Biochemistry 30:6224–6230
212. Posch KC, Burns RD, Napoli JL (1992) Biosynthesis of all-trans-retinoic acid from retinal. Recognition of retinal bound to cellular retinol binding protein (type I) as substrate by a purified cytosolic dehydrogenase. J Biol Chem 267:19676–19682
213. Quadro L, Hamberger L, Gottesman ME, Wang F, Colantuoni V, Blaner WS et al (2005) Pathways of vitamin A delivery to the embryo: insights from a new tunable model of embryonic vitamin A deficiency. Endocrinology 146:4479–4490
214. Quick TC, Ong DE (1990) Vitamin A metabolism in the human intestinal Caco-2 cell line. Biochemistry (Mosc) 29:11116–11123
215. Randolph RK, Simon M (1997) Metabolism of all-trans-retinoic acid by cultured human epidermal keratinocytes. J Lipid Res 38:1374–1383
216. Randolph RK, Winkler KE, Ross AC (1991) Fatty acyl CoA-dependent and -independent retinol esterification by rat liver and lactating mammary gland microsomes. Arch Biochem Biophys 288:500–508
217. Rask L, Peterson PA (1976) In vitro uptake of vitamin A from the retinol-binding plasma protein to mucosal epithelial cells from the monkey's small intestine. J Biol Chem 251:6360–6366
218. Ratnam KK, Feng X, Chuang PY, Verma V, Lu T-C, Wang J et al (2011) Role of the retinoic acid receptor-α in HIV-associated nephropathy. Kidney Int 79:624–634
219. Reeves PG (1997) Components of the AIN-93 diets as improvements in the AIN-76A diet. J Nutr 127:838S–841S
220. Reeves PG, Nielsen FH, Fahey GC Jr (1993) AIN-93 purified diets for laboratory rodents: final report of the American Institute of Nutrition ad hoc writing committee on the reformulation of the AIN-76A rodent diet. J Nutr 123:1939–1951
221. Rexer BN, Ong DE (2002) A novel short-chain alcohol dehydrogenase from rats with retinol dehydrogenase activity, cyclically expressed in uterine epithelium. Biol Reprod 67:1555–1564
222. Rexer BN, Zheng WL, Ong DE (2001) Retinoic acid biosynthesis by normal human breast epithelium is via aldehyde dehydrogenase 6, absent in MCF-7 cells. Cancer Res 61:7065–7070
223. Reynier M (1969) Pyrazole inhibition and kinetic studies of ethanol and retinol oxidation catalyzed by rat liver alcohol dehydrogenase. Acta Chem Scand 23:1119–1129

224. Rhinn M, Schuhbaur B, Niederreither K, Dollé P (2011) Involvement of retinol dehydrogenase 10 in embryonic patterning and rescue of its loss of function by maternal retinaldehyde treatment. Proc Natl Acad Sci U S A 108:16687–16692
225. Rieck M, Meissner W, Ries S, Müller-Brüsselbach S, Müller R (2008) Ligand-mediated regulation of peroxisome proliferator-activated receptor (PPAR) beta/delta: a comparative analysis of PPAR-selective agonists and all-trans retinoic acid. Mol Pharmacol 74:1269–1277
226. Rigtrup KM, Ong DE (1992) A retinyl ester hydrolase activity intrinsic to the brush border membrane of rat small intestine. Biochemistry (Mosc) 31:2920–2926
227. Rigtrup KM, Kakkad B, Ong DE (1994) Purification and partial characterization of a retinyl ester hydrolase from the brush border of rat small intestine mucosa: probable identity with brush border phospholipase B. Biochemistry (Mosc) 33:2661–2666
228. Rigtrup KM, McEwen LR, Said HM, Ong DE (1994) Retinyl ester hydrolytic activity associated with human intestinal brush border membranes. Am J Clin Nutr 60:111–116
229. Ross AC (1982) Retinol esterification by mammary gland microsomes from the lactating rat. J Lipid Res 23:133–144
230. Ross AC (1982) Retinol esterification by rat liver microsomes. Evidence for a fatty acyl coenzyme A: retinol acyltransferase. J Biol Chem 257:2453–2459
231. Ross AC (1993) Cellular metabolism and activation of retinoids: roles of cellular retinoid-binding proteins. FASEB J Off Publ Fed Am Soc Exp Biol 7:317–327
232. Ross AC, Goodman DS (1979) Intracellular binding proteins for retinol and retinoic acid: comparison with each other and with serum retinol-binding protein. Fed Proc 38:2515–2518
233. Ross AC, Takahashi YI, Goodman DS (1978) The binding protein for retinol from rat testis cytosol. Isolation and partial characterization. J Biol Chem 253:6591–6598
234. Ross AC, Adachi N, Goodman DS (1980) The binding protein for retinoic acid from rat testis cytosol: isolation and partial characterization. J Lipid Res 21:100–109
235. Ruff SJ, Ong DE (2000) Cellular retinoic acid binding protein is associated with mitochondria. FEBS Lett 487:282–286
236. Rühl R, Plum C, Elmazar MM, Nau H (2001) Embryonic subcellular distribution of 13-cis- and all-trans-retinoic acid indicates differential cytosolic/nuclear localization. Toxicol Sci Off J Soc Toxicol 63:82–89
237. Ruiz A, Winston A, Lim YH, Gilbert BA, Rando RR, Bok D (1999) Molecular and biochemical characterization of lecithin retinol acyltransferase. J Biol Chem 274:3834–3841
238. Ruiz A, Ghyselinck NB, Mata N, Nusinowitz S, Lloyd M, Dennefeld C et al (2007) Somatic ablation of the Lrat gene in the mouse retinal pigment epithelium drastically reduces its retinoid storage. Invest Ophthalmol Vis Sci 48:5377–5387
239. Ruiz A, Mark M, Jacobs H, Klopfenstein M, Hu J, Lloyd M et al (2012) Retinoid content, visual responses, and ocular morphology are compromised in the retinas of mice lacking the retinol-binding protein receptor, STRA6. Invest Ophthalmol Vis Sci 53:3027–3039
240. Saari JC, Crabb JW (2005) Focus on molecules: cellular retinaldehyde-binding protein (CRALBP). Exp Eye Res 81:245–246
241. Saari JC, Futterman S (1976) Separable binding proteins for retinoic acid and retinol in bovine retina. Biochim Biophys Acta 444:789–793
242. Saari JC, Futterman S, Bredberg L (1978) Cellular retinol- and retinoic acid-binding proteins of bovine retina. Purification and properties. J Biol Chem 253:6432–6436
243. Sani BP (1977) Localization of retinoic acid-binding protein in nuclei. Biochem Biophys Res Commun 75:7–12
244. Sani BP, Donovan MK (1979) Localization of retinoic acid-binding protein in nuclei and the nuclear uptake of retinoic acid. Cancer Res 39:2492–2496
245. Sani BP, Hill DL (1974) Retinoic acid: a binding protein in chick embryo metatarsal skin. Biochem Biophys Res Commun 61:1276–1282

246. Sato M, Lieber CS (1981) Hepatic vitamin A depletion after chronic ethanol consumption in baboons and rats. J Nutr 111:2015–2023
247. Schmidt CK, Brouwer A, Nau H (2003) Chromatographic analysis of endogenous retinoids in tissues and serum. Anal Biochem 315:36–48
248. Schug TT, Berry DC, Shaw NS, Travis SN, Noy N (2007) Opposing effects of retinoic acid on cell growth result from alternate activation of two different nuclear receptors. Cell 129:723–733
249. Schug TT, Berry DC, Toshkov IA, Cheng L, Nikitin AY, Noy N (2008) Overcoming retinoic acid-resistance of mammary carcinomas by diverting retinoic acid from PPARbeta/delta to RAR. Proc Natl Acad Sci U S A 105:7546–7551
250. Shaw N, Elholm M, Noy N (2003) Retinoic acid is a high affinity selective ligand for the peroxisome proliferator-activated receptor beta/delta. J Biol Chem 278:41589–41592
251. Shih MYS, Kane MA, Zhou P, Yen CLE, Streeper RS, Napoli JL et al (2009) Retinol esterification by DGAT1 is essential for retinoid homeostasis in murine skin. J Biol Chem 284:4292–4299
252. Shingleton JL, Skinner MK, Ong DE (1989) Characteristics of retinol accumulation from serum retinol-binding protein by cultured Sertoli cells. Biochemistry (Mosc) 28:9641–9647
253. Shingleton JL, Skinner MK, Ong DE (1989) Retinol esterification in Sertoli cells by lecithin-retinol acyltransferase. Biochemistry (Mosc) 28:9647–9653
254. Siegenthaler G, Saurat JH, Ponec M (1990) Retinol and retinal metabolism. Relationship to the state of differentiation of cultured human keratinocytes. Biochem J 268:371–378
255. Sima A, Parisotto M, Mader S, Bhat PV (2009) Kinetic characterization of recombinant mouse retinal dehydrogenase types 3 and 4 for retinal substrates. Biochim Biophys Acta 1790:1660–1664
256. Song M-S, Chen W, Zhang M, Napoli JL (2003) Identification of a mouse short-chain dehydrogenase/reductase gene, retinol dehydrogenase-similar. Function of non-catalytic amino acid residues in enzyme activity. J Biol Chem 278:40079–40087
257. Ström K, Gundersen TE, Hansson O, Lucas S, Fernandez C, Blomhoff R et al (2009) Hormone-sensitive lipase (HSL) is also a retinyl ester hydrolase: evidence from mice lacking HSL. FASEB J Off Publ Fed Am Soc Exp Biol 23:2307–2316
258. Sun H (2012) Membrane receptors and transporters involved in the function and transport of vitamin A and its derivatives. Biochim Biophys Acta 1821:99–112
259. Sun G, Alexson SE, Harrison EH (1997) Purification and characterization of a neutral, bile salt-independent retinyl ester hydrolase from rat liver microsomes. Relationship To rat carboxylesterase ES-2. J Biol Chem 272:24488–24493
260. Szuts EZ, Harosi FI (1991) Solubility of retinoids in water. Arch Biochem Biophys 287:297–304
261. Takahashi Y, Moiseyev G, Farjo K, Ma J-X (2009) Characterization of key residues and membrane association domains in retinol dehydrogenase 10. Biochem J 419:113–122, 1 p following 122
262. Takase S, Ong DE, Chytil F (1986) Transfer of retinoic acid from its complex with cellular retinoic acid-binding protein to the nucleus. Arch Biochem Biophys 247:328–334
263. Thumser AE, Moore JB, Plant NJ (2014) Fatty acid binding proteins: tissue-specific functions in health and disease. Curr Opin Clin Nutr Metab Care 17:124–129
264. Tzimas G, Nau H (2001) The role of metabolism and toxicokinetics in retinoid teratogenesis. Curr Pharm Des 7:803–831
265. Van Thiel DH, Gavaler J, Lester R (1974) Ethanol inhibition of vitamin A metabolism in the testes: possible mechanism for sterility in alcoholics. Science 186:941–942
266. Venepally P, Reddy LG, Sani BP (1996) Analysis of the effects of CRABP I expression on the RA-induced transcription mediated by retinoid receptors. Biochemistry (Mosc) 35:9974–9982

267. Vogel S, Mendelsohn CL, Mertz JR, Piantedosi R, Waldburger C, Gottesman ME et al (2001) Characterization of a new member of the fatty acid-binding protein family that binds all-trans-retinol. J Biol Chem 276:1353–1360
268. Vonesch JL, Nakshatri H, Philippe M, Chambon P, Dollé P (1994) Stage and tissue-specific expression of the alcohol dehydrogenase 1 (Adh-1) gene during mouse development. Dev Dyn Off Publ Am Assoc Anat 199:199–213
269. Vreeland AC, Yu S, Levi L, de Barros RD, Noy N (2014) Transcript Stabilization by the RNA-Binding Protein HuR Is Regulated by Cellular Retinoic Acid-Binding Protein 2. Mol Cell Biol 34:2135–2146
270. Wang X-D (2005) Alcohol, vitamin A, and cancer. Alcohol Fayettev N 35:251–258
271. Wang X, Penzes P, Napoli JL (1996) Cloning of a cDNA encoding an aldehyde dehydrogenase and its expression in Escherichia coli. Recognition of retinal as substrate. J Biol Chem 271:16288–16293
272. Wang J, Bongianni JK, Napoli JL (2001) The N-terminus of retinol dehydrogenase type 1 signals cytosolic orientation in the microsomal membrane. Biochemistry (Mosc) 40:12533–12540
273. Wang C, Kane MA, Napoli JL (2011) Multiple retinol and retinal dehydrogenases catalyze all-trans-retinoic acid biosynthesis in astrocytes. J Biol Chem 286:6542–6553
274. Williams JB, Napoli JL (1985) Metabolism of retinoic acid and retinol during differentiation of F9 embryonal carcinoma cells. Proc Natl Acad Sci U S A 82:4658–4662
275. Williams JB, Napoli JL (1987) Inhibition of retinoic acid metabolism by imidazole antimycotics in F9 embryonal carcinoma cells. Biochem Pharmacol 36:1386–1388
276. Wolf G (2007) Identification of a membrane receptor for retinol-binding protein functioning in the cellular uptake of retinol. Nutr Rev 65:385–388
277. Wongsiriroj N, Piantedosi R, Palczewski K, Goldberg IJ, Johnston TP, Li E et al (2008) The molecular basis of retinoid absorption: a genetic dissection. J Biol Chem 283:13510–13519
278. Wongsiriroj N, Jiang H, Piantedosi R, Yang KJZ, Kluwe J, Schwabe RF et al (2014) Genetic dissection of retinoid esterification and accumulation in the liver and adipose tissue. J Lipid Res 55:104–114
279. Wu BX, Chen Y, Chen Y, Fan J, Rohrer B, Crouch RK et al (2002) Cloning and characterization of a novel all-trans retinol short-chain dehydrogenase/reductase from the RPE. Invest Ophthalmol Vis Sci 43:3365–3372
280. Yang Q, Graham TE, Mody N, Preitner F, Peroni OD, Zabolotny JM et al (2005) Serum retinol binding protein 4 contributes to insulin resistance in obesity and type 2 diabetes. Nature 436:356–362
281. Yen C-LE, Monetti M, Burri BJ, Farese RV (2005) The triacylglycerol synthesis enzyme DGAT1 also catalyzes the synthesis of diacylglycerols, waxes, and retinyl esters. J Lipid Res 46:1502–1511
282. Yost RW, Harrison EH, Ross AC (1988) Esterification by rat liver microsomes of retinol bound to cellular retinol-binding protein. J Biol Chem 263:18693–18701
283. Yu S, Levi L, Siegel R, Noy N (2012) Retinoic acid induces neurogenesis by activating both retinoic acid receptors (RARs) and peroxisome proliferator-activated receptor β/δ (PPARβ/δ). J Biol Chem 287:42195–42205
284. Zachman RD, Grummer MA (1998) The interaction of ethanol and vitamin A as a potential mechanism for the pathogenesis of Fetal Alcohol syndrome. Alcohol Clin Exp Res 22:1544–1556
285. Zhai Y, Higgins D, Napoli JL (1997) Coexpression of the mRNAs encoding retinol dehydrogenase isozymes and cellular retinol-binding protein. J Cell Physiol 173:36–43
286. Zhai Y, Sperkova Z, Napoli JL (2001) Cellular expression of retinal dehydrogenase types 1 and 2: effects of vitamin A status on testis mRNA. J Cell Physiol 186:220–232
287. Zhang M, Chen W, Smith SM, Napoli JL (2001) Molecular characterization of a mouse short chain dehydrogenase/reductase active with all-trans-retinol in intact cells, mRDH1. J Biol Chem 276:44083–44090

288. Zhang M, Hu P, Napoli JL (2004) Elements in the N-terminal signaling sequence that determine cytosolic topology of short-chain dehydrogenases/reductases. Studies with retinol dehydrogenase type 1 and cis-retinol/androgen dehydrogenase type 1. J Biol Chem 279:51482–51489
289. Zhang M, Hu P, Krois CR, Kane MA, Napoli JL (2007) Altered vitamin A homeostasis and increased size and adiposity in the rdh1-null mouse. FASEB J Off Publ Fed Am Soc Exp Biol 21:2886–2896
290. Zizola CF, Frey SK, Jitngarmkusol S, Kadereit B, Yan N, Vogel S (2010) Cellular retinol-binding protein type I (CRBP-I) regulates adipogenesis. Mol Cell Biol 30:3412–3420

Chapter 3
Vitamin A Transport and Cell Signaling by the Retinol-Binding Protein Receptor STRA6

Noa Noy

Abstract Vitamin A, retinol, circulates in blood bound to retinol binding protein (RBP). In some tissues, the retinol-RBP complex (holo-RBP) is recognized by a membrane receptor, termed STRA6, which mediates uptake of retinol into cells. Recent studies have revealed that, in addition to serving as a retinol transporter, STRA6 is a ligand-activated cell surface signaling receptor that, upon binding of holo-RBP activates JAK/STAT signaling, culminating in the induction of STAT target genes. It has further been shown that retinol transport and cell signaling by STRA6 are critically interdependent and that both are coupled to intracellular vitamin A metabolism. The molecular mechanism of action of STRA6 and its associated machinery is beginning to be revealed, but further work is needed to identify and characterize the complete range of genes and associated signaling cascades that are regulated by STRA6 in different tissues. An understanding of STRA6 is clinically relevant, as for example, it has been shown to be hyper- activated in obese animals, leading to insulin resistance. A potential role for STRA6 in other pathologies, including cancer, awaits further investigation.

Keywords Retinol • Vitamin A • Stra6 • Vitamin A transport • Cell signaling • RBP • Insulin responsiveness

Abbreviations

CRBP	cellular retinol-binding protein
JAK	Janus kinases
LRAT	lecithin:retinol acyltransferase

N. Noy (✉)
Department of Cellular and Molecular Medicine, Lerner Research Institute, Cleveland Clinic Foundation, Cleveland, OH 44195, USA
e-mail: noyn@ccf.org

M.A. Asson-Batres, C. Rochette-Egly (eds.), *The Biochemistry of Retinoid Signaling II*, Subcellular Biochemistry 81, DOI 10.1007/978-94-024-0945-1_3

RA	retinoic acid
RBP	serum retinol-binding protein
RPE	retinal pigment epithelium
SOCS	suppressor of cytokine signaling
STAT	signal transducers and activators of transcription
STRA6	stimulated by retinoic acid 6
TTR	transthyretin
VAD	vitamin A-deficient

Introduction

Some biologically important compounds are hydrophilic in nature. These compounds dissolve in water and readily traverse the aqueous milieu that comprises about 60 % of the human body. Other substances are hydrophobic and, while they dissolve in lipid-like compartments such as cellular membranes and fat depots, they do not freely move in the aqueous spaces of blood and cell cytosol. Understanding how hydrophobic compounds, like vitamin A are distributed in the body and reach their sites of metabolism and action is therefore critical for understanding how their biological activities are exerted and regulated.

Vitamin A readily dissolves in the hydrophobic core of cellular membranes. However, the presence of excess vitamin A in membranes results in disruption of membrane structure and function and can lead to toxicity [1, 2]. In addition, in order to reach target tissues and sites of action inside cells, retinol must traverse the aqueous milieu of plasma and cytosol, a process that is complicated by its hydrophobic nature. These barriers are overcome by specific retinol-binding proteins that solubilize retinol, transport it through aqueous phases, and maintain its concentrations in cellular membranes below disruptive levels. Vitamin A is mobilized from its storage sites, present in various tissues and mainly in the liver [3], with the aid of a carrier that was identified in 1968 to be a 21 kDa protein which was named retinol-binding protein (RBP) [4]. Retinol-bound RBP (holo-RBP) is associated with another protein termed transthyretin (**trans**porter of **thy**roxin and **retinol**, TTR). TTR fulfils two roles in the delivery of vitamin A to peripheral tissues: it is involved in secretion of holo-RBP from the liver [5, 6], and it prevents loss of the low molecular weight RBP by glomerular filtration in the kidneys (see e.g. [7]).

Retinol leaves the holo-RBP-TTR complex prior to being taken up into cells. The lipophilic vitamin can readily dissociate from RBP and rapidly diffuse across membranes into cells. Nevertheless, it was hypothesized that a plasma membrane transporter specifically recognizes holo-RBP and mediates uptake of retinol into cells, and in 2007, such a transporter was indeed identified. This protein, encoded for by *Stimulated by Retinoic Acid* 6 (*Stra6*) [8], is expressed in various tissues and displays an exceptionally high level in retinal pigment epithelium cells in the eye.

However, although STRA6 is a retinol transporter, it is not required for maintaining proper vitamin A levels in most tissues either during embryonic development or in the adult [9, 10]. The exception is the eye, where STRA6 is necessary for supplementing free diffusion-mediated retinol uptake to satisfy the extreme vitamin A demands of this tissue. Recent studies further revealed that, in addition to mediating vitamin A transport, STRA6 is a surface signaling receptor which is activated by holo-RBP to trigger a phosphorylation cascade mediated by the Janus kinase JAK2 and its associated transcription factors STAT3/5. Holo-RBP thus serves as a cytokine that regulates gene transcription by activating STRA6. This role of holo-RBP in regulating gene transcription via interaction with STRA6 is independent of the role of the vitamin as a precursor for the transcriptionally-active metabolite retinoic acid.

History and Development of the Field

When isolated from blood, RBP (encoded for by the *Rbp4* gene) was found to be associated with approximately one molecule of retinol, suggesting that it contains a single binding site for the ligand [4]. The solved X-ray crystal structure of RBP supports this conclusion. The structure reveals that RBP is comprised of an eight-stranded, anti-parallel, β-sheet folded to form a β-barrel that binds one molecule of retinol. The amino terminus of RBP wraps around the back of the barrel, 'capping' that side of the pocket, while the front of the β-barrel is open, providing a portal for the ligand. The 3-dimensional structure of the protein shows that retinol binds to RBP with the β-ionone ring innermost and the hydroxyl head-group reaching to the protein surface where it is coordinated to a water molecule at the pocket entrance ([11, 12]), Fig. 3.1). Retinol associates with RBP to form the holo-RBP complex in the endoplasmic reticulum of hepatocytes prior to secretion into blood [13]. Notably, RBP is also expressed in extrahepatic tissues including lungs, spleen, brain, stomach, heart, and skeletal muscle [14], retina [15], and adipose tissue [16, 17]. Little information is currently available to indicate whether RBP is secreted from these tissues and, if so, whether it is secreted as a complex with retinol.

In blood, retinol also associates with a tryptophan-rich, 56 KDa protein that was originally called prealbumin [18]. Studies have shown that retinol does not directly bind to prealbumin, but instead, holo-RBP associates with prealbumin, forming a ternary complex with an approximate 1:1 mole ratio of holo-RBP : prealbumin [4]. Prealbumin was later re-named transthyretin (TTR) to denote that in addition to binding holo-RBP, it serves as a blood carrier for thyroid hormone [19]. The major sites of synthesis of TTR are the choroid plexus in the brain and the liver. The protein is found in plasma and in cerebrospinal fluid [20]. Notably, TTR binds holo- but not apo-RBP and the RBP-TTR complex dissociates following loss of retinol [21]. The reported 3-dimensional crystal structure of the holo-RBP-TTR complex [22] shows that RBP binds to TTR such that the association blocks the entrance to the ligand-binding pocket of RBP (Fig. 3.2). Studies of mice lacking RBP established

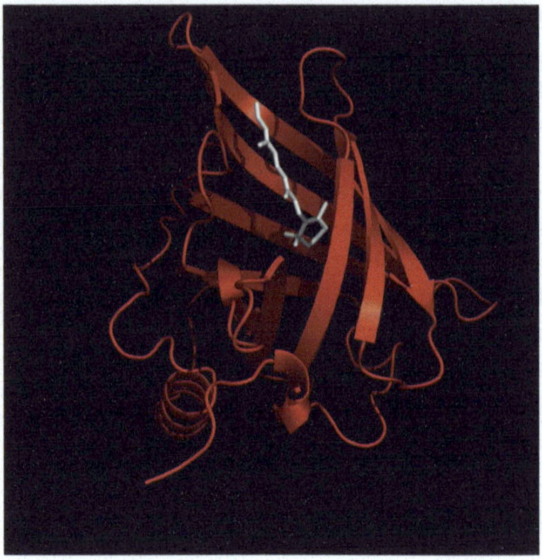

Fig. 3.1 The three dimensional crystal structure of holo-RBP. The human holo-RBP structure (PDB ID 1BRP) was rendered using Pymol (http://www.pymol.org/). The structure shows the eight stranded antiparallel β-sheet folded over itself to form a β-barrel. Retinol (*white*) is encapsulated by the barrel with the β-ionone ring buried in the binding pocket and the alcohol group is at the protein surface

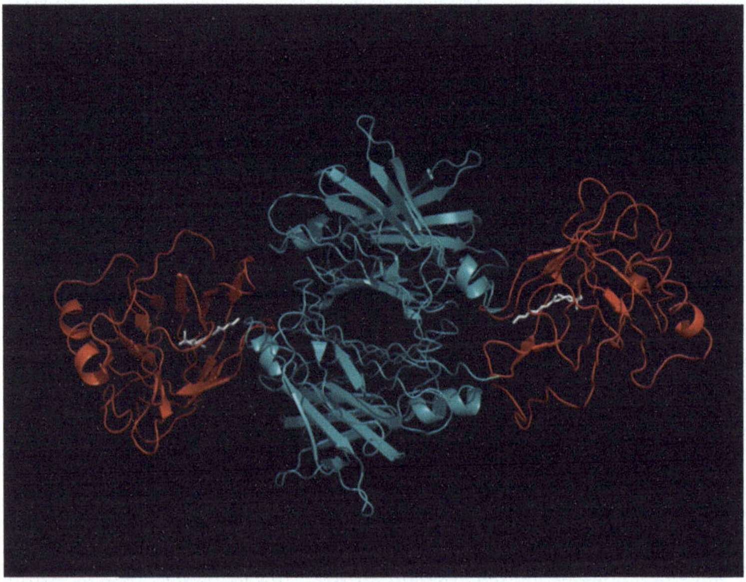

Fig. 3.2 The three dimensional crystal structure of the retinol-RBP-TTR complex. The structure of the human retinol-RBP-TTR complex (PDB ID 1QAB) was rendered using Pymol (http://www.pymol.org/). The TTR tetramer (*blue*) is comprised of a dimer of dimers with two RBP molecules (*green*) bound to the opposite dimers. Interactions between RBP and TTR are mediated by residues at the entrance to the ligand binding pocket and span across the two TTR dimers

the importance of the protein in mobilizing vitamin A from the liver [23]. *Rbp4*-null mice efficiently accumulate hepatic retinol stores but these are trapped in the liver and are not released to the circulation. Consequently, these mice display reduced blood retinol levels and markedly impaired visual function during the first months of life. Notably however, although their blood retinol levels remain low, normal vision is restored in *Rbp4*-null mice by 5 months of age if they are fed a vitamin A-sufficient diet. These observations reflect that, under these dietary conditions, retinol can be supplied to tissues directly from postprandial vitamin A-containing chylomicrons negating a need for liver stores. On the other hand, when fed a vitamin A-deficient (VAD) diet, serum retinol levels in adult *Rbp4*-null mice become undetectable, and visual function rapidly deteriorates. In accordance, feeding VAD diet to pregnant *Rbp4*-null females yields offspring suffering a pronounced fetal VAD syndrome [24]. Thus, RBP is critical for supplying vitamin A to extra-hepatic tissues in times of restricted dietary intake.

The mechanism by which retinol leaves circulating RBP and is taken up by target cells has long been controversial. While retinol readily enters cells by free diffusion, there is also evidence to support transport of retinol by a protein transporter located in the cell surface membrane. With regard to the former mechanism, it has been shown that the association of retinol with RBP is reversible, that spontaneous dissociation of retinol from RBP or the RBP-TTR complex occurs with a $t_{1/2}$ of minutes [21, 25, 26], and that retinol rapidly crosses lipid bilayers and spontaneously dissociates from this milieu at a $t_{1/2}$ of about a second [27]. These observations indicate that retinol can leave its blood carrier and move into cells by free diffusion at rates that are sufficient for physiological needs. The observations that retinol is metabolized in keratinocytes at similar rates regardless of whether retinol is provided to the cells in a free or RBP-bound form [28] further suggests that the rate of entry of retinol into cells is dictated by its metabolism and is not limited by the rate of its uptake at the plasma membrane.

With reference to the second proposed mechanism, it has been proposed that a specific receptor for RBP exists in target cells and that this receptor binds circulating RBP, releases retinol from the protein, and mediates its transfer across the plasma membrane to the cytosol [29–31]. In support of this notion, it was reported that uptake of radiolabeled retinol from holo-RBP can be inhibited by RBP complexed with non-labeled retinol [31], and that plasma membranes of some cells possess a RBP-binding activity [32]. The search for the putative RBP receptor continued for many years and took some wrong turns (e.g. [33]) but, in 2007, a receptor that binds extracellular RBP and functions as a vitamin A transporter was identified [34]. This protein, STRA6, was originally identified in P19 embryonal carcinoma cells to be a RA-responsive gene with unknown function [8]. In the adult, STRA6 is most highly expressed in retinal pigment epithelium (RPE) in the eye, is also expressed in blood organ barriers, brain, adipose tissue, spleen, kidney, testis, and female genital tract, but is undetectable in liver and intestines [8, 34]. STRA6 is a largely hydrophobic protein that lacks homology to any known protein. Computer modeling predicts that STRA6 contains 11 trans-membrane helices, a number of loops, and a large C-terminus cytosolic tail. Alternatively, it has been suggested that

STRA6 is arranged in 9 trans-membrane helices [35]. Importantly, retinol uptake by STRA6 is not affected by metabolic inhibitors [34], indicating that the receptor is not energy-driven and that instead, it mediates facilitated transport of retinol. Retinol thus enters cells following an inwardly-directed concentration gradient. A *Stra6* paralogue displaying 40 % similarity at the protein level has been identified in the mouse genome (1300002K09Rik, *Stra6.2,* also termed RBP receptor 2 RBPR2). This protein is expressed primarily in intestines and in the liver and, like STRA6, it facilitates uptake of retinol into cells [36]. *Stra6.2* is conserved amongst many mammals but is not expressed in great apes and humans where it is split in two parts with an associated break in synteny.

It has been reported that mutations in *Stra6* in humans are associated with the occurrence of a Matthew-Wood syndrome, a severe congenital disease that includes microphtalmia, pulmonary hypoplasia, heart defects and diaphragmatic hernia [37–39]. In contrast with *Stra6*, mutations in the human *Rbp4* gene only lead to mild symptoms that are traceable to vitamin A deficiency, including night blindness and a modest retinal dystrophy [40]. The lack of similarity between the impairments observed in humans with *Stra6* and *Rbp4* mutations suggest either that, in addition to its role as an RBP receptor, STRA6 may have other biological functions, or that Matthew-Wood syndrome originates from chromosomal aberrations other than *Stra6* mutations. In support of the latter possibility, it was reported that a patient suffering PAGOD syndrome, a multiple congenital anomalies syndrome with significant phenotypic overlap with Matthew-Woods syndrome, does not carry any *Stra6* mutations [41].

Importantly, while vitamin A is critical for both embryonic development and adult life, *Stra6*-null mice are obtained at Mendelian frequency, appear healthy, and do not display any histological abnormality except in the eye ([9, 10, 42]) and see http://www.kompphenotype.org/summary-tab.php?gene=Stra6). These observations indicate that STRA6 is not necessary for maintaining retinol availability to most tissues throughout life. It was reported that, with the exception of the eye, the rate of uptake of retinol from circulating holo-RBP into STRA6-expressing tissues is only modestly reduced in *Stra6*-null mice and that the retinoid content of such tissues is indistinguishable in *Stra6*-null and WT mice fed regular chow diet. Even in mice fed a VAD diet, when holo-RBP mobilized from the liver is the sole source of retinol, vitamin A levels in STRA6-expressing tissues was found to be only modestly reduced in *Stra6*-null mice as compared to WT animals [9]. Moreover, while RBP is critical for development in mice fed a VAD diet [24], *Stra6*-null fetuses born from dams fed a VAD diet do not exhibit any of the hallmarks of fetal vitamin A deficiency syndrome [9]. In contrast with most tissues, ablation of STRA6 leads to severe depletion of retinoid stores in the RPE and neurosensory retina, and *Stra6*-null mice display a shortening of rod outer- and inner- segments and a reduction in cone photoreceptor cell number and cone b-wave amplitude [10]. Eyes of *Stra6*-null mice are also characterized by the presence of vascularized structures in the vitreous humor between the posterior surface of the lens and neurosensory retina. Notably however, morphological changes and reduction in visual function in eyes of *Stra6*-null mice are mild, indicating that STRA6 is not the only pathway for retinol uptake by the RPE [10]. Hence, although STRA6 functions as a retinol transporter, the major fraction of cellular uptake of retinol from circulating holo-RBP occurs in a STRA6-independent fashion, likely by free diffusion across the plasma membrane. In the eye, diffusion-mediated retinol uptake is supplemented by STRA6-mediated transport to satisfy the extreme vitamin A demand of this tissue.

Current State of the Field

STRA6 Is a Holo-RBP-Activated Surface Signaling Receptor

The observations that STRA6 is not essential for providing vitamin A to most tissues that express it suggest that the receptor has function(s) other than to serve as a retinol transporter. This notion is supported by recent studies showing that STRA6 functions as a holo-RBP-activated signaling receptor ([9, 43–46]). Extracellular signaling polypeptides like cytokines, hormones, and growth factors are often recognized by cell surface receptors that transduce a signaling cascade mediated by tyrosine kinases called Janus kinases (JAK) and by their associated transcription factors, Signal Transducers and Activators of Transcription (STAT). Activated STATs form dimers that translocate to the nucleus where they regulate transcription of target genes. STATs thus regulate multiple aspects of cellular behavior in response to a myriad of cytokines, hormones, and growth factors [47–49]. JAK/STAT signaling can be terminated by dephosphorylation of activated receptors, JAKs, or STATs [48], and by inhibition of the transcriptional activity of STATs [50, 51]. Other key negative regulators of this pathway are encoded by the direct STAT target genes *Suppressors of Cytokine Signaling* (*Socs*). SOCS are components of negative feedback loops. They compete with STATs for binding to activated receptors, block the catalytic activity of JAK, and recruit ubiquitin ligases that facilitate the degradation of JAK by the proteasome [48, 52–54]. SOCS thus suppress multiple cytokine-initiated responses.

Inspection of the cytosolic domain of STRA6 reveals the presence of a stretch of residues resembling a consensus phospho-tyrosine motif, a sequence known to serve as a binding site for STATs [37]. It was recently demonstrated that treatment of STRA6-expressing cells with holo-RBP triggers STRA6 phosphorylation which induces recruitment and activation of JAK2 and, in a cell-specific manner, STAT3 or STAT5. It was further shown that holo-RBP-induced STRA6 activation results in upregulation of STAT target genes, including SOCS3 and the nuclear receptor PPARγ. Importantly, neither RBP nor retinol triggered JAK/STAT signaling when administered alone, and retinoic acid had no effect on this cascade [43]. Hence, holo-RBP functions like a classical cytokine to activate a STRA6/JAK2/STAT3/5 pathway. Among known cytokine receptors, STRA6 is unique in that it functions both as a transporter and as a signaling receptor. It is interesting that the two functions of this "cytokine signaling transporter" are strictly inter-dependent, i.e. STRA6

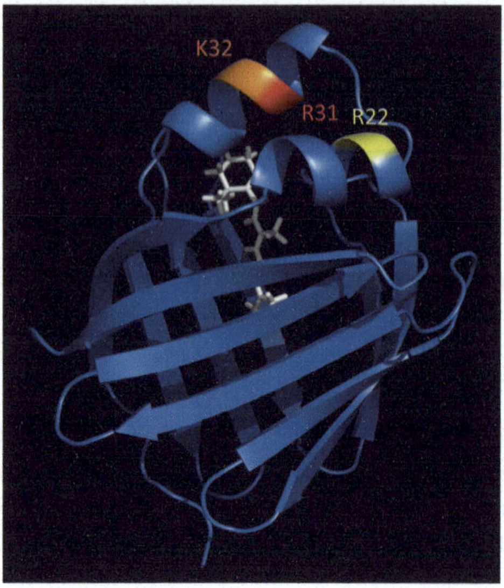

Fig. 3.3 The three dimensional crystal structure of holo-CRBP1. The structure of CRBP1 in complex with retinol (PDB accession # 1KGL) was rendered using Pymol (http://www.pymol.org/). Residues R22, R31 and K32, which serve as the ligand-controlled switch that mediates the association of CRBP1 with STRA6 are highlighted

signaling critically depends on STRA6-mediated retinol transport and, *vice versus*, retinol transport cannot proceed if STRA6 phosphorylation is impaired ([9, 45]).

Retinol transport and cell signaling by STRA6 requires expression of intracellular accessory proteins and retinol-metabolizing enzymes. Cellular retinol-binding protein 1 (CRBP1) is a member of the family of intracellular lipid-binding proteins (iLBP) that bind retinol with high affinity [55]. This protein associates with cytosolic regions of STRA6, directly accepting retinol transferred from holo-RBP by the transporter [45]. STRA6-mediated transport can only proceed if an inward-directed concentration is maintained. Some retinol-metabolizing enzymes, such as lecithin:retinol acyltransferase (LRAT), which catalyzes the conversion of retinol to its storage species retiny lesters receive retinol directly from CRBP1 [56]. Thus, STRA6 'channels' retinol from RBP to CRBP1, which, in turn, shuttles the vitamin to its sites of metabolism, bypassing the need for this lipophilic compound to traverse an aqueous milieu. LRAT unloads retinol from CRBP1 and converts it to retinyl esters, thereby maintaining a low intracellular retinol concentration as well as regenerating apo-CRBP1, enabling it to re-associate with STRA6. It was shown that LRAT supports both retinol uptake and cell signaling by STRA6 [46, 57].

Apo-CRBP1 simultaneously binds to two regions of STRA6. One of these regions is an intracellular loop of the receptor and the other is localized in the cytosolic C-terminal tail. The association with the loop is mediated by arginine-22, arginine-31 and lysine-32 located at the helix-loop-helix that caps the entrance to the retinol binding pocket of CRBP1 (Fig. 3.3). Binding of retinol induces conformational changes in these residues, leading to dissociation of CRBP1 from the loop [45]. Interestingly, these residues are conserved in several iLBPs but they are used for different purposes in different proteins. While in CRBP1, they mediate the protein's interactions with STRA6 [45], these residues comprise a ligand-controlled

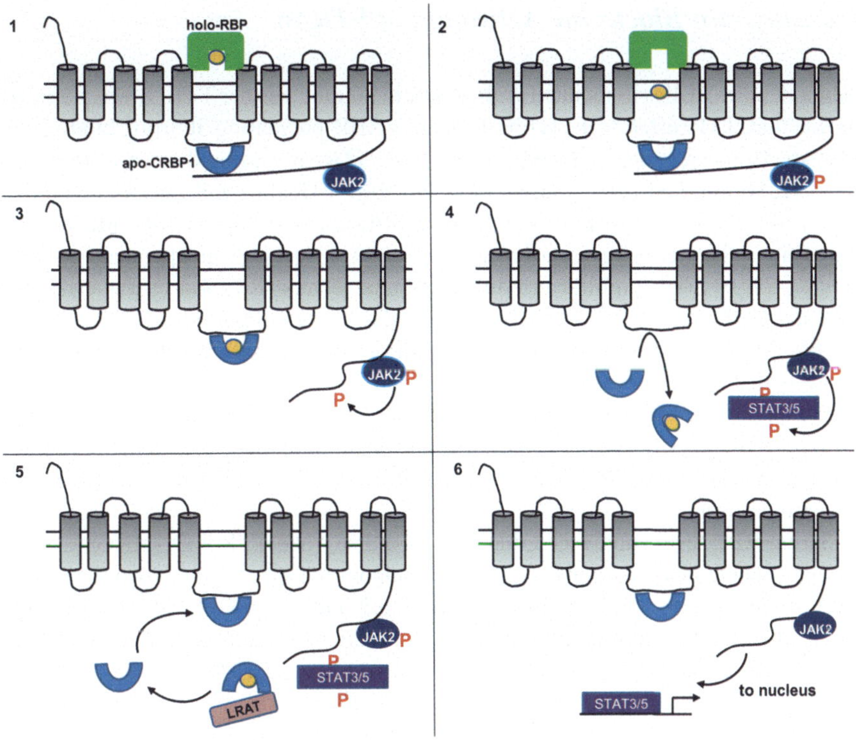

Fig. 3.4 Model of the mechanism of action of STRA6. See text for explanation

nuclear localization signal in some other iLBPs [58–60]. The other region of STRA6 that interacts with CRBP1 is a stretch in the receptor's cytoplasmic tail which contains the signaling residue of STRA6 tyrosine-643. These interactions are disrupted when the tyrosine is phosphorylated upon activation of the receptor [45].

Current understanding of the mechanism of action of STRA6 is depicted by the following model (Fig. 3.4): (**1**). In the quiescent state, STRA6 is "primed" with basal levels of bound apo-CRBP1 and JAK2. CRBP1 associates with STRA6 through interactions with both an intracellular loop and with the phospho-tyrosine-containing region in the receptor's cytoplasmic tail. (**2**). Retinol transfers through STRA6 from extracellular holo-RBP to apo-CRBP1, a process that triggers phosphorylation of JAK2. (**3**). JAK2 phosphorylates STRA6,leading to release of the C-terminus tail from CRBP1. (**4**). Following ligand-binding,-CRBP1 dissociates from STRA6, and STAT3 or STAT5 is recruited and activated. (**5**). Holo-CRBP1 delivers retinol to LRAT, and regenerated apo-CRBP1 re-associates with STRA6. (**6**). Activated STAT3/5 moves to the nucleus and induces the expressionof STAT target genes.. Thus, STRA6 couples "sensing" of serum holo-RBP levels and intracellular retinol metabolism to a signaling cascade which, in turn, regulates expression of multiple STAT target genes.

Transthyretin Blocks the Activation of STRA6

In blood, holo-RBP is associated with its serum binding partner TTR. It was recently shown that TTR effectively sequesters holo-RBP preventing it from binding to STRA6. Consequently, TTR suppresses both STRA6-mediated retinol transport and STRA6-initiated cell signaling [44]. Notably, TTR does not interfere with the ability of retinol to move into cells by free diffusion [44], further supporting early suggestions that retinol primarily enters cells by a receptor-independent process [21, 27, 28, 61]. The equilibrium dissociation constants (Kd) that characterize the association of holo-RBP with TTR and with STRA6 were reported to be 70 nM [62] and 60 nM [34], respectively. The similar affinity of holo-RBP for the two proteins suggests that STRA6 can robustly function only under circumstances in which its local expression level is high enough to allow it to successfully compete with TTR for holo-RBP, such as in the retina [10] and in some human cancers [63, 64], or when serum holo-RBP levels exceed that of TTR, allowing STRA6 to have free access to retinol.

Serum RBP level can exceed that of TTR either when TTR levels decrease, e.g. during an acute phase response (APR) to inflammation [65, 66], or when the serum level of RBP is elevated, e. g. in obese mice and humans [44, 67, 68]. Under such circumstances, the resulting high RBP:TTR ratio in blood may lead to hyperactivation of STRA6. Notably,, it has been reported that hepatic TTR expression is regulated by sex hormones [69], and that expression of RBP in brown adipose tissue and liver are regulated by cAMP-mediated pathways and by the nuclear receptors PPARα and PPARγ [70, 71]. Whether these factors regulate the RBP:TTR ratio in blood by controlling TTR or RBP expression and thus modulate STRA6 signaling remains to be clarified.

The RBP/STRA6 Signaling Pathway Regulates Insulin Responses

It has been reported that administration of recombinant RBP or transgenic overexpression of the protein in mice results in insulin resistance. Consistent with these observations, genetic deletion of *Rbp4* in mice enhances insulin sensitivity [72]. It has also been shown that treatment of cultured adipocytes with holo-RBP decreases the phosphorylation status of the insulin receptor (IR) and suppresses insulin-induced mobilization of the glucose transporter GluT4 to the plasma membranes [43]. The discovery that activation of STRA6 by holo-RBP leads to induction of STAT target genes provides a rationale for understanding the basis for the inhibitory effect of RBP on insulin signaling. STAT target genes in insulin-responsive tissues that express STRA6, such as adipose tissue and muscle, include *Suppressor of cytokine signaling 3* (*Socs3*), a potent inhibitor of the insulin receptor (IR) [53, 73–77], and it was shown that holo-RBP-induced suppression of IR signaling in these tissues is mediated by STRA6, JAK2, STAT5, and SOCS3 [43]. Demonstrating the specificity of the response, RBP did not affect JAK/STAT activation, the level of SOCS3, or the phosphorylation status of IR in liver, a tissue that does not express STRA6 [43]. The observation that *Stra6*-null mice are completely protected from insulin resistance by administration of RBP provided conclusive evidence for the critical role of STRA6 in suppressing RBP signaling [9]. Taken together, available information demonstrates that STRA6 and its associated signaling pathway mediate RBP-induced insulin resistance.

Serum RBP levels are often elevated in obese humans and mice and it was suggested that the resulting high serum RBP levels may contribute to the well-documented but incompletely understood link through which obesity leads to insulin resistance [67, 72, 78]. Recent studies support this notion. It was thus shown that *Stra6*-null mice are partially protected from insulin resistance brought about by high fat feeding [9] and that even partial reduction of STRA6 in adipose tissues improves insulin responsiveness in obese mice [79].

Relevance

It has long been thought that the parent vitamin A molecule, retinol, is biologically inert and that it's biological functions are exerted solely by active metabolites. The discovery that by activating STRA6, holo-RBP triggers a signaling cascade culminating in activation of STAT3/5 demonstrates that retinol, in conjunction with its serum carrier, is a transcriptional regulator in its own right. These findings reveal that retinol not only functions as a precursor to the transcriptionally-active metabolite retinoic acid (RA), which activates the classical RA receptors (RARs) and the alternative RA receptor PPARβ/δ but it also can activate STAT3 and STAT5. Vitamin A thus activates a larger group of transcription factors, regulates the expression of a larger cohort of genes, and controls a larger spectrum of biological activities than previously recognized (Fig. 3.5).

Holo-RBP is a unique cytokine that both binds to its receptor and provides a ligand whose transport that activates the receptor. Among cytokine-activated receptors, STRA6 is unique in that it couples ligand transport to cell signaling. This coupling obligatorily relies on intracellular processing of retinol. The observations that STRA6 "senses" both free holo-RBP levels in blood and intracellular retinol metabolism and transduces this information to cell signaling makes the receptor a founding member of a novel class of proteins that serve as "cytokine signaling transporters".

The holo-RBP/STRA6 pathway regulates insulin responses, and its hyperactivation in obese animals contributes to insulin resistance brought about by high fed feeding. These observations suggest that the path may comprise a promising novel target for therapy of insulin resistance and associated metabolic disease.

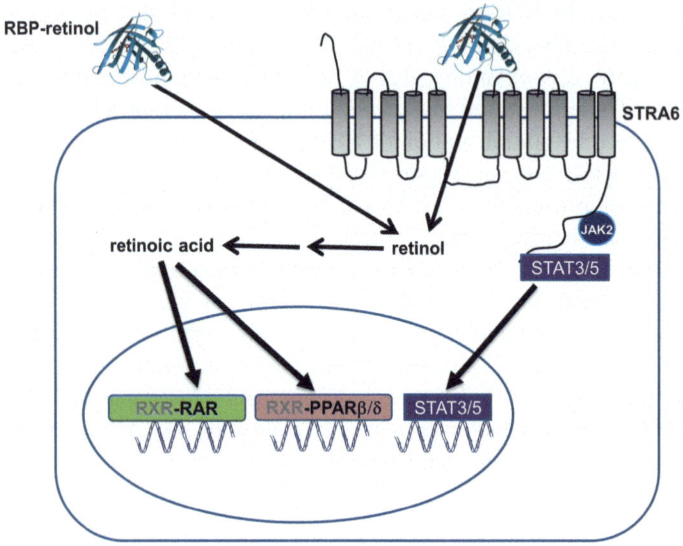

Fig. 3.5 Regulation of gene transcription by vitamin AN Retinol can enter cells from circulating holo-RBP by free diffusion through the plasma membrane, or through STRA6-mediated transport. Once in cells, retinol can be metabolically converted to RA which regulates transcription by activating the classical RA receptors RARs and the alternative receptor PPARβ/δ. Retinol transported through STRA6 regulates transcription by activating STAT3/5. Vitamin A thus controls gene expression by activating multiple transcription factors

Future Directions

The signaling activities of holo-RBP are constrained by its association with TTR and by the short half-life of uncomplexed holo-RBP in the circulation. TTR efficiently inhibits both STRA6-mediated retinol transport and STRA6-initiated signaling [44], and, due to its low molecular weight, free holo-RBP is rapidly excreted by glomerular filtration. Consequently, STRA6 is activated only in cells where its expression level is very high, or when serum holo-RBP exceeds that of TTR. For example, STRA6 mediates a large fraction of retinol uptake by the retinal pigment epithelium, where it is exceptionally highly expressed [9, 10]. STRA6 is also activated in obese animals where the blood level of RBP is elevated, resulting in an increase in the RBP:TTR ratio in plasma. Under these circumstances, hyperactivation of the receptor leads to insulin resistance [43, 44]. Other physiological circumstances under which the path is activated remain to be identified.

It has been reported that STRA6 is upregulated in several human cancers [63]. The functional significance of the increased expression of STRA6 in carcinoma cells is currently unknown but the discovery that STRA6 activates a JAK2/STAT3/5 cascade may provide a clue. These STATs are associated with inflammation, cellular transformation, survival, proliferation, invasion, and angiogenesis and they are considered to be oncogenes [80–83]. Hence, an intriguing possibility is that STRA6 and its associated machinery may be involved in oncogenic activities. Notably, while obesity is a risk factor for multiple types of cancers [84–86], the molecular mechanisms that underlie the association are incompletely understood. The possibility that activation of STAT3/5 by STRA6 may comprise an important molecular mechanism through which obesity promotes cancer development remains to be investigated.

Available information indicates that activation of STRA6 is supported by the cellular retinol-binding protein CRBP1 and by the retinol-metabolizing enzyme LRAT [45, 46]. Other CRBPs and retinol-metabolizing enzymes that may enable STRA6-mediated vitamin A transport and signaling remain to be identified.

Finally, the complete range of genes regulated by STRA6 and its associated signaling cascade, and the biological activities controlled by the path in the different tissues that express the receptor also remain to be elucidated.

Acknowledgements Work in the author's laboratory was supported by NIH grants RO1 DK060684, RO1 DK088669, and R01 CA166955

References

1. Jetten AM, Meeks RG, De Luca LM (1981) Specific and nonspecific alterations in membrane microviscosity induced by retinoids in embryonal carcinoma and fibroblast cells. Ann N Y Acad Sci 359:398–400
2. Meeks RG, Zaharevitz D, Chen RF (1981) Membrane effects of retinoids: possible correlation with toxicity. Arch Biochem Biophys 207:141–147
3. Harrison EH (2005) Mechanisms of digestion and absorption of dietary vitamin a. Annu Rev Nutr 25:87–103
4. Kanai M, Raz A, Goodman DS (1968) Retinol-binding protein: the transport protein for vitamin a in human plasma. J Clin Invest 47:2025–2044
5. Wei S, Episkopou V, Piantedosi R, Maeda S, Shimada K, Gottesman ME, Blaner WS (1995) Studies on the metabolism of retinol and retinol-binding protein in transthyretin-deficient mice produced by homologous recombination. J Biol Chem 270:866–870
6. Bellovino D, Morimoto T, Tosetti F, Gaetani S (1996) Retinol binding protein and transthyretin are secreted as a complex formed in the endoplasmic reticulum in hepg2 human hepatocarcinoma cells. Exp Cell Res 222:77–83
7. van Bennekum AM, Wei S, Gamble MV, Vogel S, Piantedosi R, Gottesman M, Episkopou V, Blaner WS (2001) Biochemical basis for depressed serum retinol levels in transthyretin-deficient mice. J Biol Chem 276:1107–1113
8. Bouillet P, Sapin V, Chazaud C, Messaddeq N, Decimo D, Dolle P, Chambon P (1997) Developmental expression pattern of stra6, a retinoic acid-responsive gene encoding a new type of membrane protein. Mech Dev 63:173–186
9. Berry DC, Jacobs H, Marwarha G, Gely-Pernot A, O'Byrne SM, DeSantis D, Klopfenstein M, Feret B, Dennefeld C, Blaner WS, Croniger CM, Mark M, Noy N, Ghyselinck NB (2013) The stra6 receptor is essential for retinol-binding protein-induced insulin resistance but not for maintaining vitamin a homeostasis in tissues other than the eye. J Biol Chem 288:24528–24539
10. Ruiz A, Mark M, Jacobs H, Klopfenstein M, Hu J, Lloyd M, Habib S, Tosha C, Radu RA, Ghyselinck NB, Nusinowitz S, Bok D (2012) Retinoid content, visual responses and ocular

morphology are compromised in the retinas of mice lacking the retinol-binding protein receptor, stra6. Invest Ophthalmol Vis Sci 53:3027–3039
11. Newcomer ME, Jones TA, Aqvist J, Sundelin J, Eriksson U, Rask L, Peterson PA (1984) The three-dimensional structure of retinol-binding protein. EMBO J 3:1451–1454
12. Zanotti G, Berni R, Monaco HL (1993) Crystal structure of liganded and unliganded forms of bovine plasma retinol-binding protein. J Biol Chem 268:10728–10738
13. Smith JE, DeMoor LM, Handler CE, Green EL, Ritter SJ (1998) The complex between retinol and retinol-binding protein is formed in the rough microsomes of liver following repletion of vitamin a-depleted rats. Biochim Biophys Acta 1380:10–20
14. Soprano DR, Soprano KJ, Goodman DS (1986) Retinol-binding protein messenger rna levels in the liver and in extrahepatic tissues of the rat. J Lipid Res 27:166–171
15. Ong DE (1984) A novel retinol-binding protein from rat. Purification and partial characterization. J Biol Chem 259:1476–1482
16. Makover A, Soprano DR, Wyatt ML, Goodman DS (1989) Localization of retinol-binding protein messenger rna in the rat kidney and in perinephric fat tissue. J Lipid Res 30:171–180
17. Tsutsumi C, Okuno M, Tannous L, Piantedosi R, Allan M, Goodman DS, Blaner WS (1992) Retinoids and retinoid-binding protein expression in rat adipocytes. J Biol Chem 267:1805–1810
18. Alvsaker JO, Haugli FB, Laland SG (1967) The presence of vitamin a in human tryptophan-rich prealbumin. Biochem J 102:362–366
19. Schussler GC (2000) The thyroxine-binding proteins. Thyroid 10:141–149
20. Schreiber G (2002) The evolutionary and integrative roles of transthyretin in thyroid hormone homeostasis. J Endocrinol 175:61–73
21. Noy N, Xu ZJ (1990) Interactions of retinol with binding proteins: implications for the mechanism of uptake by cells. Biochemistry 29:3878–3883
22. Naylor HM, Newcomer ME (1999) The structure of human retinol-binding protein (rbp) with its carrier protein transthyretin reveals an interaction with the carboxy terminus of rbp. Biochemistry 38:2647–2653
23. Quadro L, Blaner WS, Salchow DJ, Vogel S, Piantedosi R, Gouras P, Freeman S, Cosma MP, Colantuoni V, Gottesman ME (1999) Impaired retinal function and vitamin a availability in mice lacking retinol-binding protein. EMBO J 18:4633–4644
24. Quadro L, Hamberger L, Gottesman ME, Wang F, Colantuoni V, Blaner WS, Mendelsohn CL (2005) Pathways of vitamin a delivery to the embryo: insights from a new tunable model of embryonic vitamin a deficiency. Endocrinology 146:4479–4490
25. Fex G, Johannesson G (1990) Transfer of retinol from retinol-binding protein complex to liposomes and across liposomal membranes. Methods Enzymol 189:394–402
26. Noy N, Blaner WS (1991) Interactions of retinol with binding proteins: studies with rat cellular retinol-binding protein and with rat retinol-binding protein. Biochemistry 30:6380–6386
27. Noy N, Xu ZJ (1990) Kinetic parameters of the interactions of retinol with lipid bilayers. Biochemistry 29:3883–3888
28. Hodam JR, Creek KE (1998) Comparison of the metabolism of retinol delivered to human keratinocytes either bound to serum retinol-binding protein or added directly to the culture medium. Exp Cell Res 238:257–264
29. Bavik CO, Eriksson U, Allen RA, Peterson PA (1991) Identification and partial characterization of a retinal pigment epithelial membrane receptor for plasma retinol-binding protein. J Biol Chem 266:14978–14985
30. Ward SJ, Chambon P, Ong DE, Bavik C (1997) A retinol-binding protein receptor-mediated mechanism for uptake of vitamin a to postimplantation rat embryos. Biol Reprod 57:751–755
31. Rask L, Peterson PA (1976) In vitro uptake of vitamin a from the retinol-binding plasma protein to mucosal epithelial cells from the monkey's small intestine. J Biol Chem 251:6360–6366
32. Sivaprasadarao A, Findlay JB (1988) The interaction of retinol-binding protein with its plasma-membrane receptor. Biochem J 255:561–569

33. Hinterhuber G, Cauza K, Brugger K, Dingelmaier-Hovorka R, Horvat R, Wolff K, Foedinger D (2004) Rpe65 of retinal pigment epithelium, a putative receptor molecule for plasma retinol-binding protein, is expressed in human keratinocytes. J Invest Dermatol 122:406–413
34. Kawaguchi R, Yu J, Honda J, Hu J, Whitelegge J, Ping P, Wiita P, Bok D, Sun H (2007) A membrane receptor for retinol binding protein mediates cellular uptake of vitamin a. Science 315:820–825
35. Kawaguchi R, Yu J, Wiita P, Ter-Stepanian M, Sun H (2008) Mapping the membrane topology and extracellular ligand binding domains of the retinol binding protein receptor. Biochemistry 47:5387–5395
36. Alapatt P, Guo F, Komanetsky SM, Wang S, Cai J, Sargsyan A, Rodriguez Diaz E, Bacon BT, Aryal P, Graham TE (2013) Liver retinol transporter and receptor for serum retinol-binding protein (rbp4). J Biol Chem 288:1250–1265
37. Pasutto F, Sticht H, Hammersen G, Gillessen-Kaesbach G, Fitzpatrick DR, Nurnberg G, Brasch F, Schirmer-Zimmermann H, Tolmie JL, Chitayat D, Houge G, Fernandez-Martinez L, Keating S, Mortier G, Hennekam RC, von der Wense A, Slavotinek A, Meinecke P, Bitoun P, Becker C, Nurnberg P, Reis A, Rauch A (2007) Mutations in stra6 cause a broad spectrum of malformations including anophthalmia, congenital heart defects, diaphragmatic hernia, alveolar capillary dysplasia, lung hypoplasia, and mental retardation. Am J Hum Genet 80:550–560
38. Golzio C, Martinovic-Bouriel J, Thomas S, Mougou-Zrelli S, Grattagliano-Bessieres B, Bonniere M, Delahaye S, Munnich A, Encha-Razavi F, Lyonnet S, Vekemans M, Attie-Bitach T, Etchevers HC (2007) Matthew-wood syndrome is caused by truncating mutations in the retinol-binding protein receptor gene stra6. Am J Hum Genet 80:1179–1187
39. Chassaing N, Golzio C, Odent S, Lequeux L, Vigouroux A, Martinovic Bouriel J, Tiziano FD, Masini L, Piro F, Maragliano G, Delezoide AL, Attie-Bitach T, Manouvrier-Hanu S, Etchevers HC, Calvas P (2009) Phenotypic spectrum of stra6 mutations: from matthew-wood syndrome to non-lethal anophthalmia. Hum Mutat 30:E673–E681
40. Biesalski HK, Frank J, Beck SC, Heinrich F, Illek B, Reifen R, Gollnick H, Seeliger MW, Wissinger B, Zrenner E (1999) Biochemical but not clinical vitamin a deficiency results from mutations in the gene for retinol binding protein. Am J Clin Nutr 69:931–936
41. Gavrilova R, Babovic N, Lteif A, Eidem B, Kirmani S, Olson T, Babovic-Vuksanovic D (2009) Vitamin a deficiency in an infant with pagod syndrome. Am J Med Genet A 149A:2241–2247
42. Terra R, Wang X, Hu Y, Charpentier T, Lamarre A, Zhong M, Sun H, Mao J, Qi S, Luo H, Wu J (2013) To investigate the necessity of stra6 upregulation in t cells during t cell immune responses. PLoS One 8, e82808
43. Berry DC, Jin H, Majumdar A, Noy N (2011) Signaling by vitamin a and retinol-binding protein regulates gene expression to inhibit insulin responses. Proc Natl Acad Sci U S A 108:4340–4345
44. Berry DC, Croniger CM, Ghyselinck NB, Noy N (2012) Transthyretin blocks retinol uptake and cell signalling by the holo-retinol-binding protein receptor stra6. Mol Cell Biol 32:3851–3859
45. Berry DC, O'Byrne SM, Vreeland AC, Blaner WS, Noy N (2012) Cross talk between signaling and vitamin a transport by the retinol-binding protein receptor stra6. Mol Cell Biol 32:3164–3175
46. Marwarha G, Berry DC, Croniger CM, Noy N (2014) The retinol esterifying enzyme lrat supports cell signaling by retinol-binding protein and its receptor stra6. FASEB J 28:26–34
47. Aaronson DS, Horvath CM (2002) A road map for those who don't know jak-stat. Science 296:1653–1655
48. O'Shea JJ, Gadina M, Schreiber RD (2002) Cytokine signaling in 2002: new surprises in the jak/stat pathway. Cell 109(Suppl):S121–S131
49. Darnell JE Jr, Kerr IM, Stark GR (1994) Jak-stat pathways and transcriptional activation in response to ifns and other extracellular signaling proteins. Science 264:1415–1421

50. Liu B, Yang Y, Chernishof V, Loo RR, Jang H, Tahk S, Yang R, Mink S, Shultz D, Bellone CJ, Loo JA, Shuai K (2007) Proinflammatory stimuli induce ikkalpha-mediated phosphorylation of pias1 to restrict inflammation and immunity. Cell 129:903–914
51. Rytinki MM, Kaikkonen S, Pehkonen P, Jaaskelainen T, Palvimo JJ (2009) Pias proteins: pleiotropic interactors associated with sumo. Cell Mol Life Sci 66:3029–3041
52. Kubo M, Hanada T, Yoshimura A (2003) Suppressors of cytokine signaling and immunity. Nat Immunol 4:1169–1176
53. Starr R, Willson TA, Viney EM, Murray LJ, Rayner JR, Jenkins BJ, Gonda TJ, Alexander WS, Metcalf D, Nicola NA, Hilton DJ (1997) A family of cytokine-inducible inhibitors of signalling. Nature 387:917–921
54. Croker BA, Kiu H, Nicholson SE (2008) Socs regulation of the jak/stat signalling pathway. Semin Cell Dev Biol 19:414–422
55. Noy N (2000) Retinoid-binding proteins: mediators of retinoid action. Biochem J 348(Pt 3):481–495
56. Herr FM, Ong DE (1992) Differential interaction of lecithin-retinol acyltransferase with cellular retinol binding proteins. Biochemistry 31:6748–6755
57. Amengual J, Golczak M, Palczewski K, von Lintig J (2012) Lecithin: retinol acyltransferase is critical for cellular uptake of vitamin a from serum retinol-binding protein. J Biol Chem 287:24216–24227
58. Sessler RJ, Noy N (2005) A ligand-activated nuclear localization signal in cellular retinoic acid binding protein-ii. Mol Cell 18:343–353
59. Gillilan RE, Ayers SD, Noy N (2007) Structural basis for activation of fatty acid-binding protein 4. J Mol Biol 372:1246–1260
60. Armstrong EH, Goswami D, Griffin PR, Noy N, Ortlund EA (2014) Structural basis for ligand regulation of the fatty acid binding protein 5, peroxisome proliferator activated receptor beta/delta (fabp5-pparbeta/delta) signaling pathway. J Biol Chem 289:14941–14954
61. Fex G, Johannesson G (1988) Retinol transfer across and between phospholipid bilayer membranes. Biochim Biophys Acta 944:249–255
62. Noy N, Slosberg E, Scarlata S (1992) Interactions of retinol with binding proteins: Studies with retinol-binding protein and with transthyretin. Biochemistry 31:11118–11124
63. Szeto W, Jiang W, Tice DA, Rubinfeld B, Hollingshead PG, Fong SE, Dugger DL, Pham T, Yansura DG, Wong TA, Grimaldi JC, Corpuz RT, Singh JS, Frantz GD, Devaux B, Crowley CW, Schwall RH, Eberhard DA, Rastelli L, Polakis P, Pennica D (2001) Overexpression of the retinoic acid-responsive gene stra6 in human cancers and its synergistic induction by wnt-1 and retinoic acid. Cancer Res 61:4197–4205
64. Berry DC, Levi L, Noy N (2014) Holo-retinol-binding protein and its receptor stra6 drive oncogenic transformation. Cancer Res 74:6341–6351
65. Baumann H, Gauldie J (1994) The acute phase response. Immunol Today 15:74–80
66. Pepys MB, Baltz ML (1983) Acute phase proteins with special reference to c-reactive protein and related proteins (pentaxins) and serum amyloid a protein. Adv Immunol 34:141–212
67. Yang Q, Graham TE, Mody N, Preitner F, Peroni OD, Zabolotny JM, Kotani K, Quadro L, Kahn BB (2005) Serum retinol binding protein 4 contributes to insulin resistance in obesity and type 2 diabetes. Nature 436:356–362
68. Berry DC, Soltanian H, Noy N (2010) Repression of cellular retinoic acid-binding protein ii during adipocyte differentiation. J Biol Chem 285:15324–15332
69. Goncalves I, Alves CH, Quintela T, Baltazar G, Socorro S, Saraiva MJ, Abreu R, Santos CR (2008) Transthyretin is up-regulated by sex hormones in mice liver. Mol Cell Biochem 317:137–142
70. Bianconcini A, Lupo A, Capone S, Quadro L, Monti M, Zurlo D, Fucci A, Sabatino L, Brunetti A, Chiefari E, Gottesman ME, Blaner WS, Colantuoni V (2009) Transcriptional activity of the murine retinol-binding protein gene is regulated by a multiprotein complex containing hmga1, p54 nrb/nono, protein-associated splicing factor (psf) and steroidogenic factor 1 (sf1)/liver receptor homologue 1 (lrh-1). Int J Biochem Cell Biol 41:2189–2203

71. Rosell M, Hondares E, Iwamoto S, Gonzalez FJ, Wabitsch M, Staels B, Olmos Y, Monsalve M, Giralt M, Iglesias R, Villarroya F (2012) Peroxisome proliferator-activated receptors-alpha and -gamma, and camp-mediated pathways, control retinol-binding protein-4 gene expression in brown adipose tissue. Endocrinology 153:1162–1173
72. Graham TE, Yang Q, Bluher M, Hammarstedt A, Ciaraldi TP, Henry RR, Wason CJ, Oberbach A, Jansson PA, Smith U, Kahn BB (2006) Retinol-binding protein 4 and insulin resistance in lean, obese, and diabetic subjects. N Engl J Med 354:2552–2563
73. Emanuelli B, Peraldi P, Filloux C, Sawka-Verhelle D, Hilton D, Van Obberghen E (2000) Socs-3 is an insulin-induced negative regulator of insulin signaling. J Biol Chem 275:15985–15991
74. Emanuelli B, Peraldi P, Filloux C, Chavey C, Freidinger K, Hilton DJ, Hotamisligil GS, Van Obberghen E (2001) Socs-3 inhibits insulin signaling and is up-regulated in response to tumor necrosis factor-alpha in the adipose tissue of obese mice. J Biol Chem 276:47944–47949
75. Krebs DL, Hilton DJ (2003) A new role for socs in insulin action. Suppressor of cytokine signaling. Sci STKE 2003:PE6
76. Pirola L, Johnston AM, Van Obberghen E (2004) Modulation of insulin action. Diabetologia 47:170–184
77. Dong F, Ren J (2007) Fitness or fatness--the debate continues for the role of leptin in obesity-associated heart dysfunction. Curr Diabetes Rev 3:159–164
78. Kloting N, Graham TE, Berndt J, Kralisch S, Kovacs P, Wason CJ, Fasshauer M, Schon MR, Stumvoll M, Bluher M, Kahn BB (2007) Serum retinol-binding protein is more highly expressed in visceral than in subcutaneous adipose tissue and is a marker of intra-abdominal fat mass. Cell Metab 6:79–87
79. Zemany L, Kraus BJ, Norseen J, Saito T, Peroni OD, Johnson RL, Kahn BB (2014) Downregulation of stra6 in adipocytes and adipose stromovascular fraction in obesity and effects of adipocyte-specific stra6 knockdown in vivo. Mol Cell Biol 34:1170–1186
80. Koptyra M, Gupta S, Talati P, Nevalainen MT (2011) Signal transducer and activator of transcription 5a/b: biomarker and therapeutic target in prostate and breast cancer. Int J Biochem Cell Biol 43:1417–1421
81. Aggarwal BB, Kunnumakkara AB, Harikumar KB, Gupta SR, Tharakan ST, Koca C, Dey S, Sung B (2009) Signal transducer and activator of transcription-3, inflammation, and cancer: how intimate is the relationship? Ann N Y Acad Sci 1171:59–76
82. Ferbeyre G, Moriggl R (1815) The role of stat5 transcription factors as tumor suppressors or oncogenes. Biochim Biophys Acta 2011:104–114
83. Quesnelle KM, Boehm AL, Grandis JR (2007) Stat-mediated egfr signaling in cancer. J Cell Biochem 102:311–319
84. Park J, Euhus DM, Scherer PE (2011) Paracrine and endocrine effects of adipose tissue on cancer development and progression. Endocrine Rev 32:550–570
85. Demark-Wahnefried W, Platz EA, Ligibel JA, Blair CK, Courneya KS, Meyerhardt J, Ganz PA, Rock CL, Schmitz KH, Wadden T, Philip EJ, Wolfe B, Gapstur SM, Ballard-Barbash R, McTiernan A, Minasian L, Nebeling LC, Goodwin PJ (2012) The role of obesity in cancer survival and recurrence. Cancer Epidemiol Biomark Prevent 21(8):1244–1259
86. Vainio H, Bianchini F (eds) (2002) Weight control and physical activity, vol 6. IARCPress, Lyon

Chapter 4
Vitamin A Absorption, Storage and Mobilization

William S. Blaner, Yang Li, Pierre-Jacques Brun, Jason J. Yuen, Seung-Ah Lee, and Robin D. Clugston

Abstract It is well established that chylomicron remnant (dietary) vitamin A is taken up from the circulation by hepatocytes, but more than 80 % of the vitamin A in the liver is stored in hepatic stellate cells (HSC). It presently is not known how vitamin A is transferred from hepatocytes to HSCs for storage. Since retinol-binding protein 4 (RBP4), a protein that is required for mobilizing stored vitamin A, is synthesized solely by hepatocytes and not HSCs, it similarly is not known how vitamin A is transferred from HSCs to hepatocytes. Although it has long been thought that RBP4 is absolutely essential for delivering vitamin A to tissues, recent research has proven that this notion is incorrect since total RBP4-deficiency is not lethal. In addition to RBP4, vitamin A is also found in the circulation bound to lipoproteins and as retinoic acid bound to albumin. It is not known how these different circulating pools of vitamin A contribute to the vitamin A needs of different tissues. In our view, better insight into these three issues is required to better understand vitamin A absorption, storage and mobilization. Here, we provide an up to date synthesis of current knowledge regarding the intestinal uptake of dietary vitamin A, the storage of vitamin A within the liver, and the mobilization of hepatic vitamin A stores, and summarize areas where our understanding of these processes is incomplete.

Keywords Vitamin A • RBP4 • Retinoid • Chylomicrons • Retinyl ester storage • Hepatic stellate cells • Adipose tissue

---

W.S. Blaner (✉) • Y. Li • P.-J. Brun • J.J. Yuen • S.-A. Lee • R.D. Clugston
Department of Medicine, College of Physicians and Surgeons, Columbia University, 650 W 168th St., New York, NY 10032, USA
e-mail: wsb2@columbia.edu; yl2680@columbia.edu; pb2390@cumc.columbia.edu; jy2297@caa.columbia.edu; sl3310@columbia.edu; rdc2132@columbia.edu

M.A. Asson-Batres, C. Rochette-Egly (eds.), *The Biochemistry of Retinoid Signaling II*, Subcellular Biochemistry 81, DOI 10.1007/978-94-024-0945-1_4

Abbreviations

apoE	apolipoprotein E
ARAT	acyl-CoA:retinol acyltransferase
ATRA	all-*trans*-retinoic acid
ATCA1	ATP-binding cassette, sub-family A, member 1
BCO1	β-carotene-15,15′-oxygenase
BCO2	β-carotene-9′,10′-oxygenase
CEL	carboxyl ester lipase
DGAT1	diacylglycerol acyltransferase 1
DGAT2	diacylglycerol acyltransferase 2
HDL	high density lipoprotein
HSC	hepatic stellate cell
HSL	hormone sensitive lipase
LDL	low density lipoprotein
LpL	lipoprotein lipase
LRAT	lecithin:retinol acyltransferase
$Lrat^{-/-}$	lecithin:retinol acyltransferase-deficient
LRP	low density lipoprotein receptor-related protein
PLRP2	pancreatic lipase-related protein 2
PTL	pancreatic triglyceride lipase
RBP1	cellular retinol-binding protein, type I
$Rbp1^{-/-}$	cellular retinol-binding protein, type I-deficient
RBP2	cellular retinol-binding protein, type 2
RBP3	interphotoreceptor retinoid-binding protein
RBP4	retinol-binding protein
$Rbp4^{-/-}$	retinol-binding protein-deficient
REH	retinyl ester hydrolase
SR-B1	scavenger receptor class B
siRNA	small inhibitory RNA
STRA6	stimulated by retinoic acid 6
TTR	transthyretin
VLDL	very low density lipoprotein.

Introduction

Vitamin A is an essential micronutrient and hence must be acquired from the diet [69]. Dietary vitamin A comprises both preformed retinoids and proretinoid carotenoids, for which β-carotene is the prime example. The two most abundant preformed retinoids in the diet are retinol and retinyl esters. Some retinoic acid is also present, but normally this represents well under 0.01 % of the total retinoid content of a healthy diet [9]. Dietary proretinoid carotenoids, including the relatively

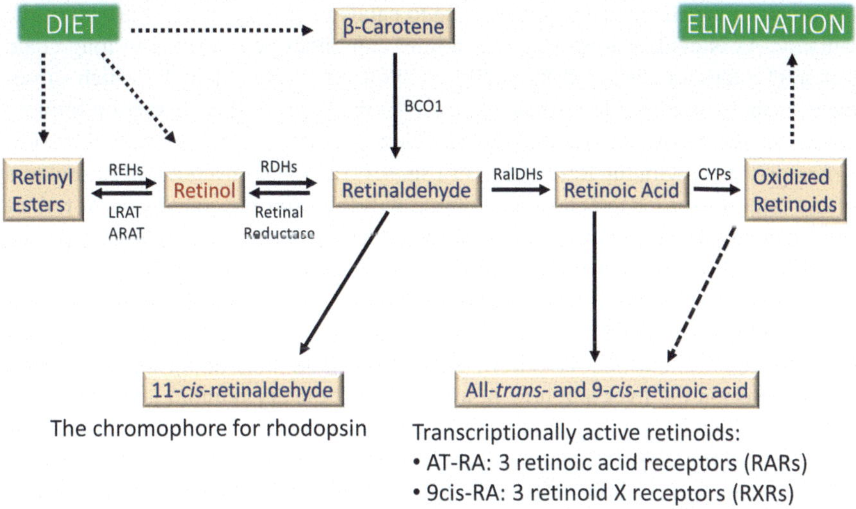

Fig. 4.1 Generalized scheme for retinoid metabolism within cells. Proretinoid carotenoids (for example, β-carotene) or preformed retinoids (primarily retinyl esters and retinol) are taken up by cells where they can undergo enzyme catalyzed metabolic transformations. These proretinoid and preformed retinoid species are also the most abundant ones present in the diet. Retinol can either be esterified to retinyl ester via the actions of LRAT or ARATs, or alternatively oxidized to retinaldehyde via the actions of retinol dehydrogenases(RDHs). Retinyl esters, the retinoid storage form, undergo hydrolysis, catalyzed by retinyl ester hydrolases (REHs) liberating retinol. β-Carotene is cleaved via the actions of β-carotene-15,15′-oxygenase (BCO1) forming retinaldehyde. Retinaldehyde undergoes irreversible oxidation to retinoic acid via the actions of retinaldehyde dehydrogenases (RalDHs) or reduction catalyzed by retinaldehyde reductases to retinol. Retinoic acid is the physiologically most important retinoid form, regulating transcription of more than 500 genes. Retinoic acid can be converted to oxidized forms through the actions of a number of different cytochrome P450 (CYP) species. These oxidized retinoic acid species may have transcriptional activity but most are destined for elimination from the body. 11 *Cis*-retinaldehyde is the visual chromophore and is required for vision. Unlike all-*trans*-retinal which is formed upon oxidation of all-*trans*-retinol, because of the unfavorable energetics needed to form 11-*cis*-retinaldehyde, this transformation requires coupled all-*trans*-retinyl ester hydrolysis and isomerization of the all-*trans*-retinoid to 11-*cis*-retinaldehyde. The isomerohydrolase reaction is catalyzed specifically and solely within the retinal pigmented epithelium of the eye by RPE65

abundant β-carotene, α-carotene and β-cryptoxanthin, either can be converted to retinoid within the intestine or absorbed intact and subsequently converted to retinoid by a number of tissues within the body. It should be noted that not all dietary carotenoids are metabolic precursors for retinoids. This is true for the common dietary carotenoids lycopene, leutein, and zeaxanthin, which may have functions within the body that do not involve retinoid-dependent pathways.

This chapter focuses on retinoid and proretinoid carotenoid absorption from the diet, uptake into the body, hepatic storage, and mobilization of retinoid from the liver. Figure 4.1 provides a general pathway for retinoid metabolism. Although there may be some differences in this scheme across different tissues and cell types, most tissues and many cell types are able to catalyze at least some, if not all of the pathway. As is evident from Fig. 4.1, retinol can either be esterified to long chain fatty acyl esters for storage within cells, or oxidized to retinaldehyde, which subsequently can be oxidized to retinoic acid. Given the focus of this chapter on retinoid dietary uptake, tissue storage, and mobilization from hepatic stores, we will be concerned primarily with the left portion of Fig. 4.1, the interconversion of retinol and retinyl ester. Other Chapters in this volume will focus on the conversion of retinol to retinaldehyde, on the conversion of retinaldehyde to retinoic acid, and on the oxidative metabolism of retinoic acid.

Although a number of key issues for understanding retinoid absorption, storage and mobilization remain to be established, the key events important to retinoid absorption in the intestine have been worked out. The most important unresolved issues center on the liver and how newly absorbed dietary retinoid is processed in this tissue, allowing for its storage, and its eventual mobilization from the liver as retinol bound to retinol-binding protein (RBP4). An equally important, unresolved issue is the question of how tissue retinoid needs are met by the circulation and how these needs are communicated to the liver, signaling the need for greater retinol mobilization from hepatic stores. This chapter will provide an historical overview of retinoid absorption, storage and metabolism, followed by consideration of how the field developed. We will then consider key gaps in our understanding and controversies that may surround these.

A provocative question raised by Fig. 4.1 that will not be specifically addressed in this chapter concerns whether all retinoid catabolism/elimination first involves the formation of retinoic acid. Retinoic acid is a very potent transcriptional regulator which is needed in proper amounts to ensure normal cellular processes [1, 21, 64]. Since as much as 99 % of all retinoid that is present in the body is either retinol or retinyl ester [9, 76], it is surprising to us that alternative mechanisms allowing for retinol and/or retinyl ester oxidation and elimination are not more prominent. If most or all of the retinoid in the body must undergo conversion to retinoic acid to allow for its elimination from the body, would this not pose a potential danger for excessive retinoic acid formation and accumulation? It also is our view that the cardinal rule governing all of retinoid storage and metabolism is that retinoic acid levels must be tightly maintained at a level that is dictated by the specific cell type and the physiologic context. Thus, the generation of retinoic acid as an intermediate in the metabolic routing needed for eliminating retinyl esters and retinol from the body seems dangerous. It is difficult to understand the evolutionary pressure or benefits that selected for this metabolism.

History

Vitamin A is by definition all-*trans*-retinol [69]. However, the term vitamin A is often used to refer collectively to all metabolites of retinol, including its retinyl ester storage form and its active forms, retinaldehyde and retinoic acid, or to speak of a biological activity (i.e. possesses vitamin A activity). When the geometric configuration of retinol or one of its metabolites is not indicated, this is normally taken to indicate the all-*trans*-isomers, which represent the predominant geometric isomers found throughout the body, aside from the eye. In vision, 11-*cis*-retinaldehyde acts as the visual chromophore and 11-*cis*-retinoids are relatively abundant [84, 127] .

In the mid-1970s, Sporn proposed that the term retinoid be used to refer to all natural and synthetic compounds that bear a structural resemblance to all-*trans*-retinol, with or without the biological activity of vitamin A [116]. An advantage of this nomenclature is that it acknowledges the relationships between the many synthetic retinoids that were generated for pharmacologic uses with the physiology and metabolism of natural vitamin A species. Presently, the term retinoid is somewhat more commonly used in the literature than vitamin A, especially when the molecular actions of these compounds are being considered. Although in this chapter the metabolism of only natural vitamin A species is to be considered, we will endeavor to use the term retinoid rather than vitamin A.

Early work, carried out in the 1930s, focusing primarily on the role of retinoids in vision, established that retinol undergoes reversible oxidation to retinaldehyde and that retinaldehyde can be subsequently oxidized to retinoic acid [127]. It was also established early on that retinyl esters are the most abundant retinoid species found in many tissues, including the eye and the liver [9].

In the late 1950s and early 1960s, seminal investigations carried out independently by the laboratories of James A. Olson, DeWitt S. Goodman, Normal I. Krinsky, and others established that dietary preformed retinoid and proretinoid carotenoids are absorbed like other dietary lipids as components of nascent chylomicrons. These intestinally derived lipoproteins are secreted into the lymphatic system and ultimately enter the general circulation. Investigations by Goodman and colleagues established that approximately 66–75 % of dietary (chylomicron) retinoid is taken up from the circulation by the liver, where it is stored [39].

In the late 1960s, Goodman and colleagues identified RBP4 as the protein present in the fasting circulation that accounts for the great majority of circulating retinoid [50]. This group of investigators subsequently established that RBP4 is synthesized and secreted by the liver, and proposed that RBP4 is responsible for mobilizing stored retinoid from the liver [113].

Although it was understood that the liver is the tissue that accounts for the great majority of retinoid present in the body, the cellular site of retinoid storage in the liver remained to be established. There was controversy in the literature over whether the hepatocyte or the hepatic stellate cell (HSC) is the cell type responsible for storage. The reader should note that in the older literature HSCs were also known as Ito cells, fat-storing cells or lipocytes, but HSC is now the preferred nomenclature (Hepatic Stellate Cell Nomenclature [45]). This issue was resolved in the 1980s, primarily by studies from Blomhoff and colleagues [14, 19] and Knook and colleagues [11], that demonstrated, employing primary liver cell isolates from rats, that HSCs account for the great majority of retinoid storage within the liver. This work also established that hepatocytes are the cellular site where dietary retinoid is taken up into the liver [14].

The early literature, a historically significant literature, has left us with a number of key unresolved issues for understanding retinoid biology. If dietary retinoid is taken up by hepatocytes and stored in HSCs, how is this retinoid transferred between these two cell types? How is HSC retinoid mobilized in times of peripherial tissue retinoid need and, does this involve the hepatocyte? How are peripheral tissue retinoid needs sensed by the liver and how does this stimulate mobilization of stored retinoid?

Perhaps the greatest problem with this early literature is that it has led to an overly simplistic and incorrect understanding of retinoid storage, mobilization, and delivery to tissues. The standard textbook summary for how tissues acquire retinoid needed to support retinoid-dependent functions is that stored retinol is released by the liver bound to RBP4 and delivered through the circulation to tissues where uptake of retinol is regulated by a plasma membrane receptor for retinol-RBP4. A rigorous reading of the literature indicates that this is a great oversimplification of how tissues acquire retinoid.

Development of the Field

Intestinal Absorption of Retinoids and Proretinoid Carotenoids

As mentioned above, intestinal aspects of preformed retinoid and proretinoid carotenoid absorption are generally well understood. Figure 4.2 provides a summary of the metabolic events that contribute to preformed retinoid and proretinoid carotenoid uptake and metabolism within the intestinal mucosa. The individual developments that led to this understanding are outlined below.

Dietary retinol is taken up directly by enterocytes lining the proximal small intestine [38, 42]. Dietary retinyl esters however are unable to be absorbed intact by the intestinal mucosa and must first be acted upon by a luminal retinyl ester hydrolase (REH) to yield free retinol. The molecular identity or identities of luminal REH(s) that are physiologically significant for retinol uptake has been studied systematically using both induced mutant mice and biochemical approaches [123, 131]. Weng et al. reported studies of dietary cholesteryl ester and dietary retinyl ester absorption in wild type (WT) and carboxyl ester lipase (CEL) knockout mice [131]. These authors showed that, compared to WT mice, mice totally deficient in CEL absorbed only about 50 % of the cholesterol provided as cholesteryl ester. However, although earlier published work had proposed that CEL acted importantly within the lumen to catalyze retinyl ester hydrolysis, WT and CEL-deficient mice were found to absorb similar amounts of retinol when it was provided as retinyl ester in a gavage dose in oil. Weng et al. concluded that enzymes other than CEL must participate in the hydrolysis of dietary cholesteryl esters and retinyl esters within the GI tract [131].

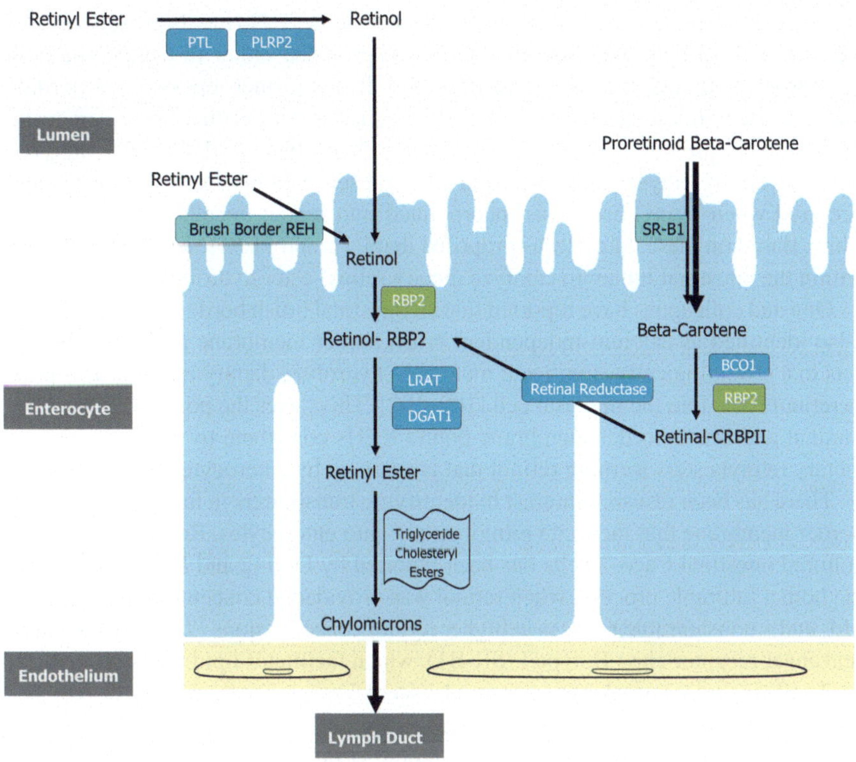

Fig. 4.2 Dietary preformed retinoid and proretinoid carotenoid uptake and metabolism by enterocytes within the intestine. Dietary retinyl ester must first be hydrolyzed to retinol, either within the intestinal lumen via the actions of pancreatic enzymes like PTL and PLRP2 or at the brush border by a membrane-bound REH. Retinol, once absorbed by the enterocyte binds RBP2 and is delivered to LRAT, and possibly ARATs like DGAT1, for re-esterification. This retinyl ester is then packaged along with other dietary lipids, including triglycerides and cholesteryl esters, in nascent chylomicrons that are secreted into the lymphatic system for uptake into the body. Dietary proretinoid carotenoids like β-carotene are taken into the enterocyte by scavenger receptors, including SR-B1. Within the enterocyte, the proretinoid carotenoid can be acted upon by BCO1 forming retinaldehyde which is bound by RBP2. Retinaldehyde bound to RBP2 is acted upon by a retinaldehyde reductase giving rise to retinol bound to RBP2. At this point, the retinol arriving as such from the diet and retinol formed through cleavage of dietary proretinoid carotenoids are metabolically indistinguishable

This group of investigators subsequently reported studies that involved the separation and partial purification of pancreatic CEL and pancreatic triglyceride lipase (PTL) by DEAE-chromatography [123]. For both rats and mice, pancreatic REH activity, measured through *in vitro* assay, was attributed mainly to PTL and, to a quantitatively lesser extent, to CEL. Purified human PTL was reported to exhibit similar enzymatic characteristics for both triglyceride hydrolysis and retinyl ester hydrolysis. Based on these biochemical data, it was proposed that PTL is the major pancreatic and hence luminal REH activity in rats and mice and is a catalytically active REH in humans, as well [123]. Purified horse PTL has been reported by Reboul et al. to hydrolyze retinyl ester when provided either in triglyceride-rich lipid droplets, mixed micelles or vesicles [93]. It was further reported that purified dog pancreatic lipase-related protein 2 (PLRP2), but not purified horse pancreatic lipase related protein 1, catalyzes retinyl ester hydrolysis [93]. PLRP2-catalyzed retinyl ester hydrolysis did not require the presence of pancreatic colipase and occurred when retinyl ester was incorporated into mixed micelles, but not emulsions. Based on these data, it was proposed that PTL and PLRP2 act synergistically within the intestinal lumen to catalyze dietary retinyl ester hydrolysis [93].

Ong and colleagues have reported that an intestinal brush border membrane REH (also identified as calcium-independent brush border membrane phospholipase B) acts in a physiologically significant manner to hydrolyze dietary retinyl esters prior to retinol entry into the mucosal cell [100–102]. This raises the possibility that both luminal and brush border membrane-bound REHs contribute to the hydrolysis of dietary retinyl esters forming retinol that is taken up by enterocytes.

There has been research interest in identifying transporters in the intestinal brush border membrane that facilitate retinol uptake into enterocytes. Retinol uptake into cultured intestinal Caco-2 cells has been reported by During and Harrison to occur via both a saturable process, when retinol was provided at concentrations below 10 μM, and a nonsaturable process at higher retinol concentrations [28]. Expression of scavenger receptor class B, type I (SR-B1), which facilitates lipid uptake into intestinal cells including uptake of carotenoids (see below), is not thought to be required for retinol uptake into Caco-2 enterocytes, since knockdown of SR-B1 expression with small inhibitory RNAs (siRNAs) failed to influence retinol uptake by the cells [28].

Interestingly, inhibition of expression of the cholesterol efflux transporter ATP-binding cassette, sub-family A, member 1 (ABCA1) in Caco-2 cells through use of either siRNAs or the drug glyburide, which inhibits transport activity of ATP-binding cassette family members including ABCA1, diminished retinol efflux from the basolateral surface of the polarized cultures of Caco-2 cells [28]. This finding led to the proposal that retinol efflux from enterocytes is probably facilitated in part by the basolateral cholesterol transporter ABCA1. However, later studies by Reboul et al. making use of both Caco-2 cells and *Abca1*-deficient mice, failed to confirm this observation [94]. Although Reboul et al. were able to convincingly demonstrate a role for ABCA1 in facilitating absorption of both α- and γ-tocopherol, their data provide no support for the notion that ABCA1 contributes to retinoid efflux, in nascent chylomicrons, from enterocytes [94]. Thus, although it is attractive to hypothesize that retinol absorption into the intestine may involve a membrane transporter, there are at present no experimental data that decisively support this hypothesis.

Unlike for retinol, there is strong experimental evidence that dietary β-carotene is taken up into enterocytes through a process that involves a specific membrane transporter. Both studies of mutant mouse models and *in vitro* cell culture experiments have established that SR-B1 is a key mediator for uptake of β-carotene from the intestinal lumen into the enterocyte [28, 63, 124]. Van Bennekum et al. studied cholesterol and β-carotene uptake by WT and SR-B1 knockout mice and concluded that SR-B1 is required for β-carotene absorption, at least for mice consuming a high fat diet [124]. SR-B1 expressed in transfected COS-7 cells [124] or in intestinal Caco-2 cells [28] was found to confer on these cells the ability to take up β-carotene from mixed bile salt micelles, phospholipid small unilamellar vesicles, and triglyceride emulsions, thus, providing further evidence for a role for SR-B1 in β-carotene absorption. In addition to facilitating mucosal uptake of the proretinoid carotenoid β-carotene, SR-B1 also acts in facilitating uptake into the enterocyte of the non-proretinoid carotenoids lycopene and lutein [71, 92].

Within the enterocyte, newly absorbed retinol must be re-esterified to retinyl ester before it can be packaged along with other dietary lipids into nascent chylomicrons [38, 125]. Although studies of lecithin:retinol acyltransferase-deficient (*Lrat*$^{-/-}$) mice have established that LRAT is the key enzyme responsible for most retinyl ester synthesis in the body [6, 62, 77], these studies also provide clear evidence that another enzyme(s) acts in the synthesis of retinyl esters in the small intestine. Specifically, O'Byrne et al. [77] reported that some retinyl ester, approximately 10 % of that of matched WT mice, was present in chylomicrons of *Lrat*$^{-/-}$ mice that had been given a physiologic oral challenge of 6 μg retinol dispersed in corn oil. It is known from *in vitro* studies that several enzymes can catalyze the acyl-CoA-dependent formation of retinyl esters. These include diacylglycerol acyltransferase 1 (DGAT1) [77, 82, 137], multifunctional acyltransferase [137], and acyl-CoA:monoacylglycerol acyltransferase [137]. Of these enzymes, the only one which has been established to act *in vivo* as an acyl-CoA:retinol acyl transferase (ARAT) is DGAT1. DGAT1 is one of two enzymes, the other being diacylglycerol acyltransferase 2 (DGAT2), which catalyzes the final step of triglyceride synthesis, transferring an acyl group from acyl-CoA to a diglyceride [103, 138]. It should be noted, based on *in vitro* studies, that DGAT2 does not possess ARAT activity [77, 137]. With regards to a role for DGAT1 as a physiologically relevant ARAT in the intestine, Wongsiriroj et al. [132] reported that mice totally lacking expression of both *Lrat* and *Dgat1* are unable to incorporate any retinyl esters in nascent chylomicrons in response to an oral challenge with a physiological dose of retinol. This is unlike *Lrat*$^{-/-}$ mice which still synthesize and incorporate some retinyl ester into nascent chylomicrons in response to a physiologic dose of retinol [77]. This implies that DGAT1 acts *in vivo* as an intestinal ARAT under physiological conditions.

Dietary proretinoid carotenoids like β-carotene can either be converted to retinaldehyde within the enterocyte or packaged unmodified into nascent chylomicrons. The intestinal enzyme responsible for the cleavage of proretinoid carotenoids to retinaldehyde is β-carotene-15,15′-oxygenase (BCO1) [126]. Cellular retinol-binding protein, type II (RBP2) is present at high concentrations in enterocytes and binds both retinaldehyde and retinol [46, 81]. Retinaldehyde, formed upon carotenoid cleavage by BCO1, binds to RBP2 and this is proposed to be the preferred substrate for reduction to retinol by an intestinal retinaldehyde reductase. Retinol bound to RBP2 is then esterified to long chain fatty acids through the action of LRAT, which utilizes retinol bound to RBP2 as a substrate for esterification [46, 74, 81]. The resulting retinyl esters are then packaged along with unmodified dietary carotenoid and other dietary lipids into chylomicrons that are secreted into the lymphatic system in route to the general circulation [9, 38, 125].

Although BCO1 is now accepted to be the sole enzyme responsible for the formation of retinoids from dietary proretinoid carotenoid precursors [126], the long history of research leading to this conclusion has been filled with controversies. It was first recognized in the early 1930s that oral β-carotene could be converted by rats into retinoid that was found in the liver [30, 68]. However, until the molecular cloning of BCO1 and a structurally-related enzyme, β-carotene-9′,10′-oxygenase (BCO2), 70 years later in the early 2000s [22, 47, 53, 59, 60, 83, 98, 134, 136], the biochemical details of how this occurs were controversial. The controversy here centered on two alternative theories for explaining the mechanism of conversion of proretinoid carotenoids to retinaldehyde that were proposed in the 1960s and 1970s. Olson [79, 80] and Goodman [40, 48] argued that β-carotene is specifically cleaved at its 15–15′ double bond by a specific enzyme(s), giving rise to two molecules of retinaldehyde. Others, originally Ganguly [111, 112] and later Krinsky, Wang and Russell [120, 128] proposed that the β-carotene molecule is eccentrically cleaved (at double bonds other than the 15–15′ central double bond), yielding β-apo-carotenoids, which in turn were proposed to be shortened stepwise, through a process analogous to β-oxidation, ultimately yielding retinol. It is now clear that BCO1 catalyzes central cleavage at the 15–15′ double bond and that BCO2 catalyzes eccentric cleavage, primarily at the 9′–10′ double bond. Recently, von Lintig and colleagues [2] reported that BCO1 can convert β-apo-10′-carotenol to retinoid and, independently, Harrison and colleagues [25] reported that BCO1 can shorten β-apo-carotenals of 22 carbons or more to retinaldehyde. Thus, it is now thought that BCO1 actions are essential for the formation of all retinoid that is derived from dietary proretinoid carotenoids. However, BCO2, to some extent, contributes to retinoid formation by generating β-apo-carotenoids that will be acted upon by BCO1 to form retinaldehyde. The quantitative extent of BCO2's *in vivo* role in retinoid formation remains to be elucidated.

The reaction mechanism of BCO1 catalysis has, over the years, also been controversial. This has created confusion with the nomenclature used for this enzyme. In the mid-1960s, Goodman and colleagues [40], employing a crude cytosol preparation from rat intestinal mucosa and doubly radiolabeled β-carotene at the 15–15′ double bond, reported findings that were consistent with the conclusion that the central cleavage of β-carotene involved a dioxygenase mechanism. Thus, the enzyme responsible for β-carotene central cleavage was referred to in the literature as β-carotene-15,15′-dioxygenase or BCDO. With the cloning of the cDNA and gene for this enzyme, Woggon and colleagues [57] reinvestigated this question by expressing purified recombinant chicken central cleavage enzyme and assessing $^{17}O_2$ or $H_2{}^{18}O$ incorporation into the product retinaldehyde. These investigators concluded that central cleavage proceeds with a monooxygenase mechanism. This led to the central cleavage enzyme being referred to in the subsequent literature as β-carotene-15,15′-monooxygenase or BCMO. Very recently however, Harrison and colleagues [26], employing the purified recombinant human protein responsible for β-carotene central cleavage and $^{16}O_2$-$H_2{}^{18}O$, concluded that the reaction mechanism is that of a dioxygenase. The reaction mechanism of BCO2 has not been systematically characterized in the literature. Since research focused on this issue is probably not yet at its final stage, throughout this chapter, we refer to the central cleavage enzyme simply as β-carotene-15,15′-oxygenase or BCO1 and the eccentric cleavage enzyme as β-carotene-9′,10′-oxygenase or BCO2. We note though that the literature employs a number of different names and acronyms for these enzymes that correspond with the era when the work was undertaken and published.

Of interest, BCO1 and BCO2 are members of a protein family whose other mammalian member is RPE65, an enzyme that is expressed in the retinal pigment epithelium and which catalyzes the coupled hydrolysis and isomerization of all-*trans*-retinyl esters to 11-*cis*-retinol for use in the retinoid cycle and vision [98, 126]. Thus, all three mammalian members of this protein family act in retinoid biology. Other members of this protein family, including ones found in plants, also catalyze the metabolism of carotenoids [37].

Dietary carotenoid uptake by the intestine is regulated by nutritional retinoid status [63]. Studies by Seino et al. identified that the intestine-specific transcription factor *Isx* (intestine specific homeobox) plays a key role in regulating expression of *Bco1* in the mouse intestine [108]. These investigators, who generated and studied *Isx*-deficient mice through knock-in of LacZ, convincingly established that mRNA levels for both *Bco1* and *SR-B1* are greatly increased in the intestines of *Isx*-knockout mice. They further showed that severe retinoid-deficiency markedly decreased *Isx* expression and that this was accompanied by an increase in *Bco1* expression in both duodenum and jejunum. Based on their data, Seino et al. suggested that *Isx* participates in the maintenance of retinoid metabolism by regulating *Bco1* expression in intestine [108]. Lobo et al. carried this idea further by showing that retinoic acid, acting through retinoic acid receptors, induces *Isx* expression [63]. This effect of retinoic acid on *Isx* expression resulted in repression of both the *Bco1* and *SR-B1* genes. Through study of *Bco1*-deficient mice, Lobo et al. were also able to demonstrate that increased *SR-B1* expression and systemic β-carotene accumulation could be prevented through administration of dietary retinoid, which induced *Isx* expression, resulting in a downregulation of *SR-B1* expression and β-carotene uptake and systemic accumulation. Thus, the work of Lobo et al. established the existence of a diet-responsive regulatory network that controls β-carotene absorption and retinoid production through negative feedback regulation of *Isx* [63].

Chylomicron Transport and Metabolism

After traversing the lymphatic system and entering the general circulation, chylomicrons undergo a process of remodeling that involves primarily the hydrolysis of triglycerides by lipoprotein lipase (LpL) and the acquisition of new apolipoprotein components, especially apolipoprotein E (apoE) [24, 95, 125]. This results in the formation of much smaller and relatively triglyceride-poor chylomicron remnants.

The hepatocyte is the cellular site within the liver where chylomicron remnants are cleared from the circulation [24, 95]. Uptake of the remnant particle by the hepatocyte involves cell surface receptors that recognize apoE [24, 95]. The topic of receptor-mediated remnant uptake has been extensively and well reviewed by both Cooper and Redgrave, and the reader is referred to these works for more details [24, 95]. Two distinct cell surface receptors are thought to facilitate chylomicron remnant uptake. One receptor-mediated pathway involves direct uptake by the low density lipoprotein (LDL) receptor, which has a high affinity for the apoE-rich chylomicron remnant particles, and internalization via endocytosis. If the LDL receptor is absent, down-regulated or saturated, the remnants may be sequestered in the space of Disse by binding to heparin sulfate proteoglycans, mediated by apoE. The remnants may also be sequestered through binding to hepatic lipase, which is enhanced by the presence of apolipoprotein B. If not immediately internalized, the remnants may eventually be transferred to LDL receptors as they become available or, if the remnants acquire enough apoE, transferred to an alternative receptor, the LDL receptor-related protein (LRP). LpL, acquired by the remnants during their formation, can facilitate uptake of remnants by LRP. Though the relative extent to which each receptor contributes to chylomicron remnant-retinyl ester removal has not been established, this likely depends on the metabolic state of the animal, which in turn can influence the amount of apoE being secreted.

During chylomicron remodeling in the circulation, some chylomicron retinyl ester is also hydrolyzed by LpL and the resulting retinol is taken up by extrahepatic tissues [13, 122]. It was first reported in the mid-1960s that approximately 66–75 % of dietary retinoid is delivered to the liver with the remainder being distributed to other tissues throughout the body [39]. The actions of LpL on chylomicrons and their remnants and the retinyl esters they contain helps explain this distribution of dietary retinoid between the liver and peripheral tissues [13, 78, 122].

Hepatic Retinoid Storage and Metabolism

Based on measures of tissue retinoid levels, Blomhoff and coauthors [19] concluded that for a nutritionally retinoid-sufficient rat, probably greater than 90 % of the total retinoid present in the whole body is found in the liver. They went on to conclude that this percentage will undoubtedly be linked to retinoid nutritional status, with a substantially lower percentage present in the livers of animals experiencing insufficient dietary retinoid intake. This conclusion is supported by compartmental modeling studies carried out in rats receiving different quantities of vitamin A in their diets [23, 41, 58].

A more recent study by Kane et al. one systematically exploring total retinoid levels in tissues of 2- to 4-month old male sv129 mice fed the AIN-93 M diet containing 4 IU vitamin A/g diet from the time of weaning, provides data which are consistent with the notion that approximately 90 % of whole body retinoid is present in the livers of these mice [51]. However, Kane et al. did not report total retinoid

Table 4.1 Comparison of retinoid-related parameters in hepatocytes and hepatic stellate cells

	Hepatocytes	Hepatic stellate cells
% total retinoid in body[a]	9 %	76 %
Cellular retinoid levels[b] (retinol eq./10^6 cells)	1.5 μg	84 μg
Retinyl ester storage?	Some	Yes; in lipid droplets
RBP4 synthesis?	Yes	No
Chylomicron uptake?	Yes	No

[a]Assuming that 85 % of all retinoid present in the body is stored in the liver estimated from levels measured for isolated rat liver cell preparations [11]

[b]Levels are give in terms of retinol equivalents per 10^6 isolated cells in the rat [11]

levels for lungs and their data perhaps somewhat overestimated the relative percentage of retinoid present in the liver. O'Byrne et al. reported tissue total retinoid levels for 3–4 month old male mixed genetic background (C57BL/6/sv129) mice fed from the time of weaning a chow diet containing 25 IU vitamin A/g diet [77]. When corrected for organ weight, the lungs of these mixed background mice were estimated to contain approximately 11 % of the total retinoid that is present in the liver. Thus, the combined data provided by Kane et al. and O'Byrne et al. suggest that, for the mouse, approximately 80–85 % of the total retinoid in the body may be found in liver. This estimate based on data obtained from mice is only slightly lower than the 90 % estimated by Blomhoff and coauthors two decades ago for the rat [19]. Thus, for these two species, there is good agreement on this point.

As noted in the previous section, the literature indicates that approximately 66–75 % of post-prandial retinoid is delivered to the liver [39]. This value is somewhat discordant with the projections of hepatic retinoid accumulation reported in the two paragraphs above. Why is this? It is well established that extrahepatic tissues are able to synthesize RBP4 [113, 115, 121] and it has been proposed based on compartmental modeling studies that this extrahepatic RBP4 recycles retinol from the periphery to the liver [19, 23, 41, 58, 113, 115]. Possibly this role for extrahepatically synthesized RBP4 provides an explanation for the discordance between these values. That is, extrahepatically synthesized RBP4 may be delivering dietary (chylomicron) retinold that was taken up in the periphery back to the liver for storage. Never-the-less, it is clear from both types of measurements that the liver is the central storage organ for retinoid within the body.

Within the liver, two distinct hepatic cell types play key roles in the storage and metabolism of retinoids, hepatocytes and the non-parenchymal hepatic stellate cells (HSCs) [11, 16]. There is no evidence that the resident macrophages of the liver, the Kupffer cells, have a role in hepatic retinoid physiology [11, 16]. There are published reports indicating that hepatic endothelial cells express LRAT protein [73]. However, there is no evidence that these cells accumulate significant concentrations of retinyl esters [11, 16]. Possibly, the hepatic endothelial cells in the future will be found to have an important role in these processes, but this remains to be established. Table 4.1 summarizes some similarities and differences among retinoid-related parameters for hepatocytes and HSCs.

The Hepatocyte

Shortly after the retinyl ester-containing chylomicron remnant is internalized by the hepatocyte, at the stage of the early endosome, the dietary retinoid, is separated from the other lipids present in these particles. This retinoid is then either bound to newly synthesized RBP4 and secreted from the hepatocyte into the circulation in order to meet peripheral tissue needs for retinoid or alternatively transferred to HSCs for storage. The process of RBP4 secretion from hepatocytes is similar to that of other secretory proteins. RBP4 is synthesized as a pre-protein with a signal peptide that is cleaved prior to secretion [114]. Retinol is loaded into newly synthesized apo-RBP in the endoplasmic reticulum, although how loading takes place has not been established [113, 117]. When retinol is unavailable, for instance upon consumption of a retinoid-deficient diet, apo-RBP4 is not secreted and accumulates in the endoplasmic reticulum to levels that are three to tenfold higher than those of a retinoid-sufficient liver [27, 72, 113]. There is evidence that newly synthesized RBP4 also binds transthyretin (TTR) within the hepatocyte [65, 66] but binding to TTR does not appear to be required for secretion to take place since RBP4 is secreted normally from *Ttr*-deficient mouse hepatocytes [129].

At the molecular level, the processes and events required for retinol transfer from hepatocytes to HSCs are not well understood and controversial. There is agreement in the literature that dietary retinyl ester must first be hydrolyzed to retinol within the hepatocyte to allow for its transfer to the HSC for storage [12]. However, the identity or identities of the REH(s) that act to catalyze retinyl ester hydrolysis within hepatocytes has not been definitively established. Harrison and colleagues have reported that newly-delivered retinyl esters and REH activity co-localize to the same populations of hepatic plasma membranes/endosomes [43]. This suggests a role for these enzymes in the initial metabolism of chylomicron remnants and this is supported by the observation that hepatic plasma membrane/endosomal fractions can catalyze the hydrolysis of chylomicron retinyl esters *in vitro*. Purification to homogeneity of the rat liver enzymes harboring neutral and acidic REH activities (pHs that would be expected to be present in early endosomes) established the identities of these as carboxylesterases ES-10 and ES-2 [61, 119]. Since ES-2 is a soluble secreted enzyme while ES-10 is membrane-bound and associated with endosomal compartments of rat liver cells, this implies that ES-10 may be importantly involved in this process. However, the possibility that ES-10 is indeed the REH responsible for the hydrolysis of newly taken up chylomicron remnant retinyl ester needs to be established through physiologically relevant *in vivo* studies.

How newly absorbed dietary retinoid is transferred from hepatocytes, where the retinyl ester-containing chylomicron remnant is taken up into the liver, to HSCs, where retinoid is stored is not understood and controversial. More than 20 years ago, it was proposed that RBP4 was responsible for this transfer [19]. However, we now know that this cannot be the case since *Rbp4*-deficient mice display normal HSC retinoid storage [88]. The controversy surrounding this process will be discussed in section "Current state of the field" below.

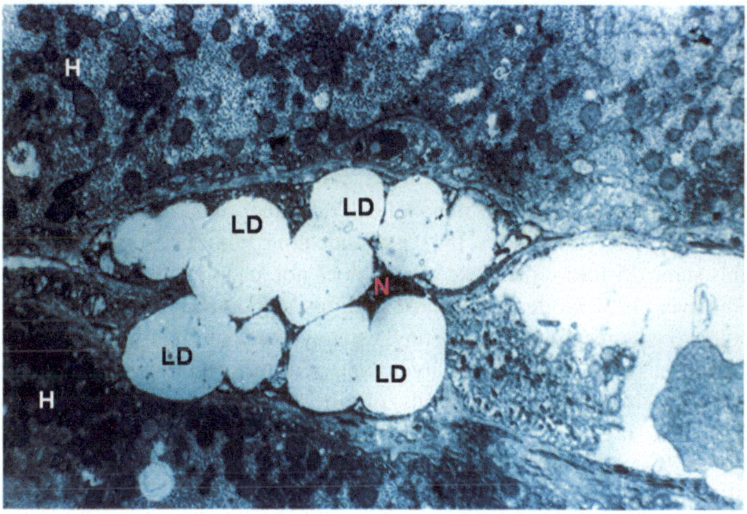

Fig. 4.3 Electron micrograph of a human hepatic stellate cell showing numerous large lipid droplets (LD) and the nucleus (N) surrounded by these lipid droplets. The cells labeled "H" immediately above and below the lipid droplet containing cell are hepatocytes

Hepatic Stellate Cells

HSCs account for approximately 8 % of the total cells present in the liver and about 1 % of hepatic protein [11, 33, 35]. However, approximately 80–90 % of the total retinoid in the liver of a healthy well nourished human or animal is present in HSCs [5, 11, 16]. Since the liver accounts for greater than 80 % of the total retinoid present in the body, the majority of the retinoid present in the entire body is localized to this small and relatively unabundant hepatic cell type. Within the HSC, retinoid is stored as retinyl ester within prominent cytoplasmic lipid droplets, which occupy most of the cytoplasm. This can be seen in Fig. 4.3, which provides a micrograph of a human hepatic stellate cell. HSC lipid droplets have an unusual lipid composition with a relatively high retinoid content, which is unlike the lipid droplets present in adipocytes or hepatocytes that consist almost entirely of triglyceride [44, 70, 135]. Moriwaki et al. reported that lipid droplets purified from HSCs isolated from rats fed a purified diet containing 8 IU vitamin A/g diet had an average lipid composition that consisted of 39.5 % retinyl ester, 31.7 % triglyceride, 15.4 % cholesteryl ester, 4.7 % cholesterol, 6.3 % phospholipids and 2.4 % free fatty acids [70]. These authors further reported that the HSC lipid droplet lipid composition was very responsive to changes in dietary retinoid intake but not to changes in dietary fat intake [70]. Based on their retinyl ester content and their responses to dietary retinoid intake, it would appear that HSC lipid droplets are highly specialized for retinoid storage and unlike those found in hepatocytes or adipocytes, not for storage of triglyceride.

HSCs express relatively high levels of LRAT and consequently have considerable enzymatic capacity for retinyl ester synthesis [55]. They also express DGAT1 but, unlike in the intestine, DGAT1 does not appear to catalyze retinyl ester formation in HSCs. Even when *Lrat*$^{-/-}$ mice are fed a diet containing a 25-fold excess of vitamin A over that which is present in a conventional chow diet, no retinyl esters accumulate in the liver [133]. Moreover, when cellular retinol-binding protein, type 1 (RBP1) (which is proposed to metabolically channel retinol to LRAT preventing esterification by ARAT activities, [81]) is absent along with LRAT in *Lrat*$^{-/-}$/*Rbp1*$^{-/-}$ mice, this same 25-fold retinol excess diet does not lead to retinyl ester accumulation in the liver [133]. Collectively, these findings support the conclusion that LRAT is the sole enzyme responsible for retinyl ester synthesis in the liver.

As is the case for retinyl ester hydrolysis within hepatocytes, the identity of physiologically important REHs catalyzing the hydrolysis of stored lipid droplet retinyl esters in HSCs is not definitively established. At least four enzymes demonstrated to possess REH activity upon *in vitro* assay are expressed in HSCs: ES-10, LpL, PLRP2, and hormone sensitive lipase (HSL) [67, 85], but it remains to be established whether these or other REHs act *in vivo* in catalyzing retinyl ester hydrolysis in HSCs. There is a clear need for new studies to establish if any of these enzymes, or possibly other enzymes, may be responsible for the hydrolysis of stored retinyl esters within the HSC lipid droplets.

Another major question still confronting us for gaining a true understanding of hepatic retinoid storage and metabolism concerns how HSC retinoid stores are mobilized. It is clear that the great preponderance, if not all, of the retinol present in the fasting circulation is bound to RBP4 [87]. HSC retinoid stores accumulate but cannot be mobilized by *Rbp4*$^{-/-}$ mice [87–89]. Thus, RBP4 is essential for retinol mobilization from the liver. But the literature is controversial as to whether HSCs synthesize and secrete significant levels of RBP4 [76]. Some argue that HSCs do account for substantial RBP4 synthesis and secretion, whereas others have refuted these findings with published data in the literature. This remains the major, and possibly the most controversial issue related to hepatic retinoid storage and mobilization that remains to be resolved. This too will be discussed in more detail in section "Current state of the field" below.

Current State of the Field

As noted above, there are a number of very important but controversial issues regarding hepatic retinoid storage and mobilization, and retinoid delivery to tissues that still need to be resolved. We will consider below the three issues that we view as the most important. For each, we have strong opinions that are based on our own work and our reading of the literature. Undoubtedly, others will hold different opinions that may be based on their work and experiences.

First, How Is Dietary Retinoid that Is Taken Up by Hepatocytes Transferred to HSCs for Storage?

Twenty years ago, the field thought that this question had been resolved – via nascent RBP4. But *Rbp4*$^{-/-}$ mice normally accumulate retinoid in HSCs [87–89], so RBP4 cannot be the answer. Since transfer does not take place unless retinyl ester is first hydrolyzed to retinol [12], it can be inferred that retinol is the retinoid species that is transferred. Although this seems likely, there presently are no data to establish this unequivocally. By the mid-1990s, several hypotheses were put forward to explain the transfer process [8, 17–19]. The most widely accepted one proposed that nascent RBP4 synthesized by hepatocytes delivers newly absorbed retinol to HSCs for storage [8, 17–19]. This hypothesis invoked the involvement of an HSC cell surface receptor for RBP4 and the possible internalization of RBP4 by HSCs [19]. A second hypothesis proposed that transfer is mediated via RBP1[29]. While another hypothesis proposed that retinoid transfer occurred through direct contacts between the two cell types, possibly through gap junctions that are known to exist between these cells [8]. The latter two hypotheses are not mutually exclusive since RBP1 protein is found in both hepatocytes and HSCs [11, 17, 18], and is sufficiently small to allow for its movement through gap junctions. The possibility that RBP4 mediates this process was disproven upon the generation and characterization of *Rbp4*$^{-/-}$ mice. Although these mice totally lack *Rbp4* in all tissues, the retinoid-containing HSC lipid droplets were identical to those found in age-, gender- and diet-matched WT mice [87–89]. This finding unequivocally establishes that RBP4 is not essentially involved in facilitating retinol transfer from hepatocytes to HSCs. Mice totally lacking *Rbp1* (*Rbp1*$^{-/-}$ mice) were also found to possess retinoid-containing lipid droplets, albeit smaller and less numerous [36]. This raises the possibility of RBP1 involvement in retinoid transfer, although RBP1 is not absolutely required for transfer. The possible involvement of gap junctions in the process has not been systematically explored and, as far as we are aware, there is no information addressing this possibility in the recent literature. Thus, although the identities of the molecular processes facilitating retinoid movement from hepatocytes to HSCs were first questioned more than 20 years ago, these still remain to be established.

We have wondered about the possibility that not all dietary retinoid present in chylomicron remnants is taken up by hepatocytes, or alternatively, whether a significant portion of dietary retinoid may be taken up directly by HSC bypassing hepatocytes. This idea is based on the observation that LpL facilitates chylomicron retinol uptake into peripheral tissues and that this provides a mechanism allowing for the distribution of postprandial retinoid to extrahepatic tissues. LpL mRNA is expressed by HSCs and, moreover, these cells are the major site of LpL expression within the liver [67]. It is possible that as the chylomicron remnants traverse the hepatic sinusoids, LpL located on the HSC cell surface hydrolyzes remnant retinyl ester facilitating retinol uptake into the HSC. Although we find this an attractive hypothesis, the early data establishing that chylomicron remnant retinyl ester is taken up first by hepatocytes are also very compelling. When chylomicrons containing [^{3}H]retinyl ester were injected into rats, the ^{3}H-label first appeared in hepatocytes followed by its later appearance in HSCs [14, 15]. This clearly suggests that postprandial retinoid is first cleared within the liver by hepatocytes.

As another possibility, one might propose that an extracellular retinoid-binding protein, one analogous in its actions to interphotoreceptor retinoid-binding protein (RBP3) in the eye that facilitates retinoid transfer between the retinal pigment epithelium and the photoreceptors [104], facilitates intercellular retinoid transfer in the liver. Although this is an attractive idea, such a hepatic protein has never been observed. We note that RBP3 is not expressed in the liver. Thus, even though there are a number of attractive hypotheses for explaining how postprandial retinoid may be acquired by HSCs for storage, each has its weaknesses. This research question, we believe, would benefit from an infusion of new ideas and/or experimental perspectives.

A Second Major Issue Concerns How Retinoid Stored in HSCs Is Acquired by Apo-RBP4 to Allow for Its Delivery to Tissues Throughout the Body

As noted above, any model for explaining how retinoid is mobilized from HSC stores must involve RBP4. But how much RBP4 do the HSCs synthesize and secrete? We would argue that it is the hepatocyte and not the HSC that synthesizes and secretes RBP4 [10]. Going back into the 1980s, the issue of how much RBP4 is synthesized and secreted by HSCs has been a controversial one. Some reports from that era maintained that immunoreactive RBP4 protein is found in freshly isolated primary rat HSCs [3, 17–19, 109, 110]. Others reported not detecting either RBP4 protein or mRNA in highly purified primary HSCs [11, 34, 104, 130]. These investigators argued that RBP4 protein or *Rbp4* mRNA present in isolated HSCs reflected hepatocyte-derived contamination of the HSCs [10]. In our view, the convincing evidence addressing this point, especially more recently published evidence, indicates that the hepatocyte is the major cellular site for RBP4 synthesis and secretion in the liver [10]. Thus, RBP4, which plays an essential role in hepatic retinoid mobilization and delivery to tissues, is synthesized primarily by hepatocytes, even though hepatic retinoid stores are predominantly localized to the HSC. It is currently not known how RBP4 that is synthesized primarily, or possibly solely, by hepatocytes acquires retinol from HSC stores in times of dietary retinoid-insufficiency. Possibly, apo-RBP4 present in the circulation and the intercellular space in the liver is able to draw retinol from HSC stores. However, when retinol is unavailable, hepatocytes retain nascent apo-RBP4 in the endoplasmic reticulum.

Our working model for this process is provided in Fig. 4.4. Newly absorbed retinol present in hepatocytes can be secreted from these cells bound to nascent RBP4. We further propose that HSC lipid droplet retinyl ester stores are continually accumulated via the actions of LRAT and broken down via the actions of one or more

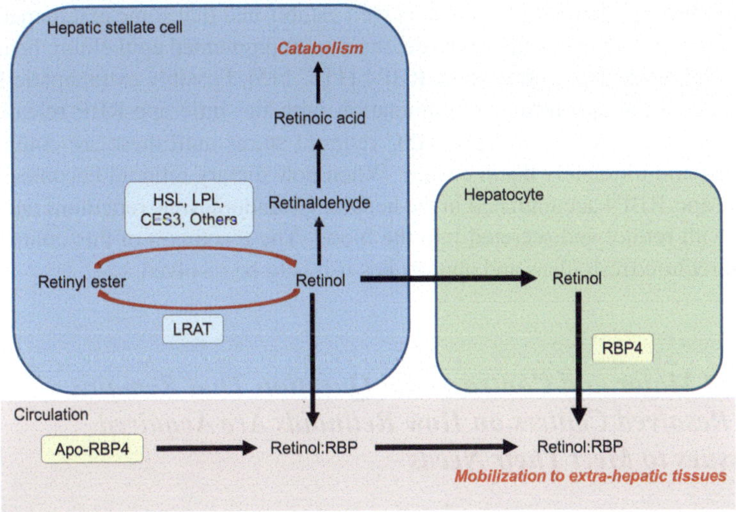

Fig. 4.4 One hypothesis for how stored retinoid within hepatic stellate cells (HSCs) is mobilized bound to RBP4. Since in our view HSCs do not synthesize or secrete physiologically significant amounts of RBP4, retinol cannot be directly mobilized from HSC stores bound to RBP4. Retinol, newly formed through the hydrolysis of retinyl ester stores (possibly catalyzed by enzymes like HSL or LpL), is either transferred to the hepatocyte where it can bind to nascent apo-RBP4 or is transferred to apo-RBP4 in the extracellular space, possibly involving an efflux transporter in the plasma membrane. The transfer of retinol from HSCs to hepatocytes could involve a reversal of the transport process that delivered newly absorbed dietary retinoid from the hepatocyte to HSCs for storage

REHs (for instance, HSL, LpL, or CES3). In times of sufficient or excessive dietary retinoid intake, the actions of LRAT in catalyzing retinyl ester accumulation will predominate. In times of insufficient dietary retinoid intake, REH actions will predominate, freeing retinol from storage. This is not unlike the continual synthesis and hydrolysis of triglycerides that is known to occur in adipocyte lipid droplets [20]. The retinol liberated will be delivered by RBP1 to the HSC plasma membrane where it crosses the membrane bilayer and is picked up on the outer leaflet of the plasma membrane by apo-RBP4 present in the extracellular space. Retinol is known to rapidly "flip-flop" between the two leaflets of the membrane bilayer, so invoking the actions of an efflux transporter may not be necessary [31, 32, 75]. However, an as yet to be identified efflux transporter may be involved in the process. Although STRA6 is known to catalyze retinol efflux from cells [49, 54], the gene for this protein is not expressed in the liver or HSCs; hence, STRA6 cannot be involved.

The weakness of our model, and what needs to be understood if the model is to be accepted as valid, is the origin of the apo-RBP4 in the extracellular space. Hepatocytes, in our view the major cellular site for RBP4 synthesis and secretion in the body, retain newly synthesized apo-RBP4, and secrete far less apo-RBP4 than holo-RBP4 [27, 72]. However, in the retinoid-deficient rat, even though blood retinol levels become undetectable, low levels of RBP4 (presumably as apo-RBP4) remain in the circulation [72]. It also is well established that some extrahepatic tissues, including kidneys, adipose tissue, the retinal pigmented epithelium, heart and others, synthesize and secrete some RBP4 [113, 115]. Possibly extrahepatic RBP4 is secreted in the apo-form, in combination with the little apo-RBP released by hepatocytes suffices to draw upon HSC retinoid stores until these are completely exhausted and ultimately death ensues. When new dietary retinoid becomes available, the apo-RBP4 accumulated in the hepatocyte endoplasmic reticulum rapidly is loaded with retinol and secreted into the blood. The resolution of this controversy too will require fresh ideas and approaches if it is to be resolved.

A Third Major and Controversial Question That Remains to Be Resolved Centers on How Retinoids Are Acquired by Tissues to Meet Their Needs

Certainly, retinol-RBP4 is central to this process. But it is generally assumed that the delivery of retinol bound to RBP4 to tissues is the sole or predominant mechanism responsible for this. This is contrary to our view. We have proposed that it is dietary retinoid obtained from chylomicrons and their remnants that accounts for why retinol-binding protein-deficient (*Rbp4*$^{-/-}$) mice and humans totally lacking *Rbp4* are viable aside from visual defects [7, 87–89, 107]. RBP4 allows for the accumulation of retinoid stores in the liver because the existence of RBP4 provides a means for mobilizing hepatic retinoid stores [87]. The ability to accumulate and mobilize retinoid stores affords the organism a great selective advantage, providing a buffer against times of low dietary retinoid intake. This is the most important function of RBP4 in the body. In our view, RBP4 is simply one of several physiologically important pathways for delivering retinoids to tissues, another important one being the post-prandial delivery of retinol described above. The literature generally does not acknowledge this point. Often, or even usually, authors propose that the existence of RBP4 and its membrane receptor STRA6 allow for controlled or regulated delivery of retinol to tissues [86, 118]. The emphasis here has always been on "controlled or regulated delivery". This ignores alternative pathways that allow tissues to acquire needed retinoids. We believe that this is far too simple a view of retinoid delivery to tissues; one focused solely on RBP4, ignoring alternative pathways, which allow *Rbp4*-deficient mice and humans to thrive with relatively mild, near normal phenotypes. We also note that retinoic acid is always present in the circulation and available for tissue uptake [56]. Moreover, retinyl esters are secreted from the liver in nascent very low density lipoprotein (VLDL) [4, 99]. Although VLDL retinyl esters are found at low levels in human and rodent circulations, lipoprotein-bound retinyl ester accounts for a large percentage of the retinoid present in the blood of some mammals, including dogs and cats [90, 91, 106]. In fact, it appears from the literature that VLDL retinyl ester is the major circulating retinoid form in some mammals, not retinol-RBP4.

Dietary proretinoid carotenoids can also be absorbed intact by the intestine, without being converted to retinoid first [40, 48, 80]. These are packaged along with retinoids and other dietary lipids in nascent chylomicrons and delivered primarily to the liver. The liver either retains these carotenoids or secretes them bound to VLDL back into the circulation. In the circulation, proretinoid carotenoids are also found in LDL, which is formed from VLDL, and in HDL [69, 97]. This results in proretinoid carotenoids being distributed to and accumulated by nearly all tissues within the body. BCO1 is present in many tissues, including the liver, lungs, testes, and eyes [22, 53, 59, 83, 98, 134, 136]. Consequently, it is now believed that proretinoid carotenoids like β-carotene present within peripheral tissues can be acted upon by BCO1 to form retinoid that will be used by tissues to maintain normal retinoid-dependent functions. However, there are no systematic studies of quantitatively how important this may be to specific tissues for generating retinoids or regarding whether this process is regulated. Never-the-less, the presence of proretinoid carotenoids in the circulation must be taken into account too when one considers how tissues can acquire retinoid needed to maintain normal cellular functions.

A more accurate way for understanding how retinoids can be acquired by peripheral tissues, and one consistent with our views, is provided in Fig. 4.5. The misunderstanding of this point, one that has been repeatedly propagated in textbooks and the literature over the last 40 years, needs to be corrected.

Relevance

The body's ability to accumulate and store retinoid as a buffer against dietary insufficiency provides a considerable selective advantage to the organism. These stores can be sufficiently large to buffer for many months against insufficient dietary intake. This is certainly unique to the retinoids and unlike other vitamins. The key unresolved issues that were considered above are all central to understanding the storage and mobilization of retinoids. The resolution of these issues is not only relevant for understanding the basic molecular mechanisms that underlie retinoid storage and ultimately actions, but also for truly understanding the evolutionary pressures that created this unique biology. Given that retinoic acid is so extensively involved in regulating cellular processes, the organism cannot go for a long period without retinoid. Hence, the accumulation of HSC retinoid stores and RBP4, and the mobilization of stores that it facilitates, underlie the maintenance of all, or nearly all, retinoid-dependent functions in the body.

We view retinoid storage and mobilization as having two very distinct arms. First, there is the primordial arm, the delivery of dietary retinoid to tissues by chylomicrons and their remnants. This allows for the delivery of newly ingested preformed retinoid and proretinoid carotenoids to tissues, for use in maintaining retinoid-dependent processes. This is not different than the situation for most macronutrients and micronutrients--uptake into the body and immediate delivery/utilization. The obvious evolutionary weakness in possessing only this pathway is

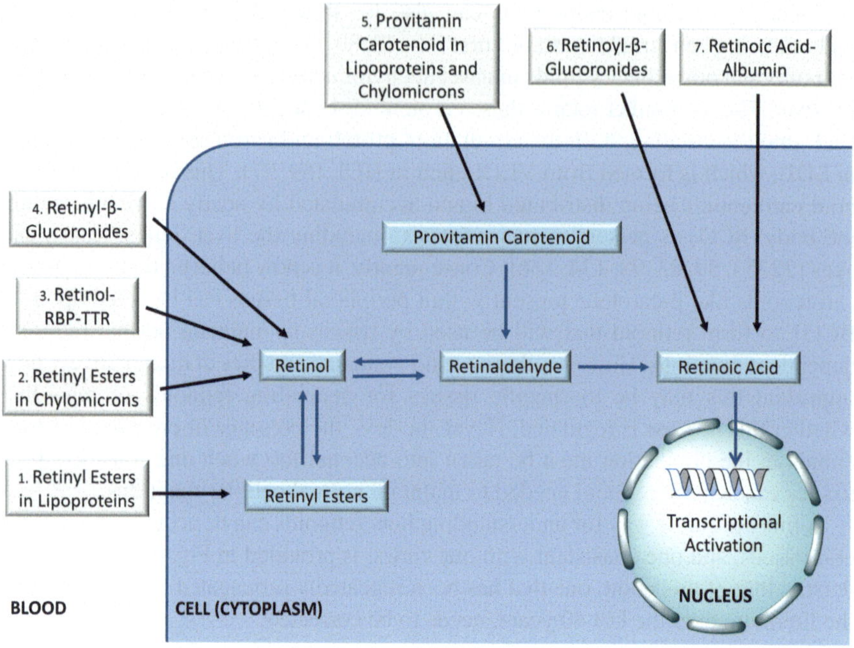

Fig. 4.5 A more rigorous way of understanding how retinoids are delivered to tissues where they are needed to support retinoid-dependent functions. Retinoids and proretinoid carotenoids are delivered to cells and tissues through a number of complementary pathways. Retinyl esters are delivered to cells and tissues either (1) bound to VLDL, LDL, or HDL (in the fasting circulation) or (2) packaged in chylomicrons/chylomicron remnants (after dietary retinoid intake). Retinol bound to the retinol-RBP4-TTR complex is taken in by cells (3), via the actions of RBP4-receptors such as STRA6 or independent of a receptor. Retinol may also be delivered to cells in the form of its water-soluble retinyl-β-glucuronide (4), although these are present only at low concentrations in the circulation. Retinoic acid is present at lower concentrations, compared to those of retinol and retinyl esters, in both the fasting and post-prandial circulations, and may be delivered to cells bound to albumin (7) or transported as its water-soluble retinoyl-β-glucuronide (6). Proretinoid carotenoids, such as β-carotene, are present in the post-prandial circulation and in circulating VLDL, LDL and HDL (5). Once inside the cell, the retinol may be esterified to retinyl ester, or oxidized to retinaldehyde. Proretinoid carotenoids may be converted to retinaldehyde. Retinaldehyde serves as an intermediate in the oxidation of retinol to retinoic acid. Retinoic acid is transported to the nucleus of the cell, where it modulates transcription of retinoid-responsive genes

that it requires very regular dietary retinoid intake in order to ensure the good health of the organism. In times of insufficient dietary retinoid intake, the health of the organism becomes vulnerable.

The second arm, which we propose was added on top of the primordial post-prandial delivery of retinoid to tissues, arose when the body acquired the capacity to accumulate/store retinoid in the liver and to synthesize and secrete RBP4 to allow for mobilization of these stores. It is not clear to us whether the capacity to accumulate/store retinoid in the liver evolved first, followed by RBP4 and its actions in retinol mobilization, or whether these evolved simultaneously. Given the research interest in RBP4 in the pathophysiology of metabolic disease development that appears to be independent of its role as a retinol transporter, it is possible that the original role of RBP4 was to signal hepatic control of metabolic responses. If this were the case, then the ability to bind and mobilize retinol may have arisen later. As we noted above, VLDL bound retinyl ester is secreted by the liver (by hepatocytes) and it may be that RBP4 evolved later to supplement this lipoprotein-based pathway for retinoid mobilization from the liver.

The ability to store retinoid in the liver and to mobilize it in times of dietary insufficiency underlies the maintenance of all retinoid-dependent process in the body. Retinoid availability ensures normal retinoid-dependent functions and good health. Thus, the understanding of how retinoid stores are accumulated and mobilized from the liver, and of how retinoids are distributed to responsive tissues is critical to all of retinoid biology within the body.

The Future – What Needs to Be Learned?

We clearly have a great deal to learn about the molecular processes that mediate and regulate hepatic retinoid storage and mobilization. Central to this is how distinct processes that occur in two different hepatic cell types are integrated and regulated. There are two key components to this. First, how is dietary retinoid that is taken up by hepatocytes transferred to HSCs for storage? And second, how is retinoid stored in HSCs acquired by apo-RBP4 to allow for its mobilization and delivery to tissues throughout the body? Overarching both of these questions is the more general and longstanding question of how these hepatic processes, ones that are central to whole body retinoid economy, are coordinated to meet body needs and whether this involves communication between peripheral tissues and the liver.

Essential to understanding whole body retinoid storage and mobilization is the longstanding and unanswered question concerning whether signals arising in the periphery modulate retinol storage in the liver, and its mobilization from the liver bound to RBP4. Investigators have long known that hepatic retinoid stores are drawn upon to defend retinoid levels in blood and peripheral tissues and many have proposed that signals arising in peripheral tissues are communicated to the liver to affect retinoid storage and mobilization. More than three decades ago, Underwood and colleagues proposed that blood levels of retinoic acid signaled retinol-RBP4 release from the liver [52]. Since hepatic *Rbp4* mRNA levels are not quantitatively different in retinoid-insufficiency versus retinoid-sufficiency [113, 115, 121], this possibility appears to be unlikely. At the time Underwood and colleagues proposed this hypothesis, it was not known that extrahepatic tissues are able to synthesize and secrete some RBP4. One possible line of communication between the periphery and liver could be the modulation of RBP synthesis and secretion by peripheral tissues. If peripheral tissues secreted apo-RBP4, this would allow for retinol mobilization from the liver and retinol could be returned from the periphery to the liver for storage and recycling in the form of holo-RBP4. Alternatively, and contrary to what has been repeatedly proposed in the literature for many decades, there may be no communication between the periphery and the liver affecting hepatic retinoid storage/mobilization. We are biased towards this latter view. We believe that the factors and processes that regulate hepatic retinoid storage and mobilization reside intrinsically within the liver, involving solely regulation of hepatocyte and HSC interactions and not RBP4 *per se*.

Why do we take this view? The origin of this question regarding peripheral tissue-liver communications resides in a faulty understanding of retinoid delivery to tissues. If retinol-RBP4 were the sole route through which peripheral tissues acquired retinoid, then such communications would be needed. However, 25–33 % of dietary retinoid is taken up by peripheral tissues [39]. Retinoic acid is always present in the circulation and available for tissue uptake [56]. Moreover, retinyl esters are secreted from the liver in nascent VLDL [4, 99]. Although VLDL retinyl esters are found at low levels in human and rodent circulations, VLDL-bound retinoid accounts for a large percentage of retinoid present in the blood of some mammals, including dogs and cats [90, 91, 106]. Because of these overlapping but complementary delivery pathways, ones which nearly all investigators and students of vitamin A ignore, there is no need for communications between the periphery and the liver to control retinoid storage or RBP4 release. There are multiple complementary pathways that ensure retinoid availability to tissues. When all of these fail, as complete retinoid-deficiency is reached, the organism can no longer thrive. However, our views on this point are simply our opinions but ones that we believe are fully grounded on published data that convincingly support our view. Should some disagree, then this remains to be resolved through future research.

References

1. Al Tanoury Z, Piskunov A, Rochette-Egly C (2013) Vitamin A and retinoid signaling: genomic and nongenomic effects. J Lipid Res 54:1761–1775
2. Amengual J, Widjaja-Adhi MA, Bodriguez-Santiago S, Hessel S, Golczak M, Palczewski K, von Lintig J (2013) Two carotenoid oxygenases contribute to mammalian provitamin A metabolism. J Biol Chem 288:34081–34096
3. Andersen KB, Nilsson A, Blomhoff HK, Oyen TB, Gabrielsen OS, Norum KR, Blomhoff R (1992) Direct mobilization of retinol from hepatic perisinusoidal stellate cells to plasma. J Biol Chem 267:1340–1344
4. Ballew C, Galuska D, Gillespie C (2001) High serum retinyl esters are not associated with reduced bone mineral density in the third national health and nutrition examination survey, 1988–1994. J Bone Mineral Res 16:2306–2312
5. Batres RO, Olson JA (1987) A marginal vitamin A status alters the distribution of vitamin A among parenchymal and stellate cells in rat liver. J Nutr 117:874–879
6. Batten ML, Imanishi Y, Maeda T, Tu DC, Moise AR, Bronson D, Possin D, Van Gelder RN, Baehr W, Palczewski K (2004) Lecithin-retinol acyltransferase is essential for accumulation of all-trans-retinyl esters in the eye and in the liver. J Biol Chem 279:10422–10432

7. Biesalski HK, Frank J, Beck SC, Heinrich F, Illek B, Reifen R, Gollnick H, Seeliger MW, Wissinger B, Zrenner E (1999) Biochemical but not clinical vitamin A deficiency results from mutations in the gene for retinol binding protein. Am J Clin Nutr 69:931–936
8. Blaner WS (1994) Retinoid (vitamin A) metabolism and the liver. In: Arias IM, Jakoby WB, Popper H, Schacter D, Shafritz DS (eds) The liver. Raven, New York, pp 529–542
9. Blaner WS, Olson JA (1994) Retinol and retinoic acid metabolism. In: Sporn MB, Roberts AB, Goodman DS (eds) The retinoids: biology, chemistry, and medicine. Raven Press Ltd., New York, pp 229–256
10. Blaner WS, O'Byrne SM, Wongsiriroj N, Kluwe J, D'Ambrosio DN, Jiang H, Schwabe RF, Hillman EM, Piantedosi R, Libien J (2009) Hepatic stellate cell lipid droplets: a specialized lipid droplet for retinoid storage. Biochim Biophys Acta 1791:467–473
11. Blaner WS, Hendriks HF, Brouwer A, de Leeuw AM, Knook DL, Goodman DS (1985) Retinoids, retinoid-binding proteins, and retinyl palmitate hydrolase distributions in different types of rat liver cells. J Lipid Res 26:1241–1251
12. Blaner WS, Dixon JL, Moriwaki H, Martino RA, Stein O, Stein Y, Goodman DS (1987) Studies on the in vivo transfer of retinoids from parenchymal to stellate cells in rat liver. Eur J Biochem 164:301–307
13. Blaner WS, Obunike JC, Kurlandsky SB, Al-Haideri M, Piantedosi R, Deckelbaum RJ, Goldberg IJ (1994) Lipoprotein lipase hydrolysis of retinyl ester. Possible implications for retinoid uptake by cells. J Biol Chem 269:16559–16565
14. Blomhoff R, Helgerud P, Rasmussen M, Berg T, Norum KR (1982) In vivo uptake of chylomicron [^{3}H]retinyl ester by rat liver: evidence for retinol transfer from parenchymal to nonparenchymal cells. Proc Natl Acad Sci U S A 79:7326–7330
15. Blomhoff R, Holte K, Naess L, Berg T (1984) Newly administered [3H]retinol is transferred from hepatocytes to stellate cells in liver for storage. Exp Cell Res 150:186–193
16. Blomhoff R, Rasmussen M, Nilsson A, Norum KR, Berg T, Blaner WS, Kato M, Mertz JR, Goodman DS, Eriksson U, Peterson PA (1985) Hepatic retinol metabolism. Distribution of retinoids, enzymes, and binding proteins in isolated rat liver cells. J Biol Chem 260:13560–13565
17. Blomhoff R, Berg T, Norum KR (1988) Distribution of retinol in rat liver cells: effect of age, sex and nutritional status. Br J Nutr 60:233–239
18. Blomhoff R, Berg T, Norum KR (1988) Transfer of retinol from parenchymal to stellate cells in liver is mediated by retinol-binding protein. Proc Natl Acad Sci U S A 85:3455–3458
19. Blomhoff R, Green MH, Green JB, Berg T, Norum KR (1991) Vitamin A metabolism: new perspectives on absorption, transport, and storage. Physiol Rev 71:951–990
20. Brasaemle DL (2007) Thematic review series: adipocyte biology. The perilipin family of structural lipid droplet proteins: stabilization of lipid droplets and control of lipolysis. J Lipid Res 48:2547–2559
21. Chambon P (1994) The retinoid signaling pathway: molecular and genetic analyses. Semin Cell Biol 5:115–125
22. Chichili GR, Nohr D, Schaffer M, von Lintig J, Biesalski HK (2005) β-Carotene conversion into vitamin A in human retinal pigment epithelial cells. Invest Ophthal Vis Sci 46:3562–3569
23. Cifelli CJ, Green JB, Green MH (2005) Dietary retinoic acid alters vitamin A kinetics in both the whole body and in specific organs of rats with low vitamin A status. J Nutr 135:746–752
24. Cooper AD (1997) Hepatic uptake of chylomicron remnants. J Lipid Res 38:2173–2192
25. dela Sena C, Narayansamy S, Riedl KM, Curley RW Jr, Schwartz SJ, Harrison EH (2013) Substrate specificity of purified recombinant human β-carotene 15,15′oxygenase (BCO1). J Biol Chem 288:37094–37103
26. dela Sena C, Riedl KM, Narayanasamy S, Curley RW Jr, Schwartz SJ, Harrison EH (2014) The human enzyme that converts dietary provitamin A carotenoids to vitamin A is a dioxygenase. J Biol Chem 289:13661–13666

27. Dixon JL, Goodman DS (1987) Studies on the metabolism of retinol-binding protein by primary hepatocytes from retinol-deficient rats. J Cell Physiol 130:14–20
28. During A, Harrison EH (2007) Mechanisms of provitamin A (carotenoid) and vitamin A (retinol) transport into and out of intestinal Caco-2 cells. J Lipid Res 48:2283–2294
29. Eriksson U, Das K, Busch C, Nordlinder H, Rask L, Sundelin J, Sallstrom J, Peterson PA (1984) Cellular retinol-binding protein. Quantitation and distribution. J Biol Chem 259:13464–13470
30. Eroglu A, Harrison EH (2013) Carotenoid metabolism in mammals, including man: formation, occurrence, and function of apocarotenoids. J Lipid Res 54:171–1730
31. Fex G, Johannesson G (1988) Retinol transfer across and between phospholipid bilayer membranes. Biochim Biophys Acta 944:249–255
32. Fex G, Johannesson G (1990) Transfer of retinol from retinol-binding protein complex to liposomes and across liposomal membranes. Methods Enzymol 189:394–402
33. Friedman SL (2008) Hepatic stellate cells: protean, multifunctional, and enigmatic cells of the liver. Physiol Rev 88:125–172
34. Friedman SL, Wei S, Blaner WS (1993) Retinol release by activated rat hepatic lipocytes: Regulation by Kupffer cell-conditioned medium and PDGF. Am J Physiol 264:G947–G952
35. Geerts A (2001) History, heterogeneity, developmental biology, and functions of quiescent hepatic stellate cells. Semin Liver Dis 21:311–335
36. Ghyselinck NB, Bavik C, Sapin V, Mark M, Bonnier D, Hindelang C, Dierich A, Nilsson CB, Hakansson H, Sauvant P, Azais-Braesco V, Frasson M, Picaud S, Chambon P (1999) Cellular retinol-binding protein 1 is essential for vitamin A homeostasis. EMBO J 18:4903–4914
37. Giuliano G, Al-Babili S, von Lintig J (2003) Carotenoid oxygenases: cleave it or leave it. Trends Plant Sci 8:145–149
38. Goodman DS, Blaner WS (1984) Biosynthesis, absorption, and hepatic metabolism of retinol. In: Sporn MB, Roberts AB, Goodman DS (eds) The retinoids. Academic Press, New York, pp 1–39
39. Goodman DW, Huang HS, Shiratori T (1965) Tissue distribution and metabolism of newly absorbed vitamin A in the rat. J Lipid Res 6:390–396
40. Goodman DW, Huang HS, Shiratori T (1966) Mechanism of the biosynthesis of vitamin A from β-carotene. J Biol Chem 241:1929–1932
41. Green MH, Green JB, Lewis KC (1987) Variation in retinol utilization rate with vitamin A status in the rat. J Nutr 117:694–703
42. Harrison EH (2012) Mechanisms involved in the intestinal absorption of dietary vitamin A and provitamin A carotenoids. Biochim Biophys Acta 1821:70–77
43. Harrison EH, Gad MZ, Ross AC (1995) Hepatic uptake and metabolism of chylomicron retinyl esters: probable role of plasma membrane/endosomal retinyl ester hydrolases. J Lipid Res 36:1498–1506
44. Hendriks HF, Brekelmans PJ, Buytenhek R, Brouwer A, de Leeuw AM, Knook DL (1987) Liver parenchymal cells differ from the fat-storing cells in their lipid composition. Lipids 22:266–273
45. Hepatic stellate cell nomenclature. Hepatology 1996;23:193
46. Herr FM, Ong DE (1992) Differential interaction of lecithin-retinol acyltransferase with cellular retinol binding proteins. Biochemistry 31:6748–6755
47. Hu KQ, Liu C, Ernst H, Krinsky NI, Russell RM, Wang XD (2006) The biochemical characterization of ferret carotene-9′,10′-monooxygenase catalyzing cleavage of carotenoids *in vitro* and *in vivo*. J Biol Chem 281:19327–19338
48. Huang H, Goodman DS (1965) Vitamin A and carotenoids. I. Intestinal absorption and metabolism of ^{14}C-labeled vitamin A alcohol, and β-carotene in the rat. J Biol Chem 240:2839–2844
49. Isken A, Golczak M, Oberhauser V, Hunzelmann S, Driever W, Imanishi Y, Palczewski K, von Lintig J (2008) Rbp4 disrupts vitamin A uptake homeostasis in a stra6-deficient animal model for matthew-wood syndrome. Cell Metab 7:258–268

50. Kanai M, Raz A, Goodman DS (1968) Retinol-binding protein: the transport protein for vitamin A in human plasma. J Clin Invest 47:2025–2044
51. Kane MA, Folias AE, Napoli JL (2008) HPLC/UV quantitation of retinal, retinol, and retinyl esters in serum and tissues. Anal Biochem 378:71–79
52. Keilson B, Underwood BA, Loerch JD (1979) Effects of retinoic acid on the mobilization of vitamin A from the liver in rats. J Nutr 109:787–795
53. Kiefer C, Hessel S, Lampert JM, Vogt K, Lederer MO, Breithaupt DE, von Lintig J (2001) Identification and characterization of a mammalian enzyme catalyzing the asymmetric oxidative cleavage of provitamin A. J Biol Chem 276:14110–14116
54. Kim YK, Wassef L, Hamberger L, Piantedosi R, Palczewski K, Blaner WS, Quadro L (2008) Retinyl ester formation by lecithin:retinol acyltransferase is a key regulator of retinoid homeostasis in mouse embryogenesis. J Biol Chem 283:5611–5621
55. Kluwe J, Wongsiriroj N, Troeger JS, Gwak GY, Dapito DH, Pradere JP, Jiang H, Siddiqi M, Piantedosi R, O'Byrne SM, Blaner WS, Schwabe RF (2011) Absence of hepatic stellate cell retinoid lipid droplets does not enhance hepatic fibrosis but decreases hepatic carcinogenesis. Gut 60:1260–1268
56. Kurlandsky SB, Gamble MV, Ramakrishnan R, Blaner WS (1995) Plasma delivery of retinoic acid to tissues in the rat. J Biol Chem 270:17850–17857
57. Leuenberger MG, Engeloch-Jarret C, Woggon WD (2001) The reaction mechanism of the enzyme-catalyzed central cleavage of β-carotene to retinal. Angew Chem Int Ed 40:2614–2617
58. Lewis KC, Green MH, Green JB, Zech LA (1990) Retinol metabolism in rats with low vitamin A status: a compartmental model. J Lipid Res 31:1535–1548
59. Lindqvist A, Andersson S (2002) Biochemical properties of purified recombinant human β-carotene 15,15′-monooxygenase. J Biol Chem 277:23942–23948
60. Lindqvist A, He YG, Andersson S (2005) Cell type-specific expression of beta-carotene 9′,10′-monooxygenase in human tissues. J Histochem Cytochem 53:1403–1412
61. Linke T, Dawson H, Harrison EH (2005) Isolation and characterization of a microsomal acid retinyl ester hydrolase. J Biol Chem 280:23287–23294
62. Liu L, Gudas LJ (2005) Disruption of the lecithin:retinol acyltransferase gene makes mice more susceptible to vitamin A deficiency. J Biol Chem 280:40226–40234
63. Lobo GP, Hessel S, Eichinger A, Noy N, Moise AR, Wyss A, Palczewski K, von Lintig J (2010) ISX is a retinoic acid-sensitive gatekeeper that controls intestinal beta, beta-carotene absorption and vitamin A production. FASEB J 24:1656–1666
64. Mangelsdorf DJ, Umesono K, Evans RM (1994) The retinoid receptors. In: Sporn MB, Roberts AB, Goodman DS (eds) The retinoids: biology, chemistry, and medicine. Raven Press Ltd., New York, pp 319–350
65. Melhus H, Nilsson T, Peterson PA, Rask L (1991) Retinol-binding protein and transthyretin expressed in HeLa cells form a complex in the endoplasmic reticulum in both the absence and the presence of retinol. Exp Cell Res 197:119–124
66. Melhus H, Laurent B, Rask L, Peterson PA (1992) Ligand-dependent secretion of rat retinol-binding protein expressed in hela cells. J Biol Chem 267:12036–12041
67. Mello T, Nakatsuka A, Fears S, Davis W, Tsukamoto H, Bosron WF, Sanghani SP (2008) Expression of carboxylesterase and lipase genes in rat liver cell-types. Biochem Biophys Res Commun 374:460–464
68. Moore T (1930) Vitamin A and carotene: the conversion of carotene to vitamin A *in vivo*. Biochem J 24:692–702
69. Moore T (1957) Vitamin A. Elsevier Publishing Company, Amsterdam
70. Moriwaki H, Blaner WS, Piantedosi R, Goodman DS (1988) Effects of dietary retinoid and triglyceride on the lipid composition of rat liver stellate cells and stellate cell lipid droplets. J Lipid Res 29:1523–1534

71. Moussa M, Landrier JF, Reboul E, Ghiringhelli O, Comera C, Collet X, Frohlich K, Bohm V, Borel P (2008) Lycopene absorption in human intestinal cells and in mice involves scavenger receptor class B type 1 but not Niemann-Pick C1-like 1. J Nutr 138:1432–1436
72. Muto Y, Smith JE, Milch PO, Goodman DS (1972) Regulation of retinol-binding protein metabolism by vitamin A status in the rat. J Biol Chem 247:2542–2550
73. Nagatsuma K, Hayashi Y, Hano H, Sagara H, Murakami K, Saito M, Masaki T, Lu T, Tanaka M, Enzan H, Aizawa Y, Tajiri H, Matsuura T (2009) Lecithin:retinol acyltransferase protein is distributed in both hepatic stellate cells and endothelial cells of normal rodent and human liver. Liver Int 29:47–54
74. Noy N (2000) Retinoid-binding proteins: mediators of retinoid action. Biochem J 348:481–495
75. Noy N, Blaner WS (1991) Interactions of retinol with binding proteins: studies with rat cellular retinol-binding protein and with rat retinol-binding protein. Biochemistry 30:6380–6386
76. O'Byrne SM, Blaner WS (2013) Retinol and retinyl esters: biochemistry and physiology. J Lipid Res 54:1731–1743
77. O'Byrne SM, Wongsiriroj N, Libien J, Vogel S, Goldberg IJ, Baehr W, Palczewski K, Blaner WS (2005) Retinoid absorption and storage is impaired in mice lacking lecithin:retinol acyltransferase (LRAT). J Biol Chem 280:35647–35657
78. O'Byrne SM, Kako Y, Deckelbaum RJ, Hansen IH, Palczewski K, Goldberg IJ, Blaner WS (2010) Multiple pathways ensure retinoid delivery to milk: studies in genetically modified mice. Am J Physiol Endocrinol Metab 298:E862–E870
79. Olson JA (1989) Provitamin A function of carotenoids: the conversion of β-carotene to vitamin A. J Nutr 119:105–108
80. Olson JA, Hayaishi O (1965) The enzymatic cleavage of beta-carotene to vitamin A by soluble enzymes of rat liver and intestine. Proc Natl Acad Sci U S A 54:1364–1370
81. Ong DE, Newcomer ME, Chytil F (1994) Cellular retinoid-binding proteins. In: Sporn MB, Roberts AB, Goodman DS (eds) The retinoids: biology, chemistry and medicine. Raven Press Ltd., New York, pp 283–318
82. Orland MD, Anwar K, Cromley D, Chu CH, Chen L, Billheimer JT, Hussain MM, Cheng D (2005) Acyl coenzyme A dependent retinol esterification by acyl coenzyme A: Diacylglycerol acyltransferase 1. Biochim Biophys Acta 1737:76–82
83. Paik J, During A, Harrison EH, Mendelsohn CL, Lai K, Blaner WS (2001) Expression and characterization of A murine enzyme able to cleave β-carotene: the formation of retinoids. J Biol Chem 276:32160–32168
84. Palczewski K (2012) Chemistry and biology of vision. J Biol Chem 287:1612–1619
85. Pang W, Zhang Y, Wang S, Jia A, Dong W, Cai C, Hua Z, Zhang J (2011) The mplrp2 and mclps genes are involved in the hydrolysis of retinyl esters in the mouse liver. J Lipid Res 52:934–941
86. Pfeffer BA, Clark VM, Flannery JG, Bok D (1986) Membrane receptors for retinol-binding protein in cultured human retinal pigment epithelium. Invest Ophthalmol Vis Sci 27:1031–1040
87. Quadro L, Blaner WS, Salchow DJ, Vogel S, Piantedosi R, Gouras P, Freeman S, Cosma MP, Colantuoni V, Gottesman ME (1999) Impaired retinal function and vitamin A availability in mice lacking retinol-binding protein. EMBO J 18:4633–4644
88. Quadro L, Blaner WS, Hamberger L, Novikoff PM, Vogel S, Piantedosi R, Gottesman ME, Colantuoni V (2004) The role of extrahepatic retinol binding protein in the mobilization of retinoid stores. J Lipid Res 45:1975–1982
89. Quadro L, Hamberger L, Gottesman ME, Wang F, Colantuoni V, Blaner WS, Mendelsohn CL (2005) Pathways of vitamin A delivery to the embryo: insights from a new tunable model of embryonic vitamin A deficiency. Endocrinology 146:4479–4490
90. Raila J, Mathews U, Schweigert FJ (2001) Plasma transport and tissue distribution of beta-carotene, vitamin A and retinol-binding protein in domestic cats. Comp Biochem Physiol A Mol Integr Physiol 130:849–856

91. Raila J, Radon R, Trupschuch A, Schweigert FJ (2002) Retinol and retinyl ester responses in the blood plasma and urine of dogs after a single oral dose of vitamin A. J Nutr 132:1673S–1675S
92. Reboul E, Abou L, Mikail C, Ghiringhelli O, Andre M, Portugal H, Jourdheuil-Rahmani D, Amiot MJ, Lairon D, Borel P (2005) Lutein transport by Caco-2 TC-7 cells occurs partly by a facilitated process involving the scavenger receptor class B type 1 (SR-B1). Biochem J 387:455–461
93. Reboul E, Berton A, Moussa M, Kreuzer C, Crenon I, Borel P (2006) Pancreatic lipase and pancreatic lipase-related protein 2, but not pancreatic lipase-related protein 1, hydrolyze retinyl palmitate in physiological conditions. Biochim Biophys Acta 1761:4–10
94. Reboul E, Trompier D, Moussa M, Klein A, Landrier JF, Chimini G, Borel P (2009) ATP-binding cassette transporter A1 is significantly involved in the intestinal absorption of alpha- and gamma-tocopherol but not in that of retinyl palmitate in mice. Am J Clin Nutr 89:177–184
95. Redgrave TG (2004) Chylomicron metabolism. Biochem Soc Trans 32:79–82
96. Redlich CA, Grauer JN, van Bennekum AM, Clever SL, Ponn RB, Blaner WS (1996) Characterization of carotenoid, vitamin A, and alpha-tocopheral levels in human lung tissue and pulmonary macrophages. Am J Res Crit Care Med 154:1436–1443
97. Redlich CA, Chung JS, Cullen MR, Blaner WS, van Bennekum AM, Berglund L (1999) Effect of long-term beta-carotene and vitamin A on serum cholesterol and triglyceride levels among participants in the Carotene and Retinol Efficacy Trial (CARET). Atherosclerosis 143:427–434
98. Redmond TM, Gentleman S, Duncan T, Yu S, Wiggert B, Gantt E, Cunningham FX Jr (2001) Identification, expression, and substrate specificity of a mammalian β-carotene 15,15′-dioxygenase. J Biol Chem 276:6560–6565
99. Relas H, Gylling H, Miettinen TA (2000) Effect of stanol ester on postabsorptive squalene and retinyl palmitate. Metabolism 49:473–478
100. Rigtrup KM, Ong DE (1992) A retinyl ester hydrolase activity intrinsic to the brush border membrane of rat small intestine. Biochemistry 31:2920–2926
101. Rigtrup KM, Kakkad B, Ong DE (1994) Purification and partial characterization of a retinyl ester hydrolase from the brush border of rat small intestine mucosa: Probable identity with brush border phospholipase B. Biochemistry 33:2661–2666
102. Rigtrup KM, McEwen LR, Said HM, Ong DE (1994) Retinyl ester hydrolytic activity associated with human intestinal brush border membranes. Am J Clin Nutr 60:111–116
103. Ruggles KV, Turkish A, Sturley SL (2013) Making, baking, and breaking: the synthesis, storage, and hydrolysis of neutral lipids. Annu Rev Nutr 33:413–451
104. Saari JC (1999) Retinoids in mammalian vision. In: Nau H, Blaner WS (eds) The handbook of experimental pharmacology, the retinoids. Springer Verlag, Berlin, pp 563–588
105. Sauvant P, Sapin V, Alexandre-Gouabau MC, Dodeman I, Delpal S, Quadro L, Partier A, Abergel A, Colantuoni V, Rock E, Azais-Braesco V (2001) Retinol mobilization from cultured rat hepatic stellate cells does not require retinol binding protein synthesis and secretion. Int J Biochem Cell Biol 33:1000–1012
106. Schweigert FJ (1988) Insensitivity of dogs to the effects of nonspecific bound vitamin A in plasma. Int J Vitam Nutr Res 58:23–25
107. Seeliger MW, Biesalski HK, Wissinger B, Gollnick H, Gielen S, Frank J, Beck S, Zernner E (1999) Phenotype in retinol deficiency due to a hereditary defect in retinol binding protein synthesis. Invest Ophthalmol Vis Sci 40:3–11
108. Seino Y, Miki T, Kiyonari H, Abe T, Fujimoto W, Kimura K, Takeuchi A, Takahashi Y, Oiso Y, Iwanaga T, Seino S (2008) ISX participates in the maintenance of vitamin A metabolism by regulation of β-carotene 15,15′-monooxygenase (Bcmo1) expression. J Biol Chem 283:4905–4911

109. Senoo H, Stang E, Nilsson A, Kindberg GM, Berg T, Roos N, Norum KR, Blomhoff R (1990) Internalization of retinol-binding protein in parenchymal and stellate cells of rat liver. J Lipid Res 31:1229–1239
110. Senoo H, Smeland S, Malaba L, Bjerknes T, Stang E, Roos N, Berg T, Norum KR, Blomhoff R (1993) Transfer of retinol-binding protein from HepG2 human hepatoma cells to cocultured rat stellate cells. Proc Natl Acad Sci U S A 90:3616–3620
111. Sharma RV, Mathur SN, Ganguly J (1976) Studies on the relative biopotencies and intestinal absorption of different apo-β-carotenoids in rats and chickens. Biochem J 158:377–383
112. Sharma RV, Mathur SN, Dmitrovskii AA, Das RC (1976) Ganguly J:Studies on the metabolism of beta-carotene and apo-beta-carotenoids in rats and chickens. Biochem Biophys Acta 486:183–194
113. Soprano DR, Blaner WS (1994) Plasma retinol-binding protein. In: Sporn MB, Roberts AB, Goodman DS (eds) The retinoids: biology, chemistry, and medicine. Raven Press Ltd., New York, pp 257–281
114. Soprano DR, Pickett CB, Smith JE, Goodman DS (1981) Biosynthesis of plasma retinol-binding protein in liver as a larger molecular weight precursor. J Biol Chem 256:8256–8258
115. Soprano DR, Soprano KJ, Goodman DS (1986) Retinol-binding protein messenger RNA levels in the liver and in extrahepatic tissues of the rat. J Lipid Res 27:166–171
116. Sporn MB, Dunlop NM, Newton DL, Smith JM (1976) Prevention of chemical carcinogenesis by vitamin A and its synthetic analogs (retinoids). Fed Proc 35:1332–1338
117. Suhara A, Kato M, Kanai M (1990) Ultrastructural localization of plasma retinol-binding protein in rat liver. J Lipid Res 31:1669–1681
118. Sun H, Kawaguchi R (2011) The membrane receptor for plasma retinol-binding protein, a new type of cell-surface receptor. Int Rev Cell Mol Biol 288:1–41
119. Sun G, Alexson SE, Harrison EH (1997) Purification and characterization of a neutral, bile salt-independent retinyl ester hydrolase from rat liver microsomes. Relationship to rat carboxylesterase ES-2. J Biol Chem 272:24488–24493
120. Tang GW, Wang XD, Russell RM, Krinsky NL (1991) Characterization of beta-apo-13-carotenone and beta-apo-14′-carotenal as enzymatic products of the excentric cleavage of beta-carotene. Biochemistry 30:9829–9834
121. Tsutsumi C, Okuno M, Tannous L, Piantedosi R, Allan M, Goodman DS, Blaner WS (1992) Retinoids and retinoid-binding protein expression in rat adipocytes. J Biol Chem 267:1805–1810
122. van Bennekum AM, Kako Y, Weinstock PH, Harrison EH, Deckelbaum RJ, Goldberg IJ, Blaner WS (1999) Lipoprotein lipase expression level influences tissue clearance of chylomicron retinyl ester. J Lipid Res 40:565–574
123. van Bennekum AM, Fisher EA, Blaner WS, Harrison EH (2000) Hydrolysis of retinyl esters by pancreatic triglyceride lipase. Biochemistry 39:4900–4906
124. van Bennekum A, Werder M, Thuahnai ST, Han CH, Duong P, Williams DL, Wettstein P, Schulthess G, Phillips MC, Hauser H (2005) Class B scavenger receptor-mediated intestinal absorption of dietary beta-carotene and cholesterol. Biochemistry 44:4517–4525
125. Vogel S, Gamble MV, Blaner WS (1999) Retinoid uptake, metabolism and transport. In: Nau H, Blaner WS (eds) Handbook of experimental pharmacology, the retinoids. Springer Verlag, Berlin/Heidelberg, New York, pp 31–95
126. von Lintig J (2010) Colors with functions: elucidating the biochemical and molecular basis of carotenoid metabolism. Annu Rev Nutr 30:35–56
127. Wald G (1968) Molecular basis of visual excitation. Science 162:230–239
128. Wang XD, Krinsky NI, Tang GW, Russell RM (1992) Retinoic acid can be produced from excentric cleavage of beta-carotene in human intestinal mucosa. Arch Biochem Biophys 293:298–304
129. Wei S, Episkopou V, Piantedosi R, Maeda S, Shimada K, Gottesman ME, Blaner WS (1995) Studies on the metabolism of retinol and retinol-binding protein in transthyretin-deficient mice produced by homologous recombination. J Biol Chem 270:866–870

130. Weiner FR, Blaner WS, Czaja MJ, Shah A, Geerts A (1992) Ito cell expression of a nuclear retinoic acid receptor. Hepatology 15:336–342
131. Weng W, Li L, van Bennekum AM, Potter SH, Harrison EH, Blaner WS, Breslow JL, Fisher EA (1999) Intestinal absorption of dietary cholesteryl ester is decreased but retinyl ester absorption is normal in carboxyl ester lipase knockout mice. Biochemistry 38:4143–4149
132. Wongsiriroj N, Piantedosi R, Palczewski K, Goldberg IJ, Johnston TP, Li E, Blaner WS (2008) The molecular basis of retinoid absorption: a genetic dissection. J Biol Chem 283:13510–13519
133. Wongsiriroj N, Jiang H, Piantedosi R, Yang KJ, Kluwe J, Schwabe RF, Ginsberg H, Goldberg IJ, Blaner WS (2014) Genetic dissection of retinoid esterification and accumulation in the liver and adipose tissue. J Lipid Res 55:104–114
134. Wyss A, Wirtz GM, Woggon WD, Brugger R, Wyss M, Friedlein A, Riss G, Bachmann H, Hunziker W (2001) Expression pattern and localization of beta, beta-carotene 15,15′-dioxygenase in different tissues. Biochem J 354:521–529
135. Yamada M, Blaner WS, Soprano DR, Dixon JL, Kjeldbye HM, Goodman DS (1987) Biochemical characteristics of isolated rat liver stellate cells. Hepatology 7:1224–1229
136. Yan W, Jang GF, Haeseleer F, Esumi N, Chang J, Kerrigan M, Campochiaro M, Campochiaro P, Palczewski K, Zack DJ (2001) Cloning and characterization of a human β, β-carotene-15, 15′-dioxygenase that is highly expressed in the retinal pigment epithelium. Genomics 72:193–202
137. Yen CL, Brown CHT, Monetti M, Farese RV Jr (2005) A human skin multifunctional o-acyltransferase that catalyzes the synthesis of acylglycerols, waxes, and retinyl esters. J Lipid Res 46:2388–2397
138. Yen CL, Stone SJ, Koliwad S, Harris C, Farese RV Jr (2008) Thematic review series: glycerolipids. DGAT enzymes and triacylglycerol biosynthesis. J Lipid Res 49:2283–2301

Chapter 5
Retinoic Acid Synthesis and Degradation

Natalia Y. Kedishvili

Abstract Retinoic acid (RA) was identified as the biologically active form of vitamin A almost 70 years ago and work on its function and mechanism of action is still of major interest both from a scientific and a clinical perspective. The currently accepted model postulates that RA is produced in two sequential oxidative steps: first, retinol is oxidized reversibly to retinaldehyde, and then retinaldehyde is oxidized irreversibly to RA. Excess RA is inactivated by conversion to hydroxylated derivatives. Much is left to learn, especially about retinoid binding proteins and the trafficking of the hydrophobic retinoid substrates between membrane bound and cytosolic enzymes. Here, background on development of the field and an update on recent advances in our understanding of the enzymatic pathways and mechanisms that control the rate of RA production and degradation are presented with a focus on the many questions that remain unanswered.

Keywords Retinol • Retinaldehyde • ADH • DHRS3 • Dehydrogenase • RDH10 • CRBP1 • RALDH • SDR • Short-chain dehydrogenases/reductases

Abbreviations

ADH	alcohol dehydrogenase
AKR	aldo-keto reductase
CRBPI	cellular retinol binding protein type I
CRABPI	cellular retinoic acid binding protein type I
CYP	cytochrome P450
MDR	medium chain dehydrogenase/reductase
NAD	nicotinamide adenine dinucleotide (diphosphopyridine nucleotide)
NADP	nicotinamide adenine dinucleotide phosphate
RA	all-*trans*-retinoic acid
RAR	retinoic acid receptor

N.Y. Kedishvili (✉)
Department of Biochemistry and Molecular Genetics, University of Alabama at Birmingham, 720 20th Street South, 35294 Birmingham, AL, USA
e-mail: nkedishvili@uab.edu

M.A. Asson-Batres, C. Rochette-Egly (eds.), *The Biochemistry of Retinoid Signaling II*, Subcellular Biochemistry 81, DOI 10.1007/978-94-024-0945-1_5

RALDH	retinaldehyde dehydrogenase
RDH or RoDH	retinol dehydrogenase
SDR	short-chain dehydrogenase reductase

Introduction

Humans obtain vitamin A from the diet in the form of retinyl esters from animal sources or β-carotene from plant sources. After absorption and transfer to tissues, the dietary forms of vitamin A are converted enzymatically into retinoic acid (RA), the derivative that was identified as the biologically active form of vitamin A almost 70 years ago [6, 168, 169]. Too much or too little RA produce harmful effects in developing embryos and postnatal animals, so levels of RA have to be tightly controlled in tissues. The exact enzymes and control mechanisms that regulate its biosynthesis and degradation are yet to be fully defined. The currently accepted model postulates that RA is produced in two sequential oxidative steps: first, retinol is oxidized reversibly to retinaldehyde, and then retinaldehyde is oxidized irreversibly to RA, which can be inactivated by conversion to hydroxylated derivatives (Fig. 5.1).

History

At first, there were doubts about the natural occurrence of RA as a normal product of retinol metabolism because initial attempts to demonstrate RA formation from retinol or retinaldehyde *in vivo* had been unsuccessful [90], and little if any RA could be found in tissues after administering milligram doses of RA [6, 145]. Other studies showed that small amounts of RA rapidly appeared in rat tissues [44, 45] and bile [49] after parenteral administration of milligram quantities of retinaldehyde. However, these observations raised a question of whether RA is formed only in systems treated with abnormally large amounts of retinaldehyde. Eventually, it was demonstrated that RA forms *in vivo* in rat liver and small intestine from microgram quantities of intrajugularly administered ^{14}C-radiolabeled retinol [53], and by the end of the 1960s, it was established that RA is a natural product that is formed

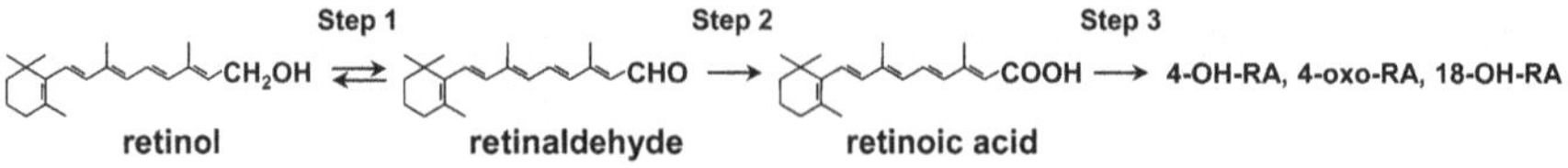

Fig. 5.1 Pathway of retinoic acid biosynthesis and degradation. Retinol is oxidized reversibly to retinaldehyde (Step 1). Retinaldehyde is oxidized irreversibly to retinoic acid (Step 2). Retinoic acid is eliminated by conversion to 4-*hydroxy* (4-OH-RA), 4-*oxo* (4-oxo-RA), and 18-*hydroxy* (18-OH-RA) intermediates (Step 3)

in very small quantities from either retinol or retinaldehyde. Using ^{14}C-labeled RA, it was also established that RA and its metabolites are rapidly excreted in the bile of bile-duct cannulated rats [50, 185].

Characterization of the Enzymes that Catalyze the Reversible Oxidation of Retinol to Retinaldehyde

Studies conducted with homogenates of frog and cattle retinas demonstrated that the vitamin A aldehyde (retinene) found in rhodopsin could be reduced to vitamin A alcohol by an enzymatic process [23, 171] (Fig. 5.1, step 1, reductive conversion of retinaldehyde to retinol) that required reduced diphosphopyridine nucleotide (NADH) as coenzyme [172]. The reduction was reversible (Fig. 5.1, step 1, oxidation of retinol to retinaldehyde). Because it could be catalyzed by crude rabbit liver extracts which contain high levels of alcohol dehydrogenases, known to interconvert alcohols and aldehydes, and also by crystalline horse liver alcohol dehydrogenase (ADH) [25], it was suggested that the reduction of retinaldehyde and oxidation of retinol were both catalyzed by an alcohol dehydrogenase [22, 24]. This notion was reinforced by the finding that the same enzyme in rat liver appeared to be responsible for both the NADH-dependent retinaldehyde-reductive and NAD-dependent ethanol-oxidizing activities [184]. Furthermore, ethanol was shown to inhibit the oxidation of retinol by human liver and cattle retina [122], providing further evidence for potential involvement of ADH in metabolism of retinol.

ADHs belong to the medium-chain dehydrogenase/reductase (MDR) superfamily of proteins and are dimeric zinc metalloproteins with ~350-residue subunits. The human and mouse ADH families each contain six classes of cytosolic ADH enzymes [75, 141] (Table 5.1). Early on, ADH isoenzymes were best known for their ability

Table 5.1 Classes of human ADH and their kinetic constants towards retinol. ADH5 and ADH6 have been identified at the gene and transcriptional levels only, and their functions are still unknown. Kinetic constants of human ADH3 activity towards retinol have not been reported. Activity assays were performed using bovine serum albumin (BSA) to solubilize retinol and high performance liquid chromatography (HPLC) to separate retinoids ([64]). ND, not determined using HPLC method with BSA as solubilizer

		Retinol as substrate		
Class	Gene name	K_m, μM	k_{cat}	k_{cat}/K_m
I	ADH1A	ND	ND	ND
	ADH1B	0.3	21	70,000
	ADH1C	ND	ND	ND
II	ADH2	0.14	4	29,000
III	ADH3	ND	ND	ND
IV	ADH4	0.3	190	640,000
V/VI	ADH5/6	ND	ND	ND

to metabolize ethanol. However, interest in retinol as a natural substrate for ADH prompted studies from several laboratories which showed that in humans, class IV ADH (ADH4, also known as stomach σσ ADH) was the most catalytically efficient retinol dehydrogenase *in vitro*, followed by ADHs of class I (ADH1) and liver-specific class II (ADH2) (27, 39, 69, 182]. ADH4 protein was proposed to catalyze the oxidation of retinol to retinaldehyde in the epithelia of the stomach, esophagus, skin, and respiratory tract [73, 74]. ADH 1 was thought to be the primary retinol dehydrogenase in mouse liver, where the enzyme constitutes 0.9 % of liver protein [4]. The ubiquitously expressed Class III ADH (ADH3) did not exhibit significant catalytic activity towards retinol and functioned primarily as a GSH-dependent formaldehyde dehydrogenase [124].

While kinetic analyses indicated that ADH4 and ADH1 isoenzymes had significant retinol dehydrogenase activity *in vitro*, the *in vivo* role of ADHs in retinoid metabolism was challenged by studies showing evidence for non-ADH retinol dehydrogenases [100]. Koen and Shaw reported a retinol dehydrogenase in liver and retina that was distinct from ethanol dehydrogenase [88]. Napoli [127] showed that 220 mM ethanol did not impair RA production from 10 μM retinol or retinaldehyde by LLC-PK1 cells, an epithelial, non-tumorigenic cell line derived from pig kidney. In addition, a well-known inhibitor of ADH activity, 4-methylpyrazole, did not inhibit RA synthesis significantly at concentrations as high as 10 mM.

These observations prompted a search for an alternative retinol dehydrogenase, unrelated to classical ethanol-oxidizing cytosolic ADH. Leo et al. [102] reported that liver of ADH-null deer mice contained a microsomal retinol dehydrogenase that had a 2.5-fold higher activity with NAD^+ than with $NADP^+$ and was not inhibited by 4-methylpyrazole, the strong inhibitor of the cytosolic retinol dehydrogenase activity. The microsomal enzyme also catalyzed the reverse direction, the reduction of retinaldehyde to retinol. Together, the cytosolic and microsomal fractions accounted for 94 % of the total NAD^+-dependent retinol dehydrogenase activity of liver homogenate, with the microsomal activity constituting only about 16 % of the total activity.

Lieber's group also showed that liver microsomes from either $ADH^{+/+}$ or $ADH^{-/-}$ deer mice did not produce RA from retinol, but the addition of cytosol to the microsomes resulted in the formation of RA [87]. The authors concluded that retinol was converted to retinaldehyde by the liver microsomes in $ADH^{-/-}$ deer mice, and then retinaldehyde was oxidized to RA by the cytosolic enzymes. The intermediate role of retinaldehyde in the production of RA was further demonstrated by the experiments where the addition of unlabeled retinaldehyde to labeled retinol resulted in isotopic dilution of the RA formed upon incubation with $ADH^{+/+}$ deer mice cytosol, and also by the finding that tritiated RA was produced from tritiated retinol in the mixture of microsomes and cytosol from ADH- deer mice but not by microsomes or cytosol alone.

Another important observation related to the conversion of retinol to RA was made by Napoli who showed that the rate-limiting step in this process was the dehydrogenation of retinol, as demonstrated using LLC-PK1 cells [127]. The oxidation of retinol to retinaldehyde was consistently 30–60-fold slower than the oxidation of retinaldehyde to RA. Overall, less than 5 % of the retinol added to the cells was converted into RA. The synthesis of RA in LLC-PK1 cells was not sensitive to inhibitors of alcohol metabolism, providing further evidence for the existence of retinol dehydrogenases distinct from ethanol dehydrogenases.

Faced with the growing number of enzymes that oxidize retinol *in vitro* and uncertainty with respect to their physiological significance, Napoli and colleagues speculated that if, as had been suggested, intermediates pass through metabolic pathways by direct transfer between proteins [17, 38, 161], rather than by diffusion through the aqueous phase, the first protein in the pathway from retinol to RA would be the ubiquitous cellular retinol binding protein type I (CRBPI), which binds cellular retinol specifically [132]. Thus, Napoli and colleagues proposed that the physiologically relevant retinol dehydrogenase should be able to recognize the CRBPI-bound retinol (holo-CRBPI) as substrate. To test this idea, Posch et al. [142] assayed the activity of microsomal fractions from rat liver, kidney, lung, and testes with holo-CRBPI as substrate. The microsomal enzymes appeared to prefer $NADP^+$ over NAD^+ as a cofactor for oxidation of holo-CRBPI to retinaldehyde, because the V_{max} with $NADP^+$ was 3.4-fold higher than that with NAD^+. Notably, the rate for the oxidation of free retinol with $NADP^+$ as cofactor (7.4 pmol/min mg) was about 20-fold higher than that with holo-CRBPI (143 pmol/min × mg). The estimated K_m values for holo-CRBPI and unbound retinol were quite similar, ~1.6 and ~1 μM, respectively. Based on the K_d value of CRBPI for binding of retinol (16 nM), the authors calculated that the amount of free retinol present in holo-CRBPI preparation was not sufficient to support the reaction rate observed using holo-CRBPI as substrate; hence, holo-CRBPI must have been the actual substrate for the microsomal retinol dehydrogenase.

Further work from this group focused on molecular identification of the microsomal $NADP^+$-dependent retinol dehydrogenase (RoDH) and characterization of its interaction with holo-CRBPI. Boerman and Napoli [26] attempted to extract and purify the microsomal enzyme based on its activity toward 2.5 μM holo-CRBPI. Total purification was not achieved because the RoDH lost activity as purity increased. The final preparation of partially purified RoDH active toward holo-CRBPI contained two major bands, one of 34 kDa and another of 54 kDa. To identify the polypeptide that interacts with CRBPI, microsomes were incubated with CRBPI that was covalently labeled with a UV-activated, cleavable cross-linking reagent. The reagent was radioiodinated such that the iodine would be transferred covalently from the CRBPI to any target protein upon activation and cleavage. A dominant radiolabeled band was observed consistently at 37 kDa but not 54 kDa. Interestingly, this band was not recognized by the antibodies raised against the 34 kDa polypeptide purified using holo-CRBPI activity assay. The authors interpreted this observation as suggesting that the antigenic determinant was contributed by limited epitopes that were altered by crosslinking.

Protein sequencing of four internal polypeptides obtained by a trypsin digest of the 34 kDa band revealed that one internal peptide (L**W**GLV**NNAG**ISVPV) had a sequence highly conserved in proteins that belong to the superfamily of short chain dehydrogenases/reductases (SDRs) [140]. SDRs have been previously shown to be involved in metabolism of sugars, steroids, and prostaglandins. The identification of the 34 kDa polypeptide as a member of the SDR superfamily suggested that SDRs are also involved in metabolism of retinoid substrates.

By the mid-90s, two major types of enzymes emerged as candidates for the role of retinol dehydrogenases: the cytosolic ADHs and the microsomal SDRs. However, whether both of these enzyme families contribute to RA biosynthesis remained unclear.

Characterization of the Enzymes that Irreversibly Oxidize Retinaldehyde to RA

In contrast to ambiguity surrounding the identification of physiologically relevant retinol dehydrogenases, there was a general consensus early on that the second step in biosynthesis of RA, the oxidation of retinaldehyde to RA, was catalyzed by the soluble enzymes present in cell cytosol [87]. Studies from several groups provided evidence that the aldehyde form of vitamin A was oxidized irreversibly to vitamin A acid, and that this reaction was catalyzed by aldehyde dehydrogenases (ALDHs) in calf and rat liver [47, 49, 52, 61, 91, 110] (Fig. 5.1, step 2). Thus, it was concluded that together with alcohol dehydrogenases, the aldehyde-oxidizing enzymes provided a pathway from the vitamin A alcohol to the acid [8, 47, 61, 110], similar to the metabolism of ethanol to acetaldehyde and further to acetic acid.

To identify the isoenzymes of ALDH family of proteins that were active towards retinaldehyde, Sladek and co-workers examined the activities of several candidate ALDHs [48, 93–95]. Their studies showed that only one of chromatographically resolved ALDHs from human liver, ALDH-1, catalyzed the oxidation of retinaldehyde with K_m value of 0.3 μM [48].

Another major contribution was provided by Dräger and coworkers whose work focused on the identification of retinaldehyde-active ALDHs that were essential for RA biosynthesis in mouse embryonic retina during development [115–118]. These researchers used a highly sensitive RA reporter cell line that allowed them to detect low levels of RA produced by ALDHs in different segments of the developing retina and spinal cord. Using a combination of immunohistochemistry, isoelectric focusing, and biochemical assays they were able to identify three separate forms of ALDH that differed in their isoelectric points, catalytic efficiency, sensitivity to inhibitors, and their spatio-temporal expression pattern during development. One of the ALDH activities confined to the dorsal part of the retina was identified as the previously characterized murine AHD2 class I ALDH (RALDH1). In addition to this previously characterized enzyme, McCaffery et al. found two novel aldehyde dehydrogenase activities, named V1 and V2. V2 was of special interest because it was the first aldehyde dehydrogenase detectable in the early mouse embryo.

Taking a different approach, Napoli and colleagues searched for the retinaldehyde-oxidizing ALDH by fractionating rat liver cytosol by anion-exchange chromatography and assaying the catalytic activities of fractions with 2 μM retinaldehyde as substrate and NAD^+ as cofactor [143]. The purified enzyme (RALDH1) was shown to oxidize retinaldehyde in the presence of CRBPI with an average K_m of 0.13 μM. By isoelectric point (~8.3), the rat retinal dehydrogenase was most similar to human ALDH IV [65] and mouse AHD-2 [154]. The exact molecular identities of the ALDHs responsible for the oxidation of retinaldehyde to RA in rat and mice remained unknown until 1996.

Characterization of the Enzymes Involved in Degradation of RA

In vivo studies using ^{14}C-labeled RA showed that administered RA disappeared quickly from the animal [145, 168, 169]. For example, all of the ^{14}C from 16 μg of ^{14}C-labeled RA injected intravenously into rats was recovered within 48 h [43]. The products of RA metabolism included retinoyl β-glucuronide ([50, 107]); decarboxylated metabolites [43], and polar metabolites [77] (Fig. 5.1, step 3). Hanni and co-workers identified 4-oxoretinoic acid as well as the all-*trans* and 9-*cis* isomers of 5'-hydroxy-RA in the feces of rats after a single intraperitoneal dose of 27.2 mg of all-*trans*-RA [70, 71].

With the advent of high-pressure liquid chromatography (HPLC) and application of this technique to the separation of retinoids, Frolik and co-workers were able to demonstrate that incubation of [^3H]RA in the presence of hamster liver 10,000 × *g* supernatant produced several metabolites that were more polar than the parent compound. Two of these metabolites were identical with synthetic all-*trans*-4-hydroxy-RA and all-*trans*-4-oxo-RA [58, 59, 119]. Napoli and colleagues identified 5,8-oxy-RA as a metabolite of administered all-*trans*-RA in the intestine of vitamin A deficient rats [130]. However, it was pointed out that the *in vivo* product was most likely the 5,6-epoxide of RA and that 5,8-oxy-RA was derived from 5,6-epoxide under the acidic conditions of the extraction procedure [120, 130].

To determine the characteristics of the enzyme systems involved in metabolism of RA, Roberts et al. [148, 149] investigated the RA metabolism by analyzing subcellular fractions of hamster intestine and liver. They found that RA metabolism was induced in the tissues of vitamin A deficient hamsters by pretreatment of the animals with oral doses of RA. The major classes of metabolites separated by HPLC were more polar than RA and retained the carboxyl carbon atom. The metabolic activity was localized in the 100,000 × g pellet, required NADPH and oxygen, and was strongly inhibited by carbon monoxide and by some cytochrome P-450 inhibitors. The authors concluded that the initial steps in the deactivation of the RA molecule were mediated by a cytochrome P-450-type enzyme system. Cytochrome P-450 enzymes are membrane-bound mono-oxygenases encoded by the *CYP* supergene family that catalyze the oxidative metabolism of both endogenous and exogenous compounds, including carcinogens, drugs, steroids, fatty acids, and prostaglandins [186]. Interestingly, the oxidation of RA was specifically induced by the RA itself. Classic cytochrome P-450 or P-448 inducers phenobarbital or d-methylcholanthrene induced RA catabolism to only a minor extent.

Leo et al. [101] showed that feeding rats with a diet containing a hundred times the normal amount of vitamin A resulted, within 2–3 weeks, in an increase in total hepatic microsomal cytochrome P-450 content. They confirmed that this was associated with an enhanced conversion of all-*trans*-RA to polar metabolites including 4-hydroxy- and 4-oxo-RA by isolated microsomes. They also showed that purified cytochromes P-450f and b promoted conversion of RA to polar metabolites, including 4-hydroxy-RA. In addition, this group showed that purified human P450IIC8 metabolized both retinol and RA to corresponding 4-hydroxy-retinoids and other polar metabolites.

Roberts et al. [150] examined eight purified rabbit cytochrome P-450 (P-450) isozymes for their activities toward RA. Cytochrome P-450 s 2B4 and 1A2 were the most active RA 4-hydroxylases, but the maximum velocities of P-450 2B4 for hydroxylation of retinol and retinal were much greater than that with RA as substrate. None of the isozymes investigated was found to convert the 4-hydroxy derivative to the 4-oxo derivative. Additional evidence for the involvement of CYP enzymes in metabolism of RA came from studies of Van Wauwe et al. [170], who demonstrated that P-450 inhibitor liarozole potently inhibited the C-4 hydroxylation of RA.

Together these studies established that a major metabolic pathway of RA consisted of the hydroxylation at position C-4 of its cyclohexenyl moiety to form 4-hydroxy-RA, which was then oxidized further to 4-keto-RA and more polar metabolites, with glucuronylation of RA possibly being limited to liver and intestine. Both *in vitro* and *in vivo*, the C-4 hydroxylation of RA was determined to be mediated by a cytochrome P-450-dependent monooxygenase system. In addition, it was proposed that RA bound to cellular retinoic acid binding protein type I (CRABPI) could serve as substrate for RA-metabolizing cytochrome P-450s [57]. In mammals, several purified CYPs were shown to be capable of converting RA to more polar metabolites in reconstituted systems, although generally specificity for RA was not high. An open question remained whether or not there could be (RA-inducible) CYP family members specifically dedicated to the hydroxylation of RA.

Development of the Field

Oxidation of Retinol to Retinaldehyde: Re-evaluation of ADH Kinetic Constants

The comparison of the catalytic properties of the cytosolic ADH isoenzymes and the microsomal RoDH toward retinol was complicated by the methodological differences in measurements of their activities. Initially, the assays of the retinol oxidizing activity of ADH isoenzymes were conducted by a continuous spectrophotometric assay [27, 182], which measured the appearance of retinaldehyde based on its absorbance at 400 nm (ε_{400} = 29 500 M^{-1} cm^{-1}), where retinol does not absorb. However, these assays required solubilization of retinoids using detergents such as Tween-80, which acted as an apparent competitive inhibitor of retinol oxidation, increasing the K_m values of ADHs for retinoids by up to 100-fold [113]. On the other hand, the activity of microsomal RoDH was measured in a reaction buffer containing 2 mM egg yolk L-α-phosphatidylcholine and retinol bound to CRBPI or added directly to buffer from ethanol stock solution. Reaction products were extracted with hexane and separated by HPLC [26].

To exclude the inhibitory effect of Tween-80, ADH kinetics with retinoids were re-measured in a detergent-free system using bovine serum albumin as the "solubilizing" agent for retinol [64], because CRBPI-bound retinol was shown not to serve as substrate for ADH [85]. These assays produced much lower K_m values of ADH isoenzymes for retinol (0.14–0.3 μM) (Table 5.1) and retinaldehyde (0.29–0.8 μM), which were in the same range or lower than the K_m values of rat RoDH for free or CRBPI-bound retinol. The k_{cat} values of ADH isoenzymes varied from 4 to 190 min^{-1} for the oxidation of retinol and 2.3 to 300 min^{-1} for the reduction of retinaldehyde. In agreement with the previous measurements, the most catalytically efficient enzyme was class IV ADH (ADH4), followed by class I ADH (ADH1B2) and class II ADH (ADH2) (reviewed in [134]). However, the *in vivo* roles of ADH isoenzymes in RA biosynthesis remained unclear.

ADH Gene Knockouts

With the development of transgenic technologies, it became possible to assess the *in vivo* roles of the numerous retinoid-active oxidoreductases by knocking out the gene of interest. An extensive analysis of the contribution of individual ADH isoenzymes to retinoid metabolism *in vivo* was carried out by Duester and colleagues. Gene knockout studies revealed that none of the ADH genes were essential for survival during embryonic development (reviewed in [134]). When maintained on normal chow diet, *Adh4*$^{-/-}$ mice, *Adh1*$^{-/-}$ mice, *Adh3*$^{-/-}$ mice, and *Adh1/4*$^{-/-}$ double knockout mice were all viable and fertile. *Adh3*$^{-/-}$ adult mice on standard mouse chow weighed significantly less than wild-type mice (24.7 g vs. 34.3 g, respectively, at 14 weeks of age), suggesting a growth deficiency [124]. A diet supplemented with tenfold more vitamin A than standard mouse chow restored near normal growth of *Adh3*$^{-/-}$ mice.

When pregnant females were fed a vitamin A-deficient diet, the embryos of *Adh1*$^{-/-}$ mice faired no worse than wild-type mice, but the embryos of *Adh4*$^{-/-}$ mice had 34 % more stillbirths than did wild-type mice [42]. The double knockout *Adh1/4*$^{-/-}$ offspring displayed more growth and survived longer than *Adh4*$^{-/-}$ mice suggesting that the additional loss of ADH1 moderated the negative effect of a loss of ADH4 during gestational vitamin A deficiency by reducing retinol turnover.

Together, these observations supported a role for ADH3 and ADH4, but not ADH1, in survival and growth during vitamin A deficiency.

Based on its conservation among all vertebrates and ubiquitous tissue expression pattern relative to ADH4, which is absent in liver, Duester and colleagues reasoned that ADH3 was the best candidate for a retinol dehydrogenase among ADH isotypes. They re-measured the activity of mouse ADH3 using a more sensitive HPLC assay and showed that ADH3 oxidized retinol to retinaldehyde at a rate of 105 pmol/min × mg, which was comparable to the activity of the partially purified rat microsomal RoDH1 (115 pmol/min × mg). The conclusion drawn by this group was that ADH3 generates RA during vitamin A deficiency. However, others questioned this interpretation because the group did not analyze growth defects or measure RA in serum or tissues from Adh null mice [129], leaving open the possibility that effects of a lack of ADH on mouse growth and viability under conditions of vitamin A deficiency may have been brought about by a disruption of other non-retinoid metabolic or detoxifying pathways that would normally involve ADH.

An important insight into the possible roles of ADH isoenzymes in retinol metabolism was provided by dosing ADH knockout mice with retinol. These metabolic assays showed that *Adh1*$^{-/-}$ mice were unable to process a large dose of retinol as efficiently as wild-type mice. Two hours after oral administration of 50 mg/kg of retinol, the plasma levels of RA in *Adh1*$^{-/-}$ mice were 23-fold lower than in WT mice. In comparison, *Adh3*$^{-/-}$ mice had only a 3.6-fold reduction in plasma RA levels and *Adh4*$^{-/-}$ mice had RA levels similar to those in WT mice [124]. The conclusion drawn from these studies was that ADH1 plays a dominant role in clearance of an acute dose of retinol, ADH3 plays a minor role, and ADH4 plays little or no role in retinol metabolism.

The potential for CRBP to influence ADH activity was also investigated in knockout models. Whereas *CRBPI*$^{-/-}$ mice had greatly reduced levels of retinyl esters in the liver, *Adh1*$^{-/-}$ mice were observed to have significantly increased liver retinyl ester levels [125]. Interestingly, relatively normal levels of liver retinyl ester were observed in animals where both CRBPI and ADH1 were inactivated. Vitamin A deficient *CRBPI*$^{-/-}$/*Adh1*$^{-/-}$ mice were protected from an excessive loss of liver RE for the first 5 weeks of dietary deficiency and showed a greatly minimized loss of liver RE for up to 13 weeks. When *CRBPI*$^{-/-}$ mice were administered a dose of retinol, increased production of RA was observed. Such increased metabolism of retinol to RA was not observed in *CRBPI*$^{-/-}$/*Adh1*$^{-/-}$ mice. These findings indicated that CRBPI protects retinol from being oxidized by ADH [125], confirming similar predictions reported previously [85].

CRBPI binds to retinol and normally facilitates its innocuous storage as RE, but in situations where vitamin A doses are sufficiently large, CRBPI becomes saturated and the excess unbound retinol is available as a substrate for ADH. Napoli [129] proposed the excess production of retinaldehyde by this route could *contribute* to retinoid toxicity. Retinoid toxicity generated by ADH1 from high doses of retinol would be especially damaging to fetuses since studies have shown that 71 % of mice receiving a single dose of 3 mg/kg retinol develop cleft palate and 76 % of mice receiving 39 mg/kg retinol develop neural tube defects [19]. Adult animals are less impacted by retinoid toxicity since susceptible developmental processes are completed and excess RA is rapidly eliminated [43, 145, 168, 169].

In summary, the work by Duester and colleagues demonstrated that different ADH isoenzymes have different capacities for metabolizing retinol unbound to CRBPI. Their findings also indicated that the retinol dehydrogenase activities of ADH isoenzymes, especially of ADH1, play a primary role in rapid elimination of a massive dose of retinol, but that these enzymes are not essential for the tightly controlled production of RA during embryogenesis. This left open the question of the identity of the retinol dehydrogenase that does operate in RA biosynthesis during embryonic development.

Cloning of the Microsomal SDR Retinol Dehydrogenases

While other researchers focused on the roles of various ADH isoenzymes in retinoid metabolism, Napoli and coworkers investigated the 34 kDa microsomal SDR partially purified from rat liver using holo-CRBPI as substrate [26]. A PCR fragment obtained using primers based on the sequences of two internal peptides from this SDR allowed isolation of a full-length cDNA encoding 317-amino acid protein with a calculated molecular mass of 34.9 kDa [32, 33]. This protein, named RoDH1 [11, 34], shared 52 % amino acid sequence identity with another protein that was cloned earlier in 1995 and found to exhibit 11-*cis*-retinol dehydrogenase activity, but not all-*trans*-retinol dehydrogenase activity [158]. These proteins that recognized different retinoids as substrates represented newly identified members of a superfamily of SDR proteins (Table 5.2).

The enzymatic properties of recombinant rat RoDH1 were determined by expressing the protein in P19 cells [32]. The 10,000 × g supernatant of homogenized P19 cells catalyzed the production of retinaldehyde from holo-CRBPI with an average K_m value of 0.9 μM. In vivo studies showed that rat RoDH1 protein expression was restricted to liver. Since RA is produced by many different cell types and tissues, this suggested that additional retinol dehydrogenases must exist. Shortly after cloning of rat RoDH1, Napoli and colleagues discovered a second, 317-amino acid, SDR retinol dehydrogenase, RoDH2 [33]. This enzyme shared 82 % sequence identity with RoDH1 and was found to be expressed in liver, kidney, brain, lung, and testis. Similar to RoDH1, recombinant RoDH2 expressed in P19 cells produced retinaldehyde from holo-CRBPI with an average K_m value of 2 μM. The enzymes had higher activity with $NADP^+$ than with NAD^+.

In 1997, a surprising finding added a new wrinkle to the developing story of microsomal RoDH enzymes. Biswas and Russell [21] reported that both rat RoDH1 and its newly identified human homolog, which shares 62 % sequence identity with rat RoDH1, exhibited a potent 3α-hydroxysteroid dehydrogenase activity, converting weak androgen, 3α-androstanediol, into the potent androgen dihydrotestosterone. However, in the reactions with 3α-hydroxysteroids, these enzymes preferred NAD^+ over $NADP^+$, and the K_m values for androgens were much lower (~0.1 μM)

Table 5.2 Nomenclature and properties of SDR Enzymes. References to these studies are cited in the text. Nomenclature for SDR proteins can be found at http://www.sdr-enzymes.org/. To search the database, choose field "family name" and enter SDR9C or SDR7C or SDR16C to view all currently known members. Human SDR16C6 is not included in the Table because it appears to be a pseudogene

	Species	Cofactors[a]	Substrates	Nomenclature	Other names
SDR7C					
RDH11	human	NADP(H)	*at*ROL/RAL, *c*ROLs/RALs	SDR7C1	PSDR1, RalR1
	mouse	NADP(H)	*at*ROL/RAL, *c*ROLs/RALs, nonanal, nonenal	SDR7C9	SCALD
RDH12	human	NADP(H)	*at*ROL/RAL, *c*ROLs/RALs, nonanal, nonenal	SDR7C2	
	mouse		ND	SDR7C10	
RDH13	human	NADPH	*at*RAL	SDR7C3	
	mouse		ND	SDR7C11	
RDH14	human	NADP(H)	*at*RAL, 9*c*RAL, *at*ROL	SDR7C4	PAN2
	mouse		ND	SDR7C12	
SDR9C					
RoDH4	human	NAD(H)	*at*ROL/RAL, *c*ROLs/RALs, 3α-HS	SDR9C8	RDH16, RDH-E
RDH1	mouse	NAD^+	*at*ROL, 9*c*ROL, 3α-HS	SDR9C17	
RoDH1	rat	$NADP^+$, NAD(H)[b]	holoCRBPI, 3α-HS	SDR9C29	RDH7
RoDH2	rat	$NADP^+$, NAD(H)[b]	holoCRBPI, 3α-HS	SDR9C28	RDH2
RL-HSD	human	NAD(H)	*at*ROL, *at*RAL; 3α-HS	SDR9C6	HSD17B6, 3α-HSE
	mouse	NAD(H)	*at*ROL/RAL, *c*ROLs/RALs; 3α-HS, 17β-HS	SDR9C13	HSD17B6, 17β-HSD9
DHRS9	human	NAD(H)	3α-HS, may be *at*ROL	SDR9C4	RDHL, retSDR8, RDH-TBE, RoDH-E2, 3α-HSD
	mouse		ND	SDR9C12	
	rat		may be *at*ROL	SDR9C26	eRolDH2
11-*cis*-RDH	human	NAD^+	*c*ROLs; 3α-HS	SDR9C5	RDH5
	mouse	NAD(H)	*c*ROLs/RALs	SDR9C21	
HSD11B2	human	NAD(H)	11β-HS	SDR9C3	Corticosteroid 11β-HSD
	mouse			SDR9C11	
HSD17B2	human	NAD(H)	17β-HS, 20α-HS	SDR9C2	Estradiol 17β-HSD

(continued)

Table 5.2 (continued)

	Species	Cofactors[a]	Substrates	Nomenclature	Other names
	mouse			SDR9C10	
SDR16C					
retSDR1	human	NADP(H)	*at*RAL	SDR16C1	DHRS3
	mouse			SDR16C9	
RDH10	human	NAD(H)	*at*ROL/RAL, *c*ROLs/RALs	SDR16C4	
	mouse			SDR16C10	
RDHE2	human	NAD(H)	*at*ROL	SDR16C5	
	mouse	ND	*at*ROL	SDR16C11	
	frog	NAD(H)	*at*ROL	SDR16C84	rdhe2, MGC80593
RDHE2S	mouse	ND	*at*ROL	SDR16C12	
HSD17B11	human	ND	3α-androstanediol	SDR16C2	Pan1b, retSDR2
	mouse			SDR16C7	
HSD17B13	human	$NADP^+$	cortisol	SDR16C3	SCDR10B
	mouse			SDR16C8	

ND no data

[a]Some of the SDR enzymes can bind both NAD(H) and NADP(H) but their K_m values may differ by an order of magnitude. Only the preferred cofactors are listed.

[b]Rat RoDH1 and RoDH2 were reported to prefer $NADP^+$ with all-*trans*-retinol as substrate ([26, 32, 33]), but NAD^+ - with 3α-hydroxysteroids ([21, 72]).

than the reported K_m values for free or CRBP1-bound retinol (1–2 μM) [32, 33]. Interestingly, human RoDH (also known as RoDH-like-hydroxysteroid dehydrogenase or RL-HSD) was also found to possess weak 3β-hydroxysteroid dehydrogenase activity and to act as a 3(α → β)-hydroxysteroid epimerase, converting 3α-hydroxysteroids to 3β-hydroxysteroids [9, 36, 37, 76, 136]. Both the rat and human RoDHs manifested oxidative 17β-hydroxysteroid dehydrogenase activities when presented with appropriate substrates [21]. Thus, RoDH-like enzymes appeared to be multifunctional.

In humans, two more RoDH-like enzymes were identified (Table 5.2). The first enzyme, named RoDH4 (also known as RDH16), exhibited a liver-specific expression pattern, while the second one, named non-hepatic 3α-hydroxysteroid dehydrogenase (also known as DHRS9), was widely expressed. Both exhibited higher catalytic efficiencies with 3α-hydroxysteroids than with all-*trans*-retinol [36, 37, 66] but the second one had little or no activity towards all-*trans*-retinol.

In order to assess the *in vivo* contribution of RoDH enzymes to RA biosynthesis, Napoli and colleagues began cloning and characterizing the murine orthologs of rat and human RoDHs. Due to a gene duplication event, mice had many more RoDH-like genes than rats or humans [11]. Importantly, all of the mouse RoDH-like enzymes studied thus far were found to prefer NAD^+ over $NADP^+$ and recognized both 3α-hydroxysteroids and *cis*-retinoids as substrates (reviewed in [128]). In fact, only one of these mouse enzymes, RDH1, had significant activity towards all-*trans*-retinol [187]. Still, the catalytic efficiency of mouse RDH1 towards all-*trans*-retinol

(V/K_m of 3 (nmol/min/mg)/μM) was ~tenfold lower than towards 3α-androstanediol (V/K_m of 31 (nmol/min/mg)/μM). Thus, like ADH isoenzymes, the microsomal RoDH/RDH enzymes related to the original rat RoDH1 all-*trans*-retinol dehydrogenase were shown to recognize other substrates besides retinoids. Phylogenetic analysis showed that the rat, human and mouse homologs of rat RoDH1 that exhibited dual retinol/sterol dehydrogenase activity belong to the same branch of phylogenetic tree. This branch of SDRs was designated as SDR9C (Table 5.2). The individual roles of retinol/sterol dehydrogenases in the metabolism of retinol remained to be established.

Analysis of RoDH Enzymes Using In Vivo and Ex Vivo Approaches

A targeted knockout of *Rdh1* gene in mice showed that RDH1-null mice were viable [188]; hence, the function of RDH1 in retinoid or steroid metabolism was not essential for embryogenesis. However, the adult mice had an interesting phenotype - instead of being smaller when restricted in vitamin A, they grew longer and larger than wild type mice, with increased weights of multiple fat pads, liver, and kidney. There were no detectable changes in RA levels in tissues of RDH1-null mice, possibly due to a downregulation (2.5-fold) of *Cyp26a1* expression, but the amount of retinol on low vitamin A diet (0.6 IU/g) was increased in liver and kidney (1.5–2fold) relative to wild-type. Thus, RDH1 clearly had an impact on retinoid metabolism in adult mice. Whether this was due to its retinol dehydrogenase activity, and whether the contribution of RDH1 to RA biosynthesis was the only role of RDH1 in mouse metabolism remained to be established. However, it was clear that RDH1 was not the primary enzyme that produced RA during embryonic development.

The search for physiologically relevant retinol dehydrogenases appeared to have hit a roadblock when, unexpectedly, a random mutagenesis screen for novel genes essential for mouse embryogenesis linked mutations in another retinoid-active member of the SDR superfamily, Retinol Dehydrogenase 10 (RDH10), to defects in growth and patterning of the forelimb, frontonasal process, pharyngeal arches, and various organs. RDH10 was originally identified in humans based on its sequence similarity to retina SDR1 (retSDR1) (discussed below) [68, 179]. RDH10 and retSDR1 (also known as DHRS3) clustered with a different branch of SDR phylogenetic tree, SDR16C (Fig. 5.2). Initial characterization of RDH10 using large amounts (200 μg) of microsomal fraction from COS cells transfected with RDH10 expression construct and 1 μCi [^{3}H] all-*trans* retinol suggested that RDH10 acted as an $NADP^+$-preferring all-*trans*-retinol – specific dehydrogenase [68]. However, subsequent studies showed that human RDH10 (SDR16C4) recognized not only all-*trans*-retinol, but also 11-*cis* and 9-*cis* retinols as substrates and strongly preferred NAD^+ as a cofactor [14, 15], in agreement with the structural determinants of SDR cofactor specificity [80]. The essential role of RDH10

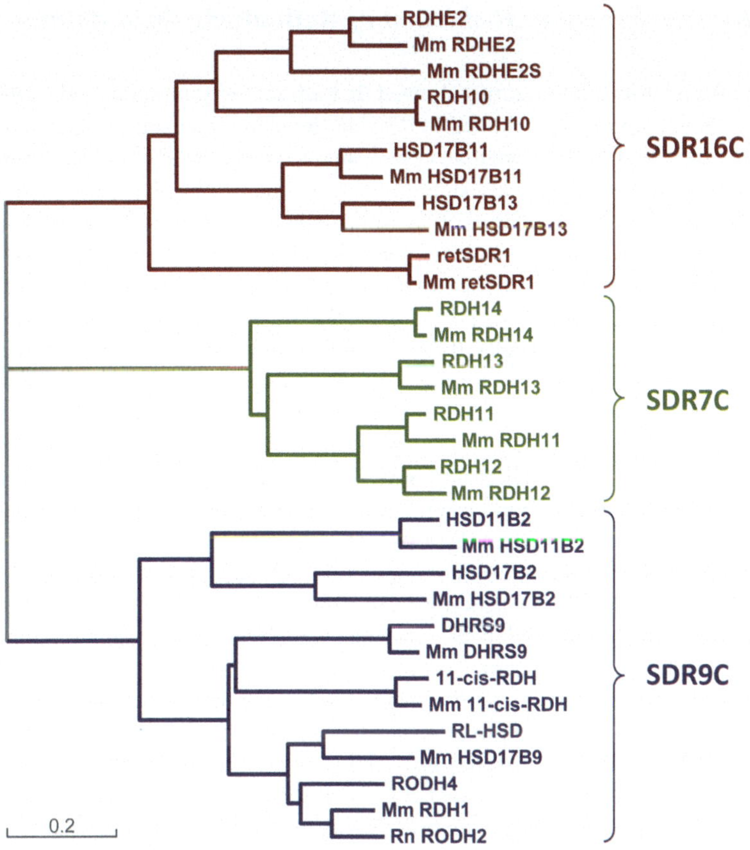

Fig. 5.2 Neighbour-joining phylogenetic tree of retinoid-active SDRs. Three branches of SDRs that include human and rodent retinoid-active enzymes implicated in retinoic acid biosynthesis are shown. SDR9C group is comprised of enzymes with preference for NAD(H) as cofactor. SDR7C group is comprised of enzymes with preference for NADP(H) as cofactor. SDR16C group includes both NAD(H) and NADP(H)-preferring enzymes. Murine enzymes are prefixed with Mm, rat enzymes are prefixed with Rn. Scale bar, 0.2 amino acid substitutions per site

in development was further confirmed by targeted inactivation of RDH10 function in mice. RDH10-null mice displayed severe malformations and embryonic lethality due to insufficient production of RA [7, 147, 156]. This phenotype could be rescued by supplementation with RA or all-*trans*-retinaldehyde [147]. Interestingly, when overexpressed in a model of human organotypic skin raft culture, RDH10, but not the SDR9C dehydrogenases, induced a phenotype consistent with overproduction of RA, which was characterized by an increased proliferation and reduced differentiation of keratinocytes [98]. Thus, at least in human epidermis, RDH10, a member of the SDR16C family appears to contribute to RA biosynthesis, while the SDR9C enzymes do not.

The Reverse Reaction: Reduction of Retinaldehyde *to Retinol*

Studies from various laboratories showed that a fraction of retinaldehyde added to intact cells in culture was oxidized to RA, but the more abundant product was retinol. This indicated that mammalian cells contained enzymes that reduce retinaldehyde to retinol. Thus the steady-state levels of cellular retinaldehyde available for RA biosynthesis could be controlled by retinaldehyde reductases. Several members of the SDR superfamily of proteins were found to catalyze the interconversion between retinol and retinaldehyde in vitro with NADP(H) as the preferred cofactor [10, 12–15, 68, 83, 86, 96] (Table 5.2, SDR7C and SDR16C family). When expressed in intact cultured cells, these NADP(H)-preferring enzymes unidirectionally converted retinaldehyde to retinol [96]. Interestingly, one of these SDRs, which was named originally retina SDR1 (retSDR1) and later renamed Dehydrogenase Reductase 3 (DHRS3, SDR16C1), was shown to be inducible by RA in human neuroblastoma cell lines, THP-1 monocytes, and rat liver [31, 191].

DHRS3 was identified at the level of cDNA in the expressed sequence database (EST) of GenBank as a photoreceptor visual cycle all-*trans*-retinol dehydrogenase [68] based on its sequence similarity to previously cloned retinoid-active SDRs [32, 158]. However, the mRNA encoding DHRS3 was found to be expressed in many human tissues, including adult heart, placenta, lung, liver, kidney, pancreas, thyroid, testis, stomach, trachea, and spinal cord, as well as fetal tissues such as kidney, liver, and lung [31, 68]. Initial assays of DHRS3 enzymatic activity using recombinant protein overexpressed in Sf9 insect cells suggested that this protein was capable of catalyzing the transfer of ^{3}H from [^{3}H]NADPH, but not from NADH, to 10 μM all-*trans*-retinaldehyde [68]. The exact rate of DHRS3-catalyzed reaction was not reported but the retinaldehyde reductase activity of the enzyme appeared to be much lower than that of other SDRs with NADPH-dependent retinaldehyde reductase activity [12, 13]. Furthermore, overexpression of DHRS3 in SK-N-AS neuroblastoma cells stimulated the accumulation of retinyl esters, but did not result in quantifiable changes in the conversion of retinol to retinaldehyde or RA [31]. Nevertheless, a knockdown of *dhrs3a* gene expression in zebrafish embryos appeared to increase the expression of *cyp26a1* gene and RARE-GFP reporter construct in the spinal cord [56], suggesting that *dhrs3a* regulated RA levels *in vivo*. In mice, targeted knockout of *Dhrs3* resulted in reduction in the levels of retinol and retinyl esters and a slight increase in RA levels, modest changes in the expression of several RA target genes, and embryonic lethality late in gestation [2, 20].

The discrepancy between the severity of the *in vivo* effects of *Dhrs3* gene knockout on retinoid metabolism and embryo survival and the seemingly negligible *in vitro* activity of DHRS3 towards retinaldehyde was puzzling. To resolve this puzzle, Adams et al. [2] undertook an in-depth study of the properties of DHRS3 as an enzyme. Unexpectedly, these studies revealed that human DHRS3 required the presence of another retinoid-active SDR, human RDH10, in order to display its full enzymatic potential. Furthermore, RDH10 was in turn activated by the presence of DHRS3. The mutual activation of DHRS3 and RDH10 did not depend on the catalytic activity of either protein and occurred equally well when wild-type proteins were substituted with the corresponding active site mutants of the partner proteins. The RDH10-activated DHRS3 was found to act as a high-affinity/high efficiency all-*trans*-retinaldehyde–specific reductase that did not recognize 11-*cis*-retinaldehyde as substrate and preferred NADPH as a cofactor. When co-expressed in HEK293 cells, the two proteins were co-localized in characteristic ring structures within the endoplasmic reticulum [2] that were described previously by two other groups [41, 78].

Importantly, the retinol dehydrogenase activity of the membrane fraction isolated from E14.5 DHRS3-null mouse embryos was about twofold lower than the activity of wild-type embryos; and mouse embryonic fibroblasts isolated from E14.5 embryos produced 1.7-fold less retinaldehyde from retinol than MEFs from wild-type embryos [2]. RDH10 was shown to be the major retinol dehydrogenase during embryonic development [147]. Hence, the results obtained with DHRS3-null mouse embryos confirmed that the mutually activating interaction between RDH10 and DHRS3 occurred *in vivo* and suggested that this mechanism was conserved across species.

Oxidation of Retinaldehyde to RA: Cloning and Characterization of Retinaldehyde Dehydrogenase Isoenzymes

Fractionation of rat tissue cytosol by anion-exchange chromatography revealed the existence of at least four retinaldehyde dehydrogenase isoenzymes [143]. The quantitatively major isoenzyme in liver, kidney, and testis, RALDH1 was purified from rat liver but cloning of the corresponding cDNA from liver was complicated by the abundance of numerous other ALDHs. To avoid abundant hepatic ALDHs, Wang et al. [173] screened rat testis cDNA library for RALDHI cDNA using PCR primers based on sequences of two highly conserved amino acid-segments in the mammalian ALDHs. As a result of this screening, a cDNA clone encoding a previously unknown ALDH, distinct from RALDH1, was identified. This isoenzyme was named RALDH2, because it catalyzed the oxidation of free retinaldehyde with the apparent K_m value of 0.7 ± 0.3 μM and the V_{max} value of 105 ± 4 nmol/min × mg of purified protein expressed in *E. coli*. The kinetic constants for the oxidation of retinaldehyde in the presence of a twofold molar excess of apo-CRBPI, which binds retinaldehyde with a relatively high K_d of 50–100 nM [103, 104], were 0.2 ± 0.06 μM (apparent K_m) and 62 ± 15 nmol/min × mg (V_{max}).

At the same time, Zhao et al. [189] succeeded in cloning a novel mouse ALDH with activity indistinguishable from that of V2 embryonic retinaldehyde dehydrogenase described in retina from mouse P19 teratocarcinoma cells. Because of the scarcity of the V2 enzyme in *vivo*, these researchers screened candidate cell lines for the presence of the V2 activity using a zymography bioassay. They found that P19 cells induced by treatment with RA expressed a new retinaldehyde dehydrogenase

Table 5.3 Kinetic properties of purified mouse retinaldehyde dehydrogenases

Parameters	ALDH1A1[a]	ALDH1A2[b]	ALDH1A3[c]
K_m, μM	11.6	0.66	3.9
V_{max}, nmol × min^{-1} × mg^{-1}	85.6	4.31	306

[a, b] Mouse ALDH1A1 (RALDH1) and ALDH1A2 (RALDH2) proteins were expressed in *E. coli* as N-terminal fusions to glutathione-S-transferase (GST) and purified using GST-affinity column ([62, 63]).
[c]RALDH3 was expressed in *E. coli* as a His-tagged protein and affinity-purified using Ni-NTA column [157]. The enzymatic products were analyzed by high pressure liquid chromatography.

activity that was absent in uninduced cells. To clone the corresponding cDNA, they designed degenerate oligonucleotide primers corresponding to two highly conserved protein motifs in ALDH sequences and employed PCR to screen cDNA synthesized from mRNA of PI9 cells, untreated and treated with retinoic acid, for aldehyde dehydrogenase cDNAs. Restriction digest of the 450-bp PCR product showed that only the RA-treated P19 cells expressed unique PCR-amplified sequence resistant to SalI digestion. The sequence of this product allowed the isolation of the full-length cDNA encoding murine RALDH2. The murine and rat RALDH2 protein sequences turned out to be 100 % identical.

Rat RALDH1 was cloned from rat kidney cDNA library [18] and testis cDNA library [138, 139]. Rat RALDH1 was found to share ~72 % amino acid sequence identity with RALDH2. A third RALDH, RALDH3, was later identified and originally referred to as ALDH6 [67]. The fourth cloned RALDH, RALDH4, was found to recognize only 9-*cis*-retinaldehyde as substrate and not all-*trans*-retinaldehyde [105, 106].

Under the currently accepted nomenclature, all all-*trans*-retinaldehyde-oxidizing ALDHs belong to ALDH1A family: ALDH1A1 (RALDH1), ALDH1A2 (RALDH2), and ALDH1A3 (RALDH3). The *in vivo* roles of individual ALDH isozymes in RA biosynthesis were summarized in several recent review articles [111, 129]. In brief, ALDH1A2 was shown to be the primary enzyme responsible for RA biosynthesis at most sites during embryogenesis. *Aldh1a2*$^{-/-}$ mice died early in embryonic development due to defects in heart morphogenesis [131]. ALDH1A3 had a more limited role during development. *Aldh1a3*$^{-/-}$ mouse embryos survived until birth but died shortly thereafter from defects in nasal development [51]. In addition, ALDH1A3 was reported to play a major role in protection of adult tissues against carcinogenesis (reviewed in [111]). ALDH1A1 was not essential for embryogenesis, but could have a role in RA biosynthesis during adulthood, because the livers of *Aldh1a1*$^{-/-}$ mice displayed reduced RA biosynthesis and increased serum retinaldehyde levels after treatment with retinol [54, 123].

To compare the catalytic properties of the retinaldehyde-active ALDH isoenzymes, Bhat and colleagues performed kinetic analysis of all three murine ALDHs under the same conditions, allowing for direct comparison of their properties (Table 5.3). ALDH1A3 was the most catalytically efficient enzyme, but it had a relatively high K_m value for all-*trans*-retinaldehyde (Table 5.3) [157]. ALDH1A1 and

ALDH1A2 had similar catalytic efficiency but the K_m value of ALDH1A2 was much lower than that of ALDH1A1 (Table 5.3) [62, 63]. Thus, ALDH1A1 was the least potent retinaldehyde dehydrogenase of the three enzymes. All three ALDHs recognized other substrates in addition to retinaldehyde, including aldehydes derived from lipid peroxidation. In fact, ALDH1A1 was shown to play a crucial role in protection of the mouse eye lens and cornea from lipid peroxidation aldehydes and cataract formation induced by oxidative stress, as demonstrated using various *Aldh1a1*$^{-/-}$ mouse models [92].

RA Degradation: Cloning and Characterization of the RA-Inducible RA-Metabolizing CYP Enzymes

It was not until the development of molecular cloning techniques that cytochrome P-450 enzymes specific for RA metabolism were identified. It was shown that in F9 cells, RAR-mediated signaling could play an important role in regulating the enzymes responsible for RA-induced self-degradation [29]. White et al. [175] used mRNA differential display to identify zebrafish genes regulated by RA during exogenous RA exposure. This approach allowed them to isolate a cDNA, P450RAI, encoding a novel member of the cytochrome P450 family. In COS-1 cells transfected with the P450RAI cDNA, all-*trans*-RA was rapidly metabolized to more polar metabolites including 4-oxo-RA and 4-OH-RA.

Subsequently, these investigators cloned the two first mammalian RA-inducible RA-metabolizing cytochromes P450 (hP450RAI and P450RAI-2) of the novel class of cytochromes (CYP26) and demonstrated that they were responsible for generation of several hydroxylated forms of all-*trans*-RA, including 4-OH-RA, 4-oxo-RA, and 18-OH-RA [176, 177]. Competition experiments with other retinoids suggested that all-*trans*-RA was the preferred substrate. The authors concluded that these enzymes (renamed CYP26A1 and CYP26B1), played a key role in RA metabolism, functioning in a feedback loop where RA levels were controlled in an autoregulatory manner. Later, a third CYP26C1 was identified that also preferred all-*trans*-RA as substrate but could metabolize 9-*cis*-RA and was much less sensitive than the other CYP26 family members to the inhibitory effects of ketoconazole [163].

Targeted gene knockout studies showed that CYP26A1-null mouse fetuses died at mid-late gestation, with multiple organ defects that were consistent with excessive RA signaling [1]. CYP26B1-null mice that were born alive died right after birth due to respiratory defects [183]. In human fetus CYP26A1 was found to be the major enzyme in brain while CYP26B1 was found in all other tissues except brain. However, in adult human tissues, the expression patterns of CYP26A1 and CYP26B1 at the level of both mRNA and protein showed a significant overlap [166]. With the exception of liver and lung, where CYP26A1 was the predominant form, all other human adult tissues contained higher levels of CYP26B1. This tissue distribution pattern was supported by another study [181], with minor discrepancies that could be potentially attributed to variability in the quality of tissue samples.

Table 5.4 Kinetic properties of purified human CYP26 enzymes. CYP26A1 and CYP26B1 were expressed as C-terminally His_6-tagged proteins using the Baculovirus expression system and purified using cobalt affinity column. The reaction products were analyzed by LC–MS/MS ([166])

Parameters	CYP26A1[a]	CYP26B1[b]
K_m, nM	50	19
V_{max}, pmol × min^{-1} × pmol $P450^{-1}$	10	0.8

Despite the low sequence identity (43 %), CYP26A1 and CYP26B1 were shown to have very similar enzymatic properties, hydroxylating RA to form 4OH-RA and 18OH-RA, and then hydroxylating these primary products further [166]. Purified recombinant CYP26A1-His_6 had a lower affinity but a ~tenfold higher rate for formation of 4OH-RA than CYP26B1-His_6 (Table 5.4). CYP26A1 also had a two to tenfold higher catalytic activity towards hydroxylated forms of RA, making it ~20-fold more efficient than CYP26B1 in the overall depletion of RA. Considering that CYP26A1 expression in liver is very sensitive to RA levels (reviewed in [152]), the high catalytic efficiency of this low affinity enzyme would enable CYP26A1 to rapidly bring down the excessive levels of RA.

CYP26C1 was found to share 45 % amino acid identity with CYP26A1 and 51 % - with CYP26B1, but it exhibited a distinctly different expression pattern and catalytic properties. CYP26C1 was expressed mainly during embryonic development, but appeared to be nonessential since CYP26C1-null mice had no apparent abnormalities [167]. In adult tissues, it was detected at low levels in adrenal gland, lung, spleen, testis, brain, adult liver and ovary [181]. Unlike the other two enzymes, CYP26C1 preferred 9-*cis*-RA as substrate and exhibited broader substrate specificity in general. Furthermore, CYP26C1 showed a different response to treatment with RA and 9-*cis*-RA, being upregulated in some tissues and cells but downregulated in others [135]. Of the three CYP26 genes, CYP26C1 is the least conserved and its physiological function remains poorly understood. A frameshift mutation in human CYP26C1 was linked to focal facial dermal dysplasia type IV, a rare syndrome characterized by facial lesions resembling aplasia cutis [159]). A potential role of other members of CYP superfamily of proteins in RA catabolism was discussed in a recent review [84].

Current State of the Field

Enzymes Involved in the Control of Retinol and RA Levels

The currently accepted model of RA biosynthesis postulates that RA is produced in two sequential oxidative steps: first, retinol is oxidized to retinaldehyde, and then retinaldehyde is oxidized to RA (Fig. 5.1). The first step is rate-limiting and reversible. The rate of retinaldehyde production is controlled by two types of SDR enzymes: the NAD^+-dependent oxidative SDR(s) that oxidize retinol to retinaldehyde (RDH10

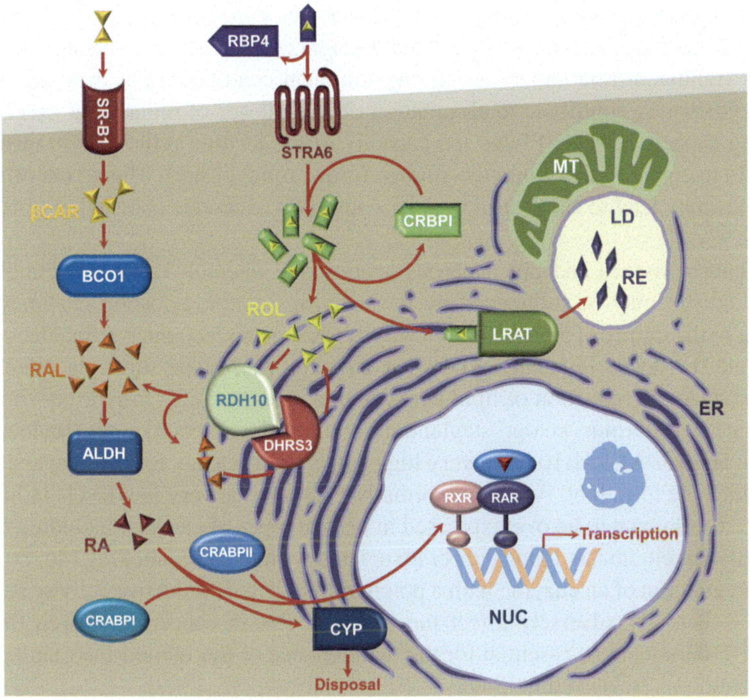

Fig. 5.3 Components of retinoid metabolic and signaling system. Retinol (ROL, depicted as yellow pyramids) is delivered to extrahepatic cells bound to plasma Retinol Binding Protein 4 (RBP4). HoloRBP4 binds to RBP4 receptor STRA6. Cellular Retinol Binding Protein type I (CRBPI) accepts retinol from STRA6 in the cytoplasm and delivers retinol to membranes of endoplasmic reticulum (ER), where retinol is either esterified by lecithin retinol acyl transferase (LRAT) to retinyl esters (RE, depicted as purple rhombus) or oxidized by Retinol Dehydrogenase 10 (RDH10) to retinaldehyde (RAL, depicted as orange pyramids). Retinaldehyde is oxidized further to retinoic acid (RA, depicted as brown pyramids) by Aldehyde Dehydrogenase (ALDH) in the cytoplasm or is reduced back to retinol by Retinaldehyde Reductase (DHRS3) in the membranes. Retinoic acid binds to Cellular Retinoic Acid Binding Proteins (CRABPs) type I or type II and is transferred by holoCRABPII to nucleus (NUC) for binding to heterodimers of Retinoic Acid Receptors (RAR and RXR) or delivered to Cytochrome P450 enzymes (CYP) by holoCRABPI for degradation. In addition, β-carotene (βCAR, depicted as duplicate of olive pyramids) is taken up by the cells through scavenger receptor class B, type 1 (SR-B1) and cleaved into two molecules of retinaldehyde by β-carotene oxygenase type 1 (BCO1). Retinaldehyde derived from β-carotene may be oxidized to retinoic acid or converted to retinol as described above. Other abbreviations: LD, lipid droplet; MT, mitochondria

and possibly others), and the NADPH-dependent reductive SDR(s) that reduce retinaldehyde back to retinol (DHRS3 and possibly others) (Fig. 5.3). Both types of SDRs are integral membrane proteins. The second step in RA biosynthesis is irreversible and is catalyzed by cytosolic ALDHs: ALDH1A1 (RALDH1), ALDH1A2 (RALDH2), and ALDH1A3 (RALDH3). RA is catabolized by the members of the CYP family of proteins, primarily by CYP26A1, CYP26B1, and CYP26C1.

It should be noted that in addition to these primary enzymes dedicated to retinoid metabolism, there are other enzymes that recognize retinoids as substrates and may play a significant role under certain physiological conditions. For example, ADH isoenzymes may contribute to elimination of large doses of retinol that exceed the binding capacity of CRBPI and the capacity of the lecithin retinol acyltransferase (LRAT) to convert retinol into the storage forms, retinyl esters. The action of ADH enzymes may lead to short-term toxicity caused by excessive conversion of retinol to RA.

In cancer tissues and precancerous lesions, the interconversion of retinol and retinaldehyde may be influenced by the activities of the cytosolic enzymes that belong to the family of aldo-keto reductases (AKRs) and recognize retinaldehyde as substrate (reviewed in [155]). AKRs are enzymes with wide substrate specificity that catalyze the reduction of lipid peroxidation products (e.g., 4-hydroxy-*trans*-2-nonenal), ketosteroids, ketoprostaglandins, and xenobiotic compounds (reviewed in [137]). Human AKR1B10 has a very high catalytic efficiency for the conversion of retinaldehyde to retinol. Although normally expressed at very low levels in human tissues, it is known to be overexpressed in certain types of cancers, including hepatocellular carcinoma and lung cancer associated with tobacco smoking [46, 60, 79]. An upregulation of an enzyme with a potent retinaldehyde reductive activity such as AKR1B10 might adversely affect the levels of RA in cancer cells even though AKR1B10 may not be essential for the maintenance of RA homeostasis under normal conditions (reviewed in [84]).

Besides CYP26 enzymes, several other members of the CYP superfamily of proteins including human CYP3A4, 3A5, 3A7, 1A1, 4A11, 2C8, 2C9, 2C22, 2C39, CYP2S1, and CYP2E1 [5, 35, 40, 55, 101, 108, 112, 121, 126, 144, 153, 160, 164, 174, 180] were shown to catabolize RA. Most of these CYPs are induced by elevated levels of RA. For example, human CYP2S1 is inducible by RA in skin of a subset of individuals and was shown to produce 4-hydroxy-RA and 5,6-epoxy-RA when expressed in *E. coli* together with NADPH cytochrome P450 reductase [160]. Human CYP2C8 and CYP2C9 genes are both induced by RA, but the less efficient CYP2C9 may have a greater impact on RA catabolism because it is expressed at about tenfold higher levels than CYP2C8 and, in addition, it's expression is significantly more responsive to RA treatment [144]. Rodent orthologs of human CYP2C8 and CYP2C9 (rat CYP2C22 and mouse CYP2C39) also 4-hydroxylate RA [5, 55, 174]. Interestingly, under normal dietary conditions, the relative level of CYP2C22 mRNA expression in rat liver exceeds that of CYP26A1 by about 100-fold [153]. Mouse CYP2C39 has a relatively high affinity for RA (K_m of 0.8 μM), but a ~40-fold lower V_{max} value compared to other CYP2C family members [5]. Nevertheless, the decreased expression of this enzyme in the liver of mice lacking the aryl hydrocarbon receptor gene has been linked to increased levels of RA, retinol, and retinyl palmitate [5]. Ethanol was shown to enhance RA metabolism into polar metabolites through the induction of CYP2E1 in the liver [40, 108]. Thus, kinetic characteristics and tissue distribution of various RA-catabolizing enzymes indicate that while CYP26A1 and CYP26B1 may be the primary CYP enzymes for the hepatic clear-

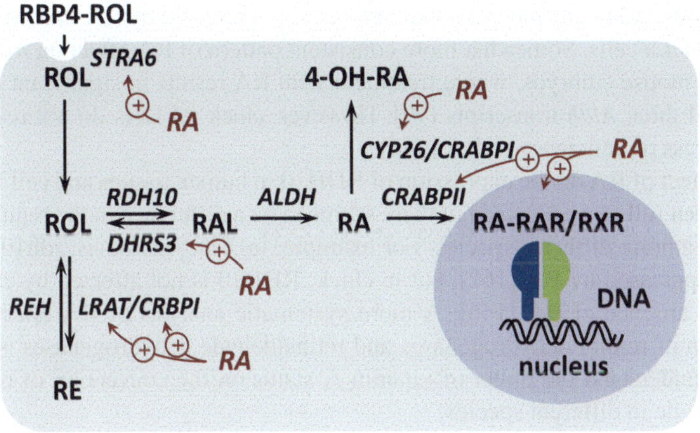

Fig. 5.4 Regulation of retinoid metabolism by retinoic acid status. Abbreviations are as in Fig. 5.3. Retinoic acid (RA) induces the expression of STRA6, LRAT, CRBPI, DHRS3, CYP26, CRABPII, and some forms of RAR

ance of RA due to their high catalytic efficiency [164, 165], other members of CYP superfamily may play an essential role in extrahepatic tissues or in liver under certain conditions.

Regulation of RA Biosynthesis and Degradation by RA Status

Some of the components of the RA biosynthetic and catabolic machinery are regulated by RA *via* a feedback regulation loop (Fig. 5.4). RA induces the expression of the retinoid enzymes and binding proteins that promote the synthesis of retinyl esters including CRBPI, LRAT [114, 151, 190], and DHRS3 [31]. At the same time, RA induces its own catabolism by up-regulating CYP26 enzymes [28, 133, 178]. STRA6, the plasma membrane receptor for retinol carrier RBP4, is known to be induced by RA in extrahepatic tissues [28, 178] while Retinol Binding Protein Receptor 2 (RBPR2), the recently discovered analogous receptor in rodent liver, is suppressed by RA, presumably to promote the uptake of retinol by extrahepatic tissues under conditions of vitamin A sufficiency [3].

The regulation of retinol dehydrogenases and retinaldehyde dehydrogenases by RA is less well understood, and it appears that the responses of the genes encoding these enzymes may be species-specific and cell context-dependent. For example, vitamin A-deficient rats have lower levels of *Aldh1a1* (*Raldh1*) mRNA in kidney and liver but elevated levels of *Aldh1a1* in testis [138, 139]. Orally administered RA restores *Aldh1a1* expression in kidney, but not in liver [138, 139]. In primary keratinocytes, a high concentration of RA (1 μM) appears to upregulate *ALDH1A3* (*RALDH3)*, but has no effect on *ALDH1A1* (*RALDH1*) or *ALDH1A2 (RALDH2)* expression [89]. *RALDH3* expression is also induced by RA in organotypic human skin cultures and in an epidermal explant, but is not affected by RA in dermal fibroblasts or HeLa cells. Somewhat more consistent pattern of RA effect on *Aldh* genes is seen in mouse embryos, where treatment with RA results in significant suppression of all three *Aldh* transcripts [99]. However, chick ALDHs do not respond to either excess or deficiency of RA [146].

The effect of RA on the expression of RDH10 in human tissues and cell lines has not yet been fully investigated but there seems to be a difference in the regulation of this gene among different species. For example, in *Xenopus laevis*, rdh10 expression is suppressed by RA [162], but in chick, RDH10 is not affected by either the excess or absence of RA [146]. A more systematic analysis of RA effect on the expression of retinol dehydrogenases and retinaldehyde dehydrogenases is needed to better understand the effect of vitamin A status on the conversion of retinol to retinaldehyde in different species.

Future Directions

Despite significant progress made since the middle of the twentieth century many unanswered questions still remain and there is much to be learned. RDH10 is the major retinol dehydrogenase responsible for RA biosynthesis during embryonic development. However, the molecular patterning defects in RDH10-null mice do not reflect a complete state of RA deficiency [147], suggesting that additional retinol dehydrogenases exist. RDH10 shares the highest sequence similarity with two other members of SDR16C family of proteins, RDHE2 and RDHE2S) (Table 5.2 and Fig. 5.2). In humans, the genes encoding RDHE2 (SDR16C5) and RDHE2S (SDR16C6) are located in close proximity to the RDH10 gene (SDR16C4), and may have originated from a common ancestor as a result of gene duplication. Proteins encoded by the mouse orthologs of human *RDHE2* and *RDHE2S* genes (SDR16C11 and SDR16C12, respectively) function as all-*trans*-retinol dehydrogenases *in vitro* and increase the rate of RA biosynthesis from retinol when expressed in living cells [16, 97]. Whether these two enzymes contribute to RA biosynthesis during development or adulthood is not yet known, but studies in frogs indicate that the frog ortholog of these enzymes is essential for frog embryonic development [16]. Frogs have a single gene, *rdhe2* (*sdr16c90*) in position orthologous to mouse *Rdhe2* (*Sdr16c11*) and *Rdhe2s* (*Sdr16c12*) genes. The enzyme encoded by the frog gene acts as a highly active retinol dehydrogenase that promotes RA biosynthesis in living cells [16]. Thus, the retinol dehydrogenase activity of RDHE2/RDHE2S enzymes is conserved in lower vertebrates. Further studies are needed in order to elucidate the roles of mammalian RDHE2 and RDHE2S in RA biosynthesis. Finally, in addition to RDH10, RDHE2, and RDHE2S, the SDR16C family includes two relatively weak steroid dehydrogenases, 17β-HSD11 (SDR16C2) [30], and 17β-HSD13 (SDR16C3) [109]. Whether the latter two enzymes also have activity towards retinoids has not yet been examined.

The recent discovery that RDH10 and DHRS3 mutually activate each other [2] raises new questions about the impact of this interaction on the cellular levels of

RA. It is not yet known whether and how the two proteins physically interact; whether their interaction is influenced by the metabolic state, RA levels, oxidative stress, etc. It is not yet known whether disruption of RDH10 and DHRS3 interaction may lead to pathophysiological changes in cell metabolism. It is also unclear whether the interaction between RDH10 and DHRS3 is limited to these two proteins or whether they have other partners among SDR proteins.

DHRS3 has now been proven to act as a physiologically relevant retinaldehyde reductase essential for embryonic development [2, 20]. However, while the retinol and retinyl ester stores in DHRS3-null mice are significantly reduced (~25 % of wild-type), they are not fully depleted. This suggests that DHRS3 is not the only retinaldehyde reductase that functions in the embryo. The molecular identity of additional retinaldehyde reductases that are responsible for the maintenance of the residual retinoid stores remains to be established.

The molecular mechanisms for trafficking of the highly hydrophobic retinoid substrates and products from one enzyme or binding protein to another remain puzzling. For example, it is not clear whether RDH10 obtains retinol directly from CRBPI or from the membranes. *In vitro* data suggest that CRBPI inhibits the activity of RDH10 towards retinol [98]; however, the effect of CRBPI on RDH10 activity *in vivo* remains unexplored, and thus far, there is no evidence of direct protein-protein interaction between RDH10 and CRBPI.

Since RDH10 and DHRS3 interact, it is possible that DHRS3 obtains retinaldehyde directly from RDH10, but experimental proof of such substrate channeling is still lacking. Furthermore, retinaldehyde is also utilized by ALDH isoenzymes, which are believed to reside in the cytosol. How the cytosolic ALDHs are able to compete for the hydrophobic retinaldehyde with the membrane-bound DHRS3, and whether CRBPI is involved in this process is currently unknown.

A recent study demonstrated that when expressed in COS7 cells [78], mouse RDH10 co-localized with mitochondria/mitochondrial associated membranes (MAM), in close proximity to CRBPI. During acyl ester biosynthesis RDH10 partially re-located to lipid droplets [78]. DHRS3 was also shown to be associated with lipid droplets [41]. The co-localization of RDH10 and DHRS3 in ring structures similar to lipid droplets was confirmed by Adams et al. [2]. Whether ALDH isoenzymes localize near the membranes or near CRBPI and whether the subcellular localization of retinoid-metabolizing enzymes and retinoid-binding proteins is affected by the cellular retinoid status or other metabolic factors remains unknown.

References

1. Abu-Abed S, Dollé P, Metzger D, Beckett B, Chambon P, Petkovich M (2001) The retinoic acid metabolizing enzyme, CYP26A1, is essential for normal hindbrain patterning, vertebral identity, and development of posterior structures. Genes Dev 15:226–240
2. Adams MK, Belyaeva OV, Wu L, Kedishvili NY (2014) The retinaldehyde reductase activity of DHRS3 is reciprocally activated by retinol dehydrogenase 10 to control retinoid homeostasis. J Biol Chem 289:14868–14880

3. Alapatt P, Guo F, Komanetsky SM, Wang S, Cai J, Sargsyan A, Rodriguez DE, Bacon BT, Aryal P, Graham TE (2013) Liver retinol transporter and receptor for serum retinol binding protein (RBP4). J Biol Chem 288:1250–1265
4. Algar EM, Seeley T-L, Holmes RS (1983) Purification and molecular properties of mouse alcohol dehydrogenase isozymes. Eur J Biochem 137:139–147
5. Andreola F, Hayhurst GP, Luo G, Ferguson SS, Gonzalez FJ, Goldstein JA, De Luca LM (2004) Mouse liver CYP2C39 is a novel retinoic acid 4-hydroxylase. Its down-regulation offers a molecular basis for liver retinoid accumulation and fibrosis in aryl hydrocarbon receptor-null mice. J Biol Chem 279:3434–3438
6. Arens JF, Van Dorp DA (1946) Synthesis of some compounds possessing vitamin A activity. Nature 157:190
7. Ashique AM, May SR, Kane MA, Folias AE, Phamluong K, Choe Y, Napoli JL, Peterson AS (2012) Morphological defects in a novel Rdh10 mutant that has reduced retinoic acid biosynthesis and signaling. Genesis 50:415–423
8. Bamji MS, Mahadevan S, Lakshmanan MR, Murthy SK (1962) Oxidation of vitamin A1 and vitamin A2 aldehydes to the corresponding acids by enzymes from pig and rat livers. Nature 196:672–673
9. Bauman DR, Steckelbroeck S, Williams MV, Peehl DM, Penning TM (2006) Identification of the major oxidative 3alpha-hydroxysteroid dehydrogenase in human prostate that converts 5alpha-androstane-3alpha,17beta-diol to 5alpha-dihydrotestosterone: a potential therapeutic target for androgen-dependent disease. Mol Endocrinol 20:444–458
10. Belyaeva OV, Kedishvili NY (2002) Human pancreas protein 2 (PAN2) has a retinal reductase activity and is ubiquitously expressed in human tissues. FEBS Lett 531:489–493
11. Belyaeva OV, Kedishvili NY (2006) Comparative genomic and phylogenetic analysis of short-chain dehydrogenases/reductases with dual retinol/sterol substrate specificity. Genomics 88:820–830
12. Belyaeva OV, Stetsenko AV, Nelson P, Kedishvili NY (2003) Properties of short-chain dehydrogenase/reductase RalR1: characterization of purified enzyme, its orientation in the microsomal membrane, and distribution in human tissues and cell lines. Biochemistry 42:14838–14845
13. Belyaeva OV, Korkina OV, Stetsenko AV, Kim T, Nelson PS, Kedishvili NY (2005) Biochemical properties of purified human retinol dehydrogenase 12 (RDH12): catalytic efficiency toward retinoids and C9 aldehydes and effects of cellular retinol-binding protein type I (CRBPI) and cellular retinaldehyde-binding protein (CRALBP) on the oxidation and reduction of retinoids. Biochemistry 44:7035–7047
14. Belyaeva OV, Johnson MP, Kedishvili NY (2008) Kinetic analysis of human enzyme RDH10 defines the characteristics of a physiologically relevant retinol dehydrogenase. J Biol Chem 283:20299–20308
15. Belyaeva OV, Korkina OV, Stetsenko AV, Kedishvili NY (2008) Human retinol dehydrogenase 13 (RDH13) is a mitochondrial short-chain dehydrogenase/reductase with a retinaldehyde reductase activity. FEBS J 275:138–147
16. Belyaeva OV, Lee SA, Adams MK, Chang C, Kedishvili NY (2012) Short chain dehydrogenase/reductase rdhe2 is a novel retinol dehydrogenase essential for frog embryonic development. J Biol Chem 287:9061–9071
17. Bernhard SA (1988) The intracellular equilibrium thermodynamic and steady-state concentrations of metabolites. Cell Biophys 12:119–132
18. Bhat PV, Labrecque J, Boutin J-M, Lacroix A, Yoshida A (1995) Cloning of a cDNA encoding rat aldehyde dehydrogenase with high activity for retinal oxidation. Gene 166:303–306
19. Biesalski HK (1989) Comparative assessment of the toxicology of vitamin A and retinoids in man. Toxicology 57:117–161
20. Billings SE, Pierzchalski K, Butler Tjaden NE, Pang XY, Trainor PA, Kane MA, Moise AR (2013) The retinaldehyde reductase DHRS3 is essential for preventing the formation of excess retinoic acid during embryonic development. FASEB J 27:4877–4889

21. Biswas MG, Russell DW (1997) Expression cloning and characterization of oxidative 17beta- and 3alpha-hydroxysteroid dehydrogenases from rat and human prostate. J Biol Chem 272:15959–15966
22. Bliss AF (1947) Enzymic formation of retinal vitamin A. Fed Proc 6(1 Pt 2):77
23. Bliss AF (1948) The mechanism of retinal vitamin A formation. J Biol Chem 172:165–178
24. Bliss AF (1950) Action of alcohol dehydrogenase on vitamin A. Biol Bull 99:330–331
25. Bliss AF (1951) The equilibrium between vitamin A alcohol and aldehyde in the presence of alcohol dehydrogenase. Arch Biochem Biophys 31:197–204
26. Boerman MH, Napoli JL (1995) Characterization of a microsomal retinol dehydrogenase: a short-chain alcohol dehydrogenase with integral and peripheral membrane forms that inter acts with holo-CRBP (type I). Biochemistry 34:7027–7037
27. Boleda MD, Saubi N, Farrés J, Parés X (1993) Physiological substrates for rat alcohol dehydrogenase classes: aldehydes of lipid peroxidation, omega-hydroxyfatty acids, and retinoids. Arch Biochem Biophys 307:85–90
28. Bouillet P, Sapin V, Chazaud C, Messaddeq N, Decimo D, Dolle P, Chambon P (1997) Developmental expression pattern of Stra6, a retinoic acid-responsive gene encoding a new type of membrane protein. Mech Dev 63:173–186
29. Boylan JF, Lufkin T, Achkar CC, Taneha R, Chambon P, Gudas LJ (1995) Targeted disruption of retinoic acid receptor alpha (RAR alpha) and RAR gamma results in receptor-specific alterations in retinoic acid-mediated differentiation and retinoic acid metabolism. Mol Cell Biol 15:843–851
30. Brereton P, Suzuki T, Sasano H, Li K, Duarte C, Obeyesekere V, Haeseleer F, Palczewski K, Smith I, Komesaroff P, Krozowski Z (2001) Pan1b (17βHSD11)-enzymatic activity and distribution in the lung. Mol Cell Endocrinol 171:111–117
31. Cerignoli F, Guo X, Cardinali B, Rinaldi C, Casaletto J, Frati L, Screpanti I, Gudas L, Gulino JA, Thiele CJ, Giannini G (2002) retSDR1, a short-chain retinol dehydrogenase/reductase, is retinoic acid-inducible and frequently deleted in human neuroblastoma cell lines. Cancer Res 62:1196–1204
32. Chai X, Boerman MH, Zhai Y, Napoli JL (1995) Cloning of a cDNA for liver microsomal retinol dehydrogenase. A tissue-specific, short-chain alcohol dehydrogenase. J Biol Chem 270:3900–3904
33. Chai X, Zhai Y, Popescu G, Napoli JL (1995) Cloning of a cDNA for a second retinol dehydrogenase type II. Expression of its mRNA relative to type I. J Biol Chem 270:28408–28412
34. Chai X, Zhai Y, Napoli JL (1996) Cloning of a rat cDNA encoding retinol dehydrogenase isozyme type III. Gene 169:219–222
35. Chen H, Fantel AG, Juchau MR (2000) Catalysis of the 4-hydroxylation of retinoic acids by cyp3a7 in human fetal hepatic tissues. Drug Metab Dispos 28:1051–1057
36. Chetyrkin SV, Belyaeva OV, Gough WH, Kedishvili NY (2001) Characterization of a novel type of human microsomal 3alpha -hydroxysteroid dehydrogenase: unique tissue distribution and catalytic properties. J Biol Chem 276:22278–22286
37. Chetyrkin SV, Hu J, Gough WH, Dumaual N, Kedishvili NY (2001) Further characterization of human microsomal 3alpha-hydroxysteroid dehydrogenase. Arch Biochem Biophys 386:1–10
38. Clegg JS (1984) Properties and metabolism of the aqueous cytoplasm and its boundaries. Am J Physiol 246:R133–R151
39. Crosas B, Allali-Hassani A, Martínez SE, Martras S, Persson B, Jörnvall H, Parés X, Farrés J (2000) Molecular basis for differential substrate specificity in class IV alcohol dehydrogenases: a conserved function in retinoid metabolism but not in ethanol oxidation. J Biol Chem 275:25180–25187
40. Dan Z, Popov Y, Patsenker E, Preimel D, Liu C, Wang XD, Seitz HK, Schuppan D, Stickel F (2005) Hepatotoxicity of alcohol-induced polar retinol metabolites involves apoptosis via loss of mitochondrial membrane potential. FASEB J 19:845–847

41. Deisenroth C, Itahana Y, Tollini L, Jin A, Zhang Y (2011) Expression of retSDR1 is activated by members of the p53 family, suggesting a potential role for retSDR1 in tumor suppression. J Biol Chem 286:28343–28356
42. Deltour L, Foglio MH, Duester G (1999) Impaired retinol utilization in *Adh4* alcohol dehydrogenase mutant mice. Dev Genet 25:1–10
43. DeLuca HF, Roberts AB (1969) Pathways of retinoic acid and retinol metabolism. Am J Clin Nutr 22:945–952
44. Deshmukh DS, Malathi P, Ganguly J (1965) Rapid conversion of retinal (vitamin A aldehyde) to retinoic acid (vitamin A acid) in the living rat. Biochim Biophys Acta 307:120
45. Deshmukh DS, Murthy SK, Mahadevan S, Ganguly J (1965) Studies on metabolism of vitamin A. Absorption of retinal (vitamin A aldehyde) in rats. Biochem J 96:377–382
46. Díez-Dacal B, Gayarre J, Gharbi S, Timms JF, Coderch C, Gago F, Pérez-Sala D (2011) Identification of aldo-keto reductase AKR1B10 as a selective target for modification and inhibition by prostaglandin A(1): implications for antitumoral activity. Cancer Res 71:4161–4171
47. Dmitrovskii AA (1961) Oxidation of vitamin A aldehyde to vitamin A acid catalyzed by aldehyde oxidase. Biokhimiya 26:126
48. Dockham PA, Lee MO, Sladek NE (1992) Identification of human liver aldehyde dehydrogenases that catalyze the oxidation of aldophosphamide and retinaldehyde. Biochem Pharmacol 43:2453–2469
49. Dunagin PE Jr, Zachman RD, Olson JA (1964) Identification of free and conjugated retinoic acid as a product of retinal (vitamin A aldehyde) metabolism in the rat in vivo. Biochim Biophys Acta 90:432
50. Dunagin PE Jr, Meadows EH Jr, Olson JA (1965) Retinoyl beta-glucuronic acid: a major metabolite of vitamin A in rat bile. Science 148:86–87
51. Dupé V, Matt N, Garnier JM, Chambon P, Mark M, Ghyselinck NB (2003) A newborn lethal defect due to inactivation of retinaldehyde dehydrogenase type 3 is prevented by maternal retinoic acid treatment. Proc Natl Acad Sci U S A 100:14036–14041
52. Elder TD, Topper YJ (1962) The oxidation of retinene (vitamin A1 aldehyde) to vitamin A acid by mammalian steroid-sensitive aldehyde dehydrogenase. Biochim Biophys Acta 64:430
53. Emerick RJ, Zile M, DeLuca HF (1967) Formation of retinoic acid from retinol in the rat. Biochem J 102:606
54. Fan X, Molotkov A, Manabe S, Donmoyer CM, Deltour L, Foglio MH, Cuenca AE, Blaner WS, Lipton SA, Duester G (2003) Targeted disruption of Aldh1a1 (Raldh1) provides evidence for a complex mechanism of retinoic acid synthesis in the developing retina. Mol Cell Biol 23:4637–4648
55. Fan LQ, Brown-Borg H, Brown S, Westin S, Mode A, Corton JC (2004) PPARalpha activators down-regulate CYP2C7, a retinoic acid and testosterone hydroxylase. Toxicology 203:41–48
56. Feng L, Hernandez RE, Waxman JS, Yelon D, Moens CB (2010) Dhrs3a regulates retinoic acid biosynthesis through a feedback inhibition mechanism. Dev Biol 338:1–14
57. Fiorella PD, Napoli JL (1994) Microsomal retinoic acid metabolism. Effects of cellular retinoic acid-binding protein (type I) and C18-hydroxylation as an initial step. J Biol Chem 269:10538–10544
58. Frolik CA, Tavela TE, Sporn MB (1978) Separation of the natural retinoids by high-pressure liquid chromatography. J Lipid Res 19:32–37
59. Frolik CA, Roberts AB, Tavela TE, Roller PP, Newton DL, Sporn MB (1979) Isolation and identification of 4-hydroxy- and 4-oxoretinoic acid. In vitro metabolites of all-trans-retinoic acid in hamster trachea and liver. Biochemistry 18:2092–2097
60. Fukumoto S, Yamauchi N, Moriguchi H, Hippo Y, Watanabe A, Shibahara J, Taniguchi H, Ishikawa S, Ito H, Yamamoto S, Iwanari H, Hironaka M, Ishikawa Y, Niki T, Sohara Y, Kodama T, Nishimura M, Fukayama M, Dosaka-Akita H, Aburatani H (2005) Overexpression

of the aldo-keto reductase family protein AKR1B10 is highly correlated with smokers' non-small cell lung carcinomas. Clin Cancer Res 11:1776–1785
61. Futterman S (1962) Enzymatic oxidation of vitamin A aldehyde to vitamin A acid. J Biol Chem 237:677
62. Gagnon I, Duester G, Bhat PV (2002) Kinetic analysis of mouse retinal dehydrogenase type-2 (RALDH2) for retinal substrates. Biochim Biophys Acta 1596:156–162
63. Gagnon I, Duester G, Bhat PV (2003) Enzymatic characterization of recombinant mouse retinal dehydrogenase type 1. Biochem Pharmacol 65:1685–1690
64. Gallego O, Belyaeva OV, Porte S, Ruiz FX, Stetsenko AV, Shabrova EV, Kostereva NV, Farrés J, Parés X, Kedishvili NY (2006) Comparative functional analysis of human medium-chain dehydrogenases, short-chain dehydrogenases/reductases and aldo-keto reductases with retinoids. Biochem J 399:101–109
65. Goedde HW, Agarwal DP (1987) Polymorphism of aldehyde dehydrogenase and alcohol sensitivity. Enzyme 37:29–44
66. Gough WH, VanOoteghem S, Sint T, Kedishvili NY (1998) cDNA cloning and characterization of a new human microsomal NAD + –dependent dehydrogenase that oxidizes all-trans-retinol and 3alpha-hydroxysteroids. J Biol Chem 273:19778–19785
67. Grün F, Hirose Y, Kawauchi S, Ogura T, Umesono K (2000) Aldehyde dehydrogenase 6, a cytosolic retinaldehyde dehydrogenase prominently expressed in sensory neuroepithelia during development. J Biol Chem 275:41210–41218
68. Haeseleer F, Huang J, Lebioda L, Saari JC, Palczewski K (1998) Molecular characterization of a novel short-chain dehydrogenase/reductase that reduces all-trans-retinal. J Biol Chem 273:21790–21799
69. Han CL, Liao CS, Wu CW, Hwong CL, Lee AR, Yin SJ (1998) Contribution to first-pass metabolism of ethanol and inhibition by ethanol for retinol oxidation in human alcohol dehydrogenase family–implications for etiology of fetal alcohol syndrome and alcohol-related diseases. Eur J Biochem 254:25–31
70. Hänni R, Bigler F (1977) Isolation and identification of three major metabolites of retinoic acid from rat feces. Helv Chim Acta 60:881–887
71. Hänni R, Bigler F, Meister W, Englert G (1976) Isolation and identification of three urinary metabolites of retinoic acid in the rat. Helv Chim Acta 59:2221–2227
72. Hardy DO, Ge RS, Catterall JF, Hou YT, Penning TM, Hardy MP (2000) Identification of the oxidative 3alpha-hydroxysteroid dehydrogenase activity of rat Leydig cells as type II retinol dehydrogenase. Endocrinology 141:1608–1617
73. Haselbeck RJ, Duester G (1997) Regional restriction of alcohol/retinol dehydrogenases along the mouse gastrointestinal epithelium. Alcohol Clin Exp Res 21:1484–1490
74. Haselbeck RJ, Ang HL, Duester G (1997) Class IV alcohol/retinol dehydrogenase localization in epidermal basal layer: potential site of retinoic acid synthesis during skin development. Dev Dyn 208:447–453
75. Höög J-O, Östberg L (2011) J. Mammalian alcohol dehydrogenases – a comparative investigation at gene and protein levels. Chem Biol Interact 191(1–3):2–7
76. Huang X-F, Luu-The V (2000) Molecular characterization of a first human 3(α → β)-hydroxysteroid epimerase. J Biol Chem 275:29452–29457
77. Ito YL, Zile M, Ahrens H, DeLuca HF (1974) Liquid-gel partition chromatography of vitamin A compounds; formation of retinoic acid from retinyl acetate in vivo. J Lipid Res 15:517–524
78. Jiang W, Napoli JL (2013) The retinol dehydrogenase Rdh10 localizes to lipid droplets during acyl ester biosynthesis. J Biol Chem 288:589–597
79. Jin J, Krishack PA, Cao D (2006) Role of aldo-keto reductases in development of prostate and breast cancer. Front Biosci 11:2767–2773
80. Kallberg Y, Oppermann U, Jörnvall H, Persson B (2002) Short-chain dehydrogenases/reductases (SDRs). Eur J Biochem 269:4409–4417

83. Kasus-Jacobi A, Ou J, Bashmakov YK, Shelton JM, Richardson JA, Goldstein JL, Brown MS (2003) Characterization of mouse short-chain aldehyde reductase (SCALD), an enzyme regulated by sterol regulatory element-binding proteins. J Biol Chem 278:32380–32389
84. Kedishvili NY (2013) Enzymology of retinoic acid biosynthesis and degradation. J Lipid Res 54:1744–1760
85. Kedishvili NY, Gough WH, Davis WI, Parsons S, Li TK, Bosron WF (1998) Effect of cellular retinol-binding protein on retinol oxidation by human class IV retinol/alcohol dehydrogenase and inhibition by ethanol. Biochem Biophys Res Commun 249:191–196
86. Kedishvili NY, Chumakova OV, Chetyrkin SV, Belyaeva OV, Lapshina EA, Lin M, Matsumura DW, Nelson PS (2002) Evidence that the human gene for prostate short-chain dehydrogenase/reductase (PSDR1) encodes a novel retinal reductase (RalR1). J Biol Chem 277:28909–28915
87. Kim CI, Leo MA, Lieber CS (1992) Retinol forms retinoic acid via retinal. Arch Biochem Biophys 294:388–393
88. Koen AL, Shaw CR (1966) Retinol and alcohol dehydrogenases in retina and liver. Biochim Biophys Acta 128:48–54
89. Koenig U, Amatschek S, Mildner M, Eckhart L, Tschachler E (2010) Aldehyde dehydrogenase 1A3 is transcriptionally activated by all-trans-retinoic acid in human epidermal keratinocytes. Biochem Biophys Res Commun 400:207–211
90. Krishnamurthy S, Bieri JG, Andrews EL (1963) Metabolism and biological activity of vitamin A acid in the chick. J Nutr 79:503–510
91. Lakshmanan MR, Vaidyanathan CS, Cama HR (1964) Oxidation of vitamin A1 aldehyde and vitamin A2 aldehyde to the corresponding acids by aldehyde oxidase from different species. Biochem J 90:569–573
92. Lassen N, Bateman JB, Estey T, Kuszak JR, Nees DW, Piatigorsky J, Duester G, Day BJ, Huang J, Hines LM, Vasiliou V (2007) Multiple and additive functions of ALDH3A1 and ALDH1A1: cataract phenotype and ocular oxidative damage in Aldh3a1(−/−)/Aldh1a1(−/−) knock-out mice. J Biol Chem 282:25668–25676
93. Lee MO, Dockham PA, Sladek NE (1990) Identification of human liver aldehyde dehydrogenases that catalyze the oxidation of retinaldehyde to retinoic acid. Pharmacologist 32:156
94. Lee MO, Manthey CL, Sladek NE (1990) Identification of mouse liver aldehyde dehydrogenases that catalyze the oxidation of retinaldehyde to retinoic acid. FASEB J 4:A659
95. Lee MO, Manthey CL, Sladek NE (1991) Identification of mouse liver aldehyde dehydrogenases that catalyze the oxidation of retinaldehyde to retinoic acid. Biochem Pharmacol 42:1279–1285
96. Lee SA, Belyaeva OV, Popov IK, Kedishvili NY (2007) Overproduction of bioactive retinoic acid in cells expressing disease-associated mutants of retinol dehydrogenase 12. J Biol Chem 282:35621–35628
97. Lee SA, Belyaeva OV, Kedishvili NY (2009) Biochemical characterization of human epidermal retinol dehydrogenase 2. Chem Biol Interact 178:182–187
98. Lee SA, Belyaeva OV, Wu L, Kedishvili NY (2011) Retinol dehydrogenase 10 but not retinol/sterol dehydrogenase(s) regulates the expression of retinoic acid-responsive genes in human transgenic skin raft culture. J Biol Chem 286:13550–13560
99. Lee LM, Leung CY, Tang WW, Choi HL, Leung YC, McCaffery PJ, Wang CC, Woolf AS, Shum AS (2012) A paradoxical teratogenic mechanism for retinoic acid. Proc Natl Acad Sci U S A 109:13668–13673
100. Leo MA, Lieber CS (1984) Normal testicular structure and reproductive function in deermice lacking retinol and alcohol dehydrogenase activity. J Clin Invest 73:593–596
101. Leo MA, Iida S, Lieber CS (1984) Retinoic acid metabolism by a system reconstituted with cytochrome P-450. Arch Biochem Biophys 234:305–312
102. Leo MA, Kim CI, Lieber CS (1987) NAD^+-dependent retinol dehydrogenase in liver microsomes. Arch Biochem Biophys 259:241–249

103. Levin MS, Locke B, Yang NC, Li E, Gordon JI (1988) Comparison of the ligand binding properties of two homologous rat apocellular retinol-binding proteins expressed in Escherichia coli. J Biol Chem 263:17715–17723
104. Li E, Qian S, Winter NS, d'Avignon A, Levin MS, Gordon JI (1991) Fluorine nuclear magnetic resonance analysis of the ligand binding properties of two homologous rat cellular retinol-binding proteins expressed in Escherichia coli. J Biol Chem 266:3622–3629
105. Lin M, Napoli JL (2000) cDNA cloning and expression of a human aldehyde dehydrogenase (ALDH) active with 9-cis-retinal and identification of a rat ortholog, ALDH12. J Biol Chem 275:40106–40112
106. Lin M, Zhang M, Abraham M, Smith SM, Napoli JL (2003) Mouse retinal dehydrogenase 4 (RALDH4), molecular cloning, cellular expression, and activity in 9-cis-retinoic acid biosynthesis in intact cells. J Biol Chem 278:9856–9861
107. Lippel K, Olson JA (1968) Origin of some derivatives of retinoic acid found in rat bile. J Lipid Res 9:580–586
108. Liu C, Russell RM, Seitz HK, Wang XD (2001) Ethanol enhances retinoic acid metabolism into polar metabolites in rat liver via induction of cytochrome P4502E1. Gastroenterology 120:179–189
109. Liu S, Huang C, Li D, Ren W, Zhang H, Qi M, Li X, Yu L (2007) Molecular cloning and expression analysis of a new gene for short-chain dehydrogenase/reductase 9. Acta Biochim Pol 54:213–218
110. Mahadevan SN, Murthy SK, Ganguly J (1962) Enzymic oxidation of vitamin A aldehyde to vitamin A acid by rat liver. Biochem J 85:326
111. Marchitti SA, Brocker C, Stagos D, Vasiliou V (2008) Non-P450 aldehyde oxidizing enzymes: the aldehyde dehydrogenase superfamily. Expert Opin Drug Metab Toxicol 4:697–720
112. Marill J, Cresteil T, Lanotte M, Chabot GG (2000) Identification of human cytochrome P450s involved in the formation of all-trans-retinoic acid principal metabolites. Mol Pharmacol 58:1341–1348
113. Martras S, Alvarez R, Gallego O, Dominguez M, de Lera AR, Farrés J, Parés X (2004) Kinetics of human alcohol dehydrogenase with ring-oxidized retinoids: effect of Tween 80. Arch Biochem Biophys 430:210–217
114. Matsuura T, Gad MZ, Harrison EH, Ross AC (1997) Lecithin:retinol acyl transferase and retinyl ester hydrolase activities are differentially regulated by retinoids and have distinct distributions between hepatocyte and nonparenchymal cell fractions of rat liver. J Nutr 127:218–224
115. McCaffery P, Dräger UC (1994) Hot spots of retinoic acid synthesis in the developing spinal cord. Proc Natl Acad Sci U S A 91:7194–7197
116. McCaffery P, Tempst P, Lara G, Dräger UC (1991) Aldehyde dehydrogenase is a positional marker in the retina. Development 112:693–702
117. McCaffery P, Lee MO, Wagner MA, Sladek NE, Dräger UC (1992) Asymmetrical retinoic acid synthesis in the dorsoventral axis of the retina. Development 115:371–382
118. McCaffrery P, Posch KC, Napoli JL, Gudas L, Dräger UC (1993) Changing patterns of the retinoic acid system in the developing retina. Dev Biol 158:390–399
119. McCormick AM, Napoli JL, Deluca HF (1978) High-pressure liquid chromatographic resolution of vitamin A compounds. Anal Biochem 86:25–33
120. McCormick AM, Napoli JL, Schnoes HK, DeLuca HF (1978) Isolation and identification of 5, 6-epoxyretinoic acid: a biologically active metabolite of retinoic acid. Biochemistry 17:4085–4090
121. McSorley LC, Daly AK (2000) Identification of human cytochrome P450 isoforms that contribute to all-trans-retinoic acid 4-hydroxylation. Biochem Pharmacol 60:517–526
122. Mezey E, Holt PR (1971) The inhibitory effect of ethanol on retinol oxidation by human liver and cattle retina. Exp Mol Pathol 15:148–156

123. Molotkov A, Duester G (2003) Genetic evidence that retinaldehyde dehydrogenase Raldh1 (Aldh1a1) functions downstream of alcohol dehydrogenase Adh1 in metabolism of retinol to retinoic acid. J Biol Chem 278:36085–36090
124. Molotkov A, Fan X, Deltour L, Foglio MH, Martras S, Farrés J, Parés X, Duester G (2002) Stimulation of retinoic acid production and growth by ubiquitously expressed alcohol dehydrogenase Adh3. Proc Natl Acad Sci U S A 99:5337–5342
125. Molotkov A, Ghyselinck NB, Chambon P, Duester G (2004) Opposing actions of cellular retinol-binding protein and alcohol dehydrogenase control the balance between retinol storage and degradation. Biochem J 383:295–302
126. Nadin L, Murray M (1999) Participation of CYP2C8 in retinoic acid 4-hydroxylation in human hepatic microsomes. Biochem Pharmacol 58:1201–1208
127. Napoli JL (1986) Retinol metabolism in LLC-PK1 Cells. Characterization of retinoic acid synthesis by an established mammalian cell line. J Biol Chem 261:13592–13597
128. Napoli JL (2001) 17beta-Hydroxysteroid dehydrogenase type 9 and other short-chain dehydrogenases/reductases that catalyze retinoid, 17beta- and 3alpha-hydroxysteroid metabolism. Mol Cell Endocrinol 171:103–109
129. Napoli JL (2012) Physiological insights into all-trans-retinoic acid biosynthesis. Biochim Biophys Acta 1821:152–167
130. Napoli JL, McCormick AM, Schnoes HK, DeLuca HF (1978) Identification of 5,8-oxyretinoic acid isolated from small intestine of vitamin A-deficient rats dosed with retinoic acid. Proc Natl Acad Sci U S A 75:2603–2605
131. Niederreither K, Subbarayan V, Dollé P, Chambon P (1999) Embryonic retinoic acid synthesis is essential for early mouse post-implantation development. Nat Genet 21:444–448
132. Ong DE, Crow JA, Chytil F (1982) Radioimmunochemical determination of cellular retinol- and cellular retinoic acid-binding proteins in cytosols of rat tissues. J Biol Chem 257:13385–13389
133. Ong DE, Newcomer ME, Chytil F (1994) Cellular retinoid-binding proteins. In: Sporn MB, Roberts AB, Goodman DS (eds) The Retinoids: biology, chemistry and medicine. Raven, New York, pp 283–318
134. Parés X, Farrés J, Kedishvili N, Duester G (2008) Medium- and short-chain dehydrogenase/reductase gene and protein families: medium-chain and short-chain dehydrogenases/reductases in retinoid metabolism. Cell Mol Life Sci 65:3936–3949
135. Pennimpede T, Cameron DA, MacLean GA, Li H, Abu-Abed S, Petkovich M (2010) The role of CYP26 enzymes in defining appropriate retinoic acid exposure during embryogenesis. Birth Defects Res A Clin Mol Teratol 88:883–894
136. Penning TM (2011) Human hydroxysteroid dehydrogenases and pre-receptor regulation: insights into inhibitor design and evaluation. J Steroid Biochem Mol Biol 125:46–56
137. Penning TM, Drury JE (2007) Human aldo-keto reductases: function, gene regulation, and single nucleotide polymorphisms. Arch Biochem Biophys 464:241–250
138. Penzes P, Wang X, Napoli JL (1997) Enzymatic characteristics of retinal dehydrogenase type I expressed in Escherichia coli. Biochim Biophys Acta 1342:175–181
139. Penzes P, Wang X, Sperkova Z, Napoli JL (1997) Cloning of a rat cDNA encoding retinal dehydrogenase isozyme type I and its expression in E. coli. Gene 191:167–172
140. Persson B, Krook M, Jörnvall H (1991) Characteristics of short-chain alcohol dehydrogenases and related enzymes. Eur J Biochem 200:537–543
141. Persson B, Hedlund J, Jörnvall H (2008) Medium- and short-chain dehydrogenase/reductase gene and protein families. Cell Mol Life Sci 65:3879–3894
142. Posch KC, Boerman MH, Burns RD, Napoli JL (1991) Holocellular retinol binding protein as a substrate for microsomal retinal synthesis. Biochemistry 30:6224–6230
143. Posch KC, Burns RD, Napoli JL (1992) Biosynthesis of all-trans-retinoic acid from retinal. Recognition of retinal bound to cellular retinol binding protein (type I) as substrate by a purified cytosolic dehydrogenase. J Biol Chem 267:19676–19682

144. Qian L, Zolfaghari R, Ross AC (2010) Liver-specific cytochrome P450 CYP2C22 is a direct target of retinoic acid and a retinoic acid-metabolizing enzyme in rat liver. J Lipid Res 51:1781–1792
145. Redfearn ER (1960) The metabolism of vitamin A acid and its C25 homolog. Arch Biochem Biophys 93:226
146. Reijntjes S, Zile MH, Maden M (2010) The expression of Stra6 and Rdh10 in the avian embryo and their contribution to the generation of retinoid signatures. Int J Dev Biol 54:1267–1275
147. Rhinn M, Schuhbaur B, Niederreither K, Dollé P (2011) Involvement of retinol dehydrogenase 10 in embryonic patterning and rescue of its loss of function by maternal retinaldehyde treatment. Proc Natl Acad Sci U S A 108:16687–16692
148. Roberts AB, Frolik CA, Nichols MD, Sporn MB (1979) Retinoid-dependent induction of the in vivo and in vitro metabolism of retinoic acid in tissues of the vitamin A-deficient hamster. J Biol Chem 254:6303–6309
149. Roberts AB, Nichols MD, Newton DL, Sporn MB (1979) In vitro metabolism of retinoic acid in hamster intestine and liver. J Biol Chem 254:6296–6302
150. Roberts ES, Vaz AD, Coon MJ (1992) Role of isozymes of rabbit microsomal cytochrome P-450 in the metabolism of retinoic acid, retinol, and retinal. Mol Pharmacol 41:427–433
151. Ross AC, Zolfaghari R (2004) Regulation of hepatic retinol metabolism: perspectives from studies on vitamin A status. J Nutr 134:2695–2755
152. Ross AC, Zolfaghari R (2011) Cytochrome P450s in the regulation of cellular retinoic acid metabolism. Annu Rev Nutr 31:65–87
153. Ross AC, Cifelli CJ, Zolfaghari R, Li N-q (2011) Multiple cytochrome P-450 genes are concomitantly regulated by vitamin A under steady state conditions and by retinoic acid during hepatic first-pass metabolism. Physiol Genomics 43:57–67
154. Rout UK, Holmes RS (1985) Isoelectric focusing studies of aldehyde dehydrogenases from mouse tissues: variant phenotypes of liver, stomach and testis isozymes. Comp Biochem Physiol 81B:647–651
155. Ruiz FX, Porté S, Parés X, Farrés J (2012) Biological role of aldo-keto reductases in retinoic acid biosynthesis and signaling. Front Pharmacol 3:58
156. Sandell LL, Sanderson BW, Moiseyev G, Johnson T, Mushegian A, Young K, Rey JP, Ma JX, Staehling-Hampton K, Trainor PA (2007) RDH10 is essential for synthesis of embryonic retinoic acid and is required for limb, craniofacial, and organ development. Genes Dev 21:1113–1124
157. Sima A, Parisotto M, Mader S, Bhat PV (2009) Kinetic characterization of recombinant mouse retinal dehydrogenase types 3 and 4 for retinal substrates. Biochim Biophys Acta 1790:1660–1664
158. Simon A, Hellman U, Wernstedt C, Eriksson U (1995) The retinal pigment epithelial-specific 11-cis retinol dehydrogenase belongs to the family of short chain alcohol dehydrogenases. J Biol Chem 270:1107–1112
159. Slavotinek AM, Mehrotra P, Nazarenko I, Tang PL, Lao R, Cameron D, Li B, Chu C, Chou C, Marqueling AL, Yahyavi M, Cordoro K, Frieden I, Glaser T, Prescott T, Morren MA, Devriendt K, Kwok PY, Petkovich M, Desnick RJ (2013) Focal facial dermal dysplasia, type IV, is caused by mutations in CYP26C1. Hum Mol Genet 22:696–703
160. Smith G, Wolf CR, Deeni YY, Dawe RS, Evans AT, Comrie MM, Ferguson J, Ibbotson SH (2003) Cutaneous expression of cytochrome P450 CYP2S1: individuality in regulation by therapeutic agents for psoriasis and other skin diseases. Lancet 361:1336–1343
161. Srivastava DK, Bernhard SA (1986) Metabolite transfer via enzyme-enzyme complexes. Science 234:1081–1086
162. Strate I, Min TH, Iliev D, Pera EM (2009) Retinol dehydrogenase 10 is a feedback regulator of retinoic acid signalling during axis formation and patterning of the central nervous system. Development 136:461–472

163. Taimi M, Helvig C, Wisniewski J, Ramshaw H, White J, Amad M, Korczak B, Petkovich M (2004) A novel human cytochrome P450, CYP26C1, involved in metabolism of 9-cis and all-trans isomers of retinoic acid. J Biol Chem 279:77–85
164. Thatcher JE, Isoherranen N (2009) The role of CYP26 enzymes in retinoic acid clearance. Expert Opin Drug Metab Toxicol 5:875–886
165. Thatcher JE, Zelter A, Isoherranen N (2010) The relative importance of CYP26A1 in hepatic clearance of all-trans retinoic acid. Biochem Pharmacol 80:903–912
166. Topletz AR, Thatcher JE, Zelter A, Lutz JD, Tay S, Nelson WL, Isoherranen N (2012) Comparison of the function and expression of CYP26A1 and CYP26B1, the two retinoic acid hydroxylases. Biochem Pharmacol 83:149–163
167. Uehara M, Yashiro K, Mamiya S, Nishino J, Chambon P, Dollé P, Sakai Y (2007) CYP26A1 and CYP26C1 cooperatively regulate anterior-posterior patterning of the developing brain and the production of migratory cranial neural crest cells in the mouse. Dev Biol 302:399–411
168. Van Dorp DA, Arens JF (1946) Biological activity of vitamin A acid. Nature 158:60
169. Van Dorp DA, Arens JF (1946) Activity of "vitamin A-acid" in the rat. Nature 158:622
170. Van Wauwe J, Van Nyen G, Coene MC, Stoppie P, Cools W, Goossens J, Borghgraef P, Janssen PA (1992) Liarozole, an inhibitor of retinoic acid metabolism, exerts retinoid-mimetic effects in vivo. J Pharmacol Exp Ther 261:773–779
171. Wald G (1950) The interconversion of the retinenes and vitamins A in vitro. Biochim Biophys Acta 4:215–228
172. Wald G, Hubbard R (1949) The reduction of retinene1 to vitamina A1 in vitro. J Gen Physiol 32:367–389
173. Wang X, Penzes P, Napoli JL (1996) Cloning of a cDNA encoding an aldehyde dehydrogenase and its expression in Escherichia coli. Recognition of retinal as substrate. J Biol Chem 271:16288–16293
174. Westin S, Sonneveld E, van der Leede BM, van der Saag PT, Gustafsson JA, Mode A (1997) CYP2C7 expression in rat liver and hepatocytes: regulation by retinoids. Mol Cell Endocrinol 129:169–179
175. White JA, Guo YD, Baetz K, Beckett-Jones B, Bonasoro J, Hsu KE, Dilworth FJ, Jones G, Petkovich M (1996) Identification of the retinoic acid-inducible all-trans-retinoic acid 4-hydroxylase. J Biol Chem 271:29922–29927
176. White JA, Beckett-Jones B, Guo YD, Dilworth FJ, Bonasoro J, Jones G, Petkovich M (1997) cDNA cloning of human retinoic acid-metabolizing enzyme (hP450RAI) identifies a novel family of cytochromes P450. J Biol Chem 272:18538–18541
177. White JA, Ramshaw H, Taimi M, Stangle W, Zhang A, Everingham S, Creighton S, Tam SP, Jones G, Petkovich M (2000) Identification of the human cytochrome P450, P450RAI-2, which is predominantly expressed in the adult cerebellum and is responsible for all-trans-retinoic acid metabolism. Proc Natl Acad Sci U S A 97:6403–6408
178. Wu L, Ross AC (2010) Acidic retinoids synergize with vitamin A to enhance retinol uptake and STRA6, LRAT, and CYP26B1 expression in neonatal lung. J Lipid Res 51:378–387
179. Wu BX, Chen Y, Chen Y, Fan J, Rohrer B, Crouch RK, Ma JX (2002) Cloning and characterization of a novel all-trans retinol short-chain dehydrogenase/reductase from the RPE. Invest Ophthalmol Vis Sci 43:3365–3372
180. Wu ZL, Sohl CD, Shimada T, Guengerich FP (2006) Recombinant enzymes overexpressed in bacteria show broad catalytic specificity of human cytochrome P450 2 W1 and limited activity of human cytochrome P450 2S1. Mol Pharmacol 69:2007–2014
181. Xi J, Yang Z (2008) Expression of RALDHs (ALDH1As) and CYP26s in human tissues and during the neural differentiation of P19 embryonal carcinoma stem cell. Gene Expr Patterns 8:438–442
182. Yang Z-N, Davis GJ, Hurley TD, Stone CL, Li T-K, Bosron WF (1994) Catalytic efficiency of human alcohol dehydrogenases for retinol oxidation and retinal reduction. Alcohol Clin Exp Res 18:587–591

183. Yashiro K, Zhao X, Uehara M, Yamashita K, Nishijima M, Nishino J, Saijoh Y, Sakai Y, Hamada H (2004) Regulation of retinoic acid distribution is required for proximodistal patterning and outgrowth of the developing mouse limb. Dev Cell 6:411–422
184. Zachman R, Olson JA (1961) A comparison of retinene reductase and alcohol dehydrogenase of rat liver. J Biol Chem 236:2309–2313
185. Zachman RD, Dunagin PE Jr, Olson JA (1966) Formation and enterohepatic circulation of metabolites of retinol and retinoic acid in bile duct-cannulated rats. Lipid Res 7:3–9
186. Zanger UM, Klein K, Thomas M, Rieger JK, Tremmel R, Kandel BA, Klein M, Magdy T (2014) Genetics, epigenetics, and regulation of drug metabolizing cytochrome p450 enzymes. Clin Pharmacol Ther 95:258–261
187. Zhang M, Chen W, Smith SM, Napoli JL (2001) Molecular characterization of a mouse short chain dehydrogenase/reductase active with all-trans-retinol in intact cells, mRDH1. J Biol Chem 276:44083–44090
188. Zhang M, Hu P, Krois CR, Kane MA, Napoli JL (2007) Altered vitamin A homeostasis and increased size and adiposity in the rdh1-null mouse. FASEB J 21:2886–2896
189. Zhao D, McCaffery P, Ivins KJ, Neve RL, Hogan P, Chin WW, Dräger UC (1996) Molecular identification of a major retinoic-acid-synthesizing enzyme, a retinaldehyde-specific dehydrogenase. Eur J Biochem 240:15–22
190. Zolfaghari R, Ross AC (2000) Lecithin:retinol acyltransferase from mouse and rat liver. cDNA cloning and liver-specific regulation by dietary vitamin A and retinoic acid. J Lipid Res 41:2024–2034
191. Zolfaghari R, Chen Q, Ross AC (2012) DHRS3, a retinal reductase, is differentially regulated by retinoic acid and lipopolysaccharide-induced inflammation in THP-1 cells and rat liver. Am J Physiol Gastrointest Liver Physiol 303:G578–G588

Chapter 6
Cellular Retinoic Acid Binding Proteins: Genomic and Non-genomic Functions and their Regulation

Li-Na Wei

Abstract Cellular retinoic acid binding proteins (CRABPs) are high-affinity retinoic acid (RA) binding proteins that mainly reside in the cytoplasm. In mammals, this family has two members, CRABPI and II, both highly conserved during evolution. The two proteins share a very similar structure that is characteristic of a "β-clam" motif built up from10-strands. The proteins are encoded by two different genes that share a very similar genomic structure. CRABPI is widely distributed and CRABPII has restricted expression in only certain tissues. The *CrabpI* gene is driven by a housekeeping promoter, but can be regulated by numerous factors, including thyroid hormones and RA, which engage a specific chromatin-remodeling complex containing either TRAP220 or RIP140 as coactivator and corepressor, respectively. The chromatin-remodeling complex binds the DR4 element in the *CrabpI* gene promoter to activate or repress this gene in different cellular backgrounds. The *CrabpII* gene promoter contains a TATA-box and is rapidly activated by RA through an RA response element. Biochemical and cell culture studies carried out in vitro show the two proteins have distinct biological functions. CRABPII mainly functions to deliver RA to the nuclear RA receptors for gene regulation, although recent studies suggest that CRABPII may also be involved in other cellular events, such as RNA stability. In contrast, biochemical and cell culture studies suggest that CRABPI functions mainly in the cytoplasm to modulate intracellular RA availability/concentration and to engage other signaling components such as ERK activity. However, these functional studies remain inconclusive because knocking out one or both genes in mice does not produce definitive phenotypes. Further studies are needed to unambiguously decipher the exact physiological activities of these two proteins.

Keywords CRABPI • CRABPII • RARs • Retinoic acid • Differentiation

---

L.-N. Wei (✉)
Department of Pharmacology, University of Minnesota Medical School,
6-120 Jackson Hall, 321 Church St. SE, 55455 Minneapolis, MN, USA
e-mail: weixx009@umn.edu

M.A. Asson-Batres, C. Rochette-Egly (eds.), *The Biochemistry of Retinoid Signaling II*, Subcellular Biochemistry 81, DOI 10.1007/978-94-024-0945-1_6

Abbreviations

AP2	Activating Protein 2
COUP	Chicken Ovalbumin Upstream Promoter
CRABP	Cellular Retinoic Acid Binding Protein
DR	Direct Repeat
ERK	Extracellular Signal-Regulated MAP kinase
ESC	Embryonic Stem Cell
GCNF	Germ Cell Nuclear Factor
LRH-1	Liver Receptor Homologue-1
PPAR	Peroxisome Proliferator Activated Receptor
RAR	Retinoic Acid Receptor
RXR	Retinoid X Receptor
RA	Retinoic Acid
SF-1	Steroidogenic Factor 1
SP1	Specific Protein 1
TR2	Testis Receptor 2
TRAP220	Thyroid hormone Receptor-Associated Protein 220
RIP140	Receptor Interacting Protein 140

Introduction

For more than a century, vitamin A and retinoids have been known to participate in crucial processes in different organisms, including humans. Among different retinoids, retinoic acid (RA) has attracted considerably more attention because it very potently and directly affects almost all types of animal tissues/cells. Early interest in the field focused on identification of RA biological targets in cells. The first such targets that were described were the cytosolic binding proteins, known as cellular RA binding protein I and II (CRABPI and CRABPII) [36]. Because gene knockout of either CRABPI or CRABPII did not produce obvious developmental or physiological defects in laboratory mice, interest in CRABPs gradually diminished after initial identification and characterization studies in the early 90s. However, the notion that CRABPs are not important because laboratory animals can survive without them is difficult to reconcile with the fact that the proteins are highly conserved throughout evolution. More recently, studies using newer techniques have begun to provide compelling evidence that CRABPs are performing important. These results have reignited enthusiasm in the field about CRABPs. Nonetheless, the decades-long query regarding the exact physiological function of CRABPs remains unsettled. Thus, the jury is still out and there remains a continuing need to study CRABPs in a physiological context.

History

Discovery of CRABPI and CRABPII

Observation that RA had a distinct binding profile in animal tissue fractions as compared with retinol-binding fractions [6] suggested there were binding proteins with specificity for RA. This was confirmed by subsequent biochemical studies in which radioisotope-labeled RA was reacted with tissue extracts. Ong and Chytil were the first to utilize classical sedimentation methods to partially purify a rat testis fraction that retained ^{3}H-RA binding activity, and determined its molecular weight to be 14,500 [35, 46]. These careful binding studies confirmed the specificity of this fraction toward RA, showing that the ligand could not be competed out even with a 200-fold molar excess of cold retinal, retinol, or oleic acid. They found the partially purified protein bound to RA has a fluorescence excitation spectrum with lambda max at 350 nm, which is different from that of cellular retinol-binding protein (CRBP) bound to retinol, which has a lambda max of 334 nm [6]. In these initial studies, RA-binding proteins were not only found in the testis, but also in several other tissues such as brain, eye, ovary, and uterus. These initial biochemical characterization studies were important in leading the way for further protein purification and, subsequently, molecular cloning of the cDNAs for CRABPs. The cloning experiments conclusively demonstrated the existence of CRABPI [50, 51, 59] and CRABPII [3, 5, 22] in rodents and humans.

CRABPs were identified long before the discovery of nuclear RA receptors (RARs). The studies performed at that time during the 1980's revealed that CRABPs are present essentially in the cytosol and would serve as shuttles transferring the ligand into the nucleus.

Genetic Analysis

Gene and amino acid sequence comparisons place CRABPs in the same family under the super family of intracellular lipid binding proteins [18, 27]. Amino acid sequence analyses have shown that the two CRABPs have a sequence conservation of approximately 72 % for the human proteins [3]. Further studies have revealed that both *crabpI* and *crabpII* genes are conserved across different vertebrate animal species, including mouse, human, rat, frog, bird and fish [40], which indicates these proteins are likely to have conserved functions.

Protein Structure

The crystal structure of bovine and mouse CRABPI was first determined for the protein complexed with its natural ligand, RA (holo-CRABP) [25]. A year later, the crystal structure of apo-CRABPI (protein without ligand) was resolved and compared with the holo-CRABPI structure [52]. The human CRABPII crystal structure was determined for the protein complexed with a synthetic retinoid [25]. The structure of CRABPII is very similar to that of CRABPI.

The CRABP structures are very similar to other intracellular lipid binding proteins in that they share a very similar *β-clam* motif consisting of two β-sheets built up from10-strands. The ligand, RA, is sandwiched inside the clam with the acidic group situated innermost [52]. Both proteins prefer binding all-*trans* RA (atRA) compared to 9-*cis*-RA, but they bind RA very differently [33]. CRABPI specifically and tightly binds atRA with a Kd in the low- to sub-nM range [20, 32, 47]. The lowest reported value is less than 0.4 nM [32]. In contrast, the Kd of CRABPII towards atRA is within the nM range and can be as high as 65 nM [5, 18, 27, 32].

Development of the Field

Although they share structural similarities, classical biochemical studies and gene expression patterns suggest that CRABPI and II behave differently and thus, likely have distinct physiological functions with regard to RA transport, metabolism, and signaling.

CRABPI

Localization of CRABPI

CRABPI is primarily localized in the cytoplasm and is found in most adult tissues [10, 18]. However, in certain cell types, CRABPI has been detected only at embryonic stages [18], and was reported to be associated with mitochondria [45]. Of note, some groups detected CRABPI in the nucleus [21], raising the hypothesis that this protein might also interfere with RAR's functions.

Biological Activities and Mechanism of Action of CRABPI

Classical biochemical studies have suggested that CRABPI mediates RA catabolism by cytochrome p450 enzymes [31]. It has been proposed that RA bound to CRABPI is a better substrate for the RA-metabolizing cytochrome p450s than free RA and that this more efficient interaction could increase the levels of polar metabolites such as 4-OH and 4-Oxo RA. However, the mechanism for the latter process is not clear, and the physiological impact remains uncertain.

In fact, mostly homogeneous distribution of CRABPI in the cytoplasm suggested the protein primarily functions in various cytosolic processes, such as cell signaling.

Therefore our group recently embarked on systematic studies aimed at carefully dissecting the potential downstream physiological targets of apo- and holo-CRABPI. Embryonic stem cells (ESCs) were used because they represent a standard model for addressing RA's pleiotropic effects. Indeed, RA is known for profoundly triggering differentiation and/or apoptosis in these cells. Moreover, CRABPI is very abundant in ESCs, while CRABPII is almost non-detectable. Our early work determined that, in ESCs, RA rapidly stimulates the post-translational phosphorylation and sumoylation of transcription factors that are important for ESC proliferation, and that this activity does not involve RARs, thus highlighting non-canonical, non-genomic effects of RA in the cytoplasm [14, 24, 38]. More recently, by employing CRABPI knock down, we identified a specific and rapid, RA-responsive process that was dependent on CRABPI, and not on RAR. This CRABPI-dependent effect involved the activation of the cytoplasmic kinase pathway ERK1/2, which is crucial for ESCs cell cycle regulation. Indeed, the RA-activated, (ERK1/2) was found to stabilize the levels of the cyclin-dependent kinase, p27, leading to a delay in cell cycle progression, while at the same time, facilitating cell differentiation.

Based on these reports, we can propose the model shown in Fig. 6.1 which posits that an ESC, when exposed to RA, exploits both immediate cytosolic (via CRABPI) and delayed nuclear (via RARs) targets to prepare its genome for proper differentiation programming [42]. The cell cycle decision is critical for stem cells because it

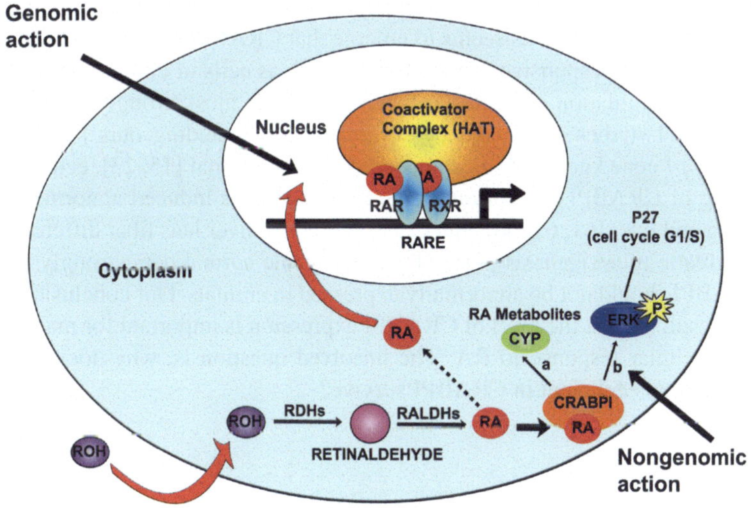

Fig. 6.1 A proposed model of CRABPI function. In cells sensitive to the genomic activity of RA such as stem cells, CRABPI functions to a) direct accidentally surged RA to its metabolic enzymes such as the cytochrome p450 system (CYP) to reduce RA toxicity to stem cells, or b) to activate cytoplasmic signaling such as ERK that in turns controls cell cycle progression regulator p27 to ensure effective cell differentiation program. In these cells, a limited amount of RA is delivered to the nucleus for its genomic function (depicted by a *broken arrow*)

indicates the difference between continuing to proliferate versus committing to a differentiation program. In the nucleus, RA regulates RA-responding genes, mediated by RARs/RXRs, to direct ESCs toward specific differentiation programs. In the cytoplasm, RA elicits rapid effects, mediated by CRABPI, to place a hold on cell cycle progression. This short window of cell cycle delay is important for a stem cell's "commitment" to proceed to proper differentiation because only when the cytosolic signaling pathway is readily activated and RA has entered the nucleus to activate RARs/RXRs, can the cell move on and commit to a specific differentiation program. How CRABPI-RA can perform such an intriguing function in the cytoplasm remains to be elucidated. Additionally, as proposed by others, CRABPI may also direct excess RA to the cytochrome p450 system for catabolism to guard against potential RA toxicity.

Alternative proposals for the actions of CRABPI have been also reported. For example, the detection of CRABPI in the nucleus [21] prompted the hypothesis that CRABPI may interact with RARs to influence their function in the nucleus (see review in [34]). Accordingly, other *in vitro* studies suggested that CRABPI establishes an RA gradient in the nucleus that passively delivers RA to RARs. This proposed mechanism of RA transfer from CRABPI to RAR is very different than the more direct mechanism of RA transfer proposed for CRABP II (see below). Intriguingly, overexpressing CRABPI failed to affect the transcriptional activity of RARs in Cos-1 cell [16] but reduced RA responses in embryonic cells [11, 67]. Thus the significance of CRABPI in the nucleus remains unclear. However, it is important to recognize that all these experiments were conducted in different cellular backgrounds with different experimental models and detection methods.

Nevertheless, a consensus seems to emerge that CRABPI expression levels matter to cells that are responsive to RA signals (such as cells of embryonic origin) in terms of gene regulation and cellular behavior. This is most strongly supported by several animal studies conducted by several groups including ours [41, 65, 66]. While *CrabpI* gene knockout mice appeared grossly normal [15, 23], ectopic overexpressing of CRABPI in two transgenic mouse models induced abnormalities in adult lung and liver [61, 65, 66], and caused defective liver lens fiber differentiation and pancreatic tumorigenesis [41]. These transgenic animal data strongly suggest that CRABPI should not be abnormally expressed in animals. Our conclusion based on these results is that the level of CRABPI expression is important for maintaining a normal cellular response to RA. The unsolved question is, why does CRABPI matter if animals deficient in CRABPI survive?

The CRABPI Gene and Its Regulation

The *crabpI* gene (Fig. 6.2) promoter does not contain the TATA box that is typically seen in housekeeping genes, but does include binding sites for the major transcription factor Specific Protein 1 (SP-1), a pair of overlapping direct repeat (DR) 4/DR5 elements (now defined as Thyroid hormone response element, or TRE) that are

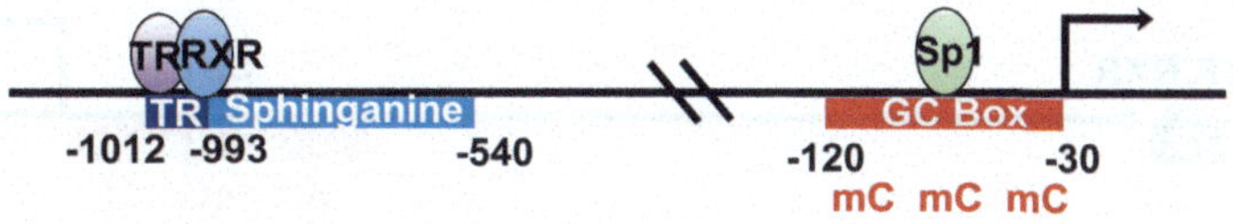

Fig. 6.2 The mouse *crabpI* gene promoter. The gene is driven by a TATA-less promoter, which contains mainly multiple GC boxes for Sp1 binding to activate basal promoter. This basal promoter is subjected to regulation by extensive DNA methylation, which accounts for basal gene repression in non-expressing cells. Approximately 1 kb upstream of this promoter lies a DR4/DR5-containing thyroid hormone response element (TRE) for regulation by hormones like T3/T4 and RA [13]. This regulatory region is also subjected to modulation by lipid signals such as sphinganine. Hormonal regulation of this gene depends upon hormone-induced chromatin juxtaposition of this TRE to GC box region (see Fig. 6.4)

responsible for its regulation by RA [60] and thyroid hormones [65, 66], a region responsive to lipid signals such as sphinganine [64] and ethanol [10], and a region enriched in DNA methylation that is known to be important in heterochromatin formation for gene silencing [58].

We have employed the mouse embryonal carcinoma cell (similar to ESC) and fibroblast-adipocyte differentiation (where RA is one important factor for commitment) models to study the physiological regulation of the *crabpI* gene. In both experimental models, the *crabpI* gene is regulated by thyroid hormones and RA through the thyroid hormone response element (TRE) [57]. These studies have exploited extensive biochemical methodology to determine whether regulation occurs at the chromatin level. We have examined the endogenous chromatin conformation of the *crabpI* gene locus, particularly its promoter and a contiguous upstream regulatory region because this segment of the mouse *crabpI* gene has been validated as physiologically functional in a transgenic reporter mouse model [62, 63]. Based on these experiments, we concluded that the *crabpI* gene is rapidly activated by thyroid hormones in undifferentiated preadipocytes, and that this activity is mediated by the Thyroid hormone Receptor-Associated Protein 220 (TRAP220)-containing activating mediator complex [37]. In this hormone-activating phase (Fig. 6.3), the enhancer region of the *crabpI* gene is looped to juxtapose its basal promoter region and, through nucleosome sliding, the transcription initiation site-spanning nucleosome falls off the chromatin, thereby opening its transcription initiation site for active gene transcription. Then, in cells stimulated to differentiate into adipocytes, the *crabpI* gene negatively responds to the same hormonal input (Fig. 6.3). In this hormone-repressive phase, chromatin remodeling on the *crabpI* gene requires the repressive remodeling machinery that contains the corepressor named Receptor Interacting Protein 140 (RIP140) whose expression level is highly elevated in differentiating adipocytes [39]. During the repressive chromatin remodeling phase, DNA methylation and heterochromatin begins to form on the *crabpI* gene promoter, rendering this gene gradually silenced in differentiated adipocytes [56]. Thus in this model, regulation of the *crabpI* gene is bimodal, depending on the differentiation status of the cells. Such a tight regulation of the *crabpI* gene is in line with the identified function of CRABPI in the stem cell model. In pre-committed

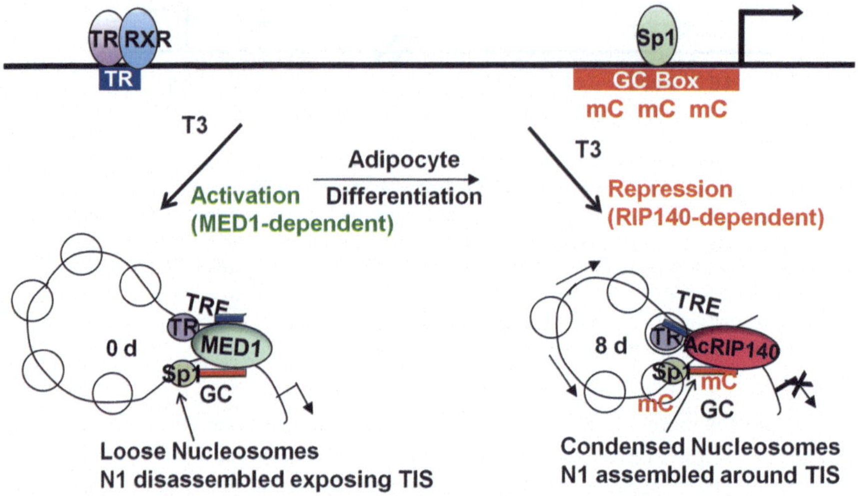

Fig. 6.3 Chromatin remodeling of *crabpI* gene in adipocyte differentiation. In adipocyte differentiation before cells are committed to adipocyte lineage, the *crabpI* gene is initially activated by thyroid hormones and RA (*left*) which induce chromatin juxtaposition, mediated by the Mediator (MED1) complex, of TRE region with the GC box region. This shortens the distance between the upstream regulatory region and the basal promoter, and simultaneously recruits chromatin remodeling machinery to cause nucleosome sliding and disassembly of N1 nucleosome from the transcription initiation site, resulting in rapid gene activation. As cells are committed to the adipocyte lineage and in the presence of T3/T4 or RA, corepressor RIP140 expression is elevated to replace MED1, which maintains the chromatin fold but recruits chromatin repressive enzymes such as histone deacetylases, heterochromatin protein 1 and DNA methyl transferases to condense this chromatin. At this stage, the CRABPI gene is completely silenced and can no longer be activated

stem cells CRABPI is needed (to reduce free RA), whereas in post-committed cells CRABPI is no longer needed and therefore, the gene is repressed.

CRABPII

Localization

During embryogenesis, CRABP II is widely expressed [44], whereas in the adult, available data suggest the protein is expressed in a more restricted pattern in tissues that are sensitive to RA, such as skin, ovary, uterus, choroid plexus, and the olfactory epithelium [2, 18, 27, 28].

CRABPII has been detected both in the cytosol and in the nucleus [21] and has been observed to traffic from the cytoplasm to the nucleus [16, 17]. Noy and colleagues reported that RA binding induces the formation of a nuclear translocation signal in the CRABPII molecule which can function as a coactivator [49, 68]. Majumbar's laboratory determined that the sumoylation state of CRABPII is also important for its nuclear localization [29].

Biological Activities and Mechanism of Action

Unfortunately, mutant mice lacking expression of CRABPII or both CRABPII and CRABPI genes display no obvious developmental or adult abnormalities [26]. However, one cannot exclude that this could be due to compensatory mechanisms in a fully protected, optimal laboratory environment. Thus, other approaches have been used to shed light on the potential functions of CRABPII, at least *in vitro*.

In vitro studies have shown that CRABPII channels RA to RAR in a "direct collisional process" that facilitates RAR transcriptional activity [12, 49]. RA-reporter assays have indicated that CRABPII (but not CRABPI), enhances the induction of RA target genes. Co-immunoprecipitation assays and gel-shift assays have further shown that CRABPII interacts directly with RAR/RXR complexes in solution [16, 17]. It was concluded from these studies that CRABPII serves as a coactivator for RARs (Fig. 6.4).

It must be stressed that most of these studies were conducted using RA-reporter assays and engineered cell lines, leaving open questions concerning the physiological relevance of the reported interpretations. One most intriguing point is that, many RA responsive cells, including ESCs, do not express CRABPII. Therefore, it would seem that a potential co-activator function of CRABPII is only needed in specific cell types.

Very recently, an interesting finding was reported that CRABPII also displays non-transcriptional activities in the absence of RA [55]. Surprisingly these studies revealed that CRABPII directly interacts with HuR, an ubiquitously expressed protein which binds mRNAs and protects them against degradation. This interaction markedly increases the affinity of HuR for some target transcripts exemplified by the apoptotic peptidase activating factor 1 (Apaf-1), which is involved in apopttic responses. Consequently, the stability and the expression levels of the transcripts are increased. Then upon RA binding, CRABPII dissociates from HuR and translocates to the nucleus, where it delivers RA to RAR. Although the spectrum of genes whose expression is regulated by HuR in cooperation with CRABPII remains to be identified, the data establish that the tumor suppressive activity of CRABPII is exerted both by its ability to deliver RA to RAR, resulting in induction of RAR-targeted growth inhibitory genes, and by its involvement in HuR-mediated stabilization of proapoptotic transcripts. In conclusion, CRABPII seems to regulate gene expression not only via regulating transcription, but also via posttranscriptional processes.

The Gene and Its Regulation

The *crabpI* and *crabpII* genes share approximately 75 % sequence homology and have a very similar gene structure in that both contain 4 exons led by a short 5'-untranscribed region. The conservation in the gene structures and the similarity

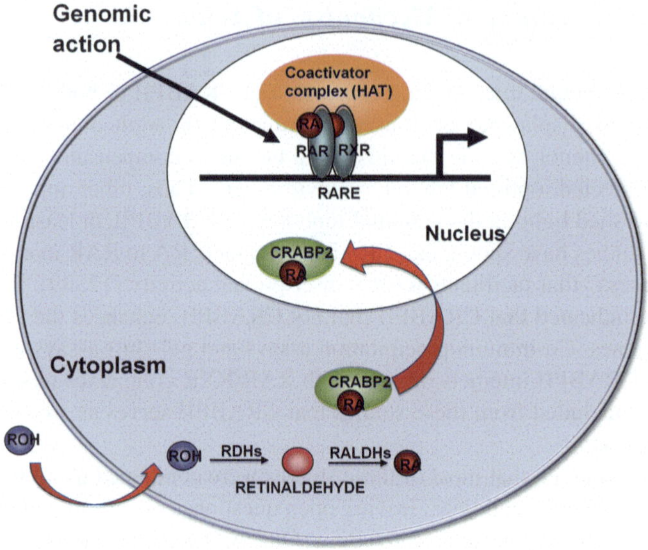

Fig. 6.4 A proposed model of CRABPII function. In cells regulated RA, CRABPII functions to capture RA from the cytoplasm, deliver RA to the nuclear RAR/RXR to activate RA target genes

of the 5' region would suggest similar regulation. Surprisingly, these two genes are regulated very differently and exhibit distinct expression profiles [36, 44].

The *crabp*II gene (Fig. 6.5) contains a typical TATA-containing promoter and is regulated primarily by RA-responding direct repeat (DR) elements [4]. In RA-induced cell differentiation models, the human *crabp*II gene is regulated by a functional DR5 located approximately 5.6 kb upstream whereas the mouse *crabp*II gene is regulated by a DR1 and a DR2 located approximately 1.1 kb upstream [19]. Studies of the *CrabpII* gene regulation utilized various cellular models including embryonal carcinoma cells [19], neuroblastoma cells [43], skin [53], uterus [28], cancer cells [30, 54] and adipocytes [8, 9]. Interestingly, these studies reported that the *crabp*II gene can be up- or down-regulated in response to various signals or hormones upon binding of transcription factors to other specific response elements located in the promoter. In this context, adipocytes differentiation is an interesting model to study the regulation of CRABPII. Indeed CRABPII is highly expressed in preadipocytes and it has been shown that it sensitizes preadipocytes to RA-induced inhibition of differentiation. Interestingly, RA inhibited differentiation if administered within a short time frame following its induction but failed to do so when administered later in the program. This failure has been attributed to the down regulation of CRABPII upon binding of the glucocorticoid receptor and of the CCAAT/enhancer-binding protein αlpha (C/EBPα) to specific response elements, during the process of adipocyte differentiation and in mature adipocytes, respectively (Fig. 6.5) [8]. Thus, down regulation of CRABPII appears critical to allow adipogenesis to proceed. In contrast, during myogenic transformation the *crabp*II gene is activated by MyoD and Sp1 binding to its basal promoter [69].

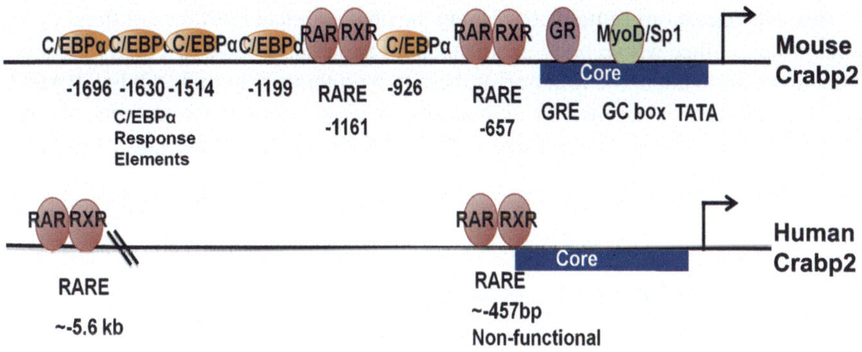

Fig. 6.5 The *crabpII* gene promoter. The basal Promoter is from −459 through −4 and contains TATA sequence and GC box. In adipocyte differentiation, this gene is repressed by glucocorticoid (through GRE) and CCAAT/enhancer-binding protein. In myogenic transformation, this gene is activated by MyoD and Sp1 binding to its basal promoter. On the human gene (*lower*), approximately 5.6 kb upstream region lies a RA-response element (RARE) that mediates major RA induction of this gene

In conclusion, a general theme emerges that the *crabpII* gene, like the *crabpI* gene, is subject to extensive positive and negative hormonal regulation. However, it has not been reported yet whether regulation of CRABPII expression involves promoter recruitment of the same positive and negative coregulators (TRAP220 and RIP240) as described for CRABPI.

Current State of the Field

As it currently stands, reports of CRABPI and CRABPII functions are primarily based upon *in vitro* experiments. Biochemical characterization of these proteins suggests their roles must be for very different aspects of RA signaling. CRABPII is thought to play an active role in gene expression via facilitating RA delivery to RARs into the nucleus and via stabilizing transcripts. In contrast, CRABPI would rather direct RA to the nucleus passively and would participate in cytosolic signaling via ERK1/2 activation.

Most importantly, at the gene level, both CRABPI and CRABPII can be regulated by several transcription factors in response to a variety of signals, indicating a need for both proteins to maintain proper expression levels in certain physiological contexts. The physiological regulation of the *crabpI* gene is better understood because specific alterations in the chromatin have been reported and considered in animal models. On the contrary, regulation of the *crabpII* gene requires further validation studies in vivo.

However, questions remain concerning the physiological relevance of these cytosolic proteins since whole body gene knockout approaches have revealed little specific information about the function of these two proteins in animals. Only CRABPI function has been examined in animals using over-expressing transgenic mouse models, and the results suggest that the level of CRABPI is important for normal animal physiology in adults.

Relevance and Future Directions

The CRABPI and II proteins were identified more than four decades ago, a long time before RARs. As both proteins are highly conserved during evolution and across species throughout the animal kingdoms, it was conceivable that they should be constrained for specific physiological needs. However, disappointingly, due to the lack of obvious phenotypes of mice deficient in either one or both genes, the functional roles of these two proteins is still debated. This has been quite discouraging and disappointing for the field.

In fact one cannot exclude that in a fully protected, optimal laboratory environment, the lack of phenotype could be due to compensatory mechanisms that spare mutant mice from lethality or deleterious effects. Thus, other approaches should be used to shed light on the physiological relevance of CRABPs.

Recent studies have begun to elucidate the functional roles of CRABPs. With regard to the function of CRABPI, it becomes increasingly evident that it mediates non-canonical effects of RA, i.e. the activation of kinase pathways. Now the question is how CRABPI performs such a function and whether this function occurs in other cell types than stem cells.

Concerning CRABPII, engineered experimental systems highlighted the role of this protein as a nuclear cofactor for RARs. The intriguing point is that, in contrast to RARs, CRABPII is not expressed in all RA-responsive cells. Such an observation raised several questions: why do only certain RA-responding cells employ CRABPII as a cofactor and is CRABPII required for only certain RA-regulated genes? In this context it is worth noting that in cells that do not express CRABPII, RA has been found to bind another type of fatty acid binding protein exemplified by FABP5 [48]. The striking observation is that FABP5 does not channel RA to RARs, rather it channels to another nuclear receptor, PPARβ/δ [1]. Consequently other genes involved in cell growth, survival and lipid metabolism are regulated.

Such a diversion from RAR to PPARβ/δ has been observed in mammary carcinoma cells with an aberrantly high FABP5/CRABPII ratio and has been correlated to RA resistance [48]. Most interestingly, *in vivo*, in a model of obese mice, RA has been shown to suppress obesity and insulin resistance via both PPARβ/δ and RARs [7]. All these data indicate that depending on the cell type, RA can bind other proteins than CRABPs and thus can regulate a wider subset of genes. All these recent data increase the complexity of the mechanism of action of RA and open new avenues in the field of the cellular retinoic acid binding proteins.

Acknowledgements This work was supported by NIH grants DK54733, and DK60521, the Dean's Commitment, and the Distinguished McKnight Professorship of University of Minnesota (LNW). I thank Shawna Persaud for preparing the figures.

References

1. Armstrong EH, Goswami D, Griffin PR, Noy N, Ortlund EA (2014) Structural basis for ligand regulation of the fatty acid-binding protein 5, peroxisome proliferator-activated receptor β/δ (FABP5-PPARβ/δ) signaling pathway. J Biol Chem 289:14941–14954
2. Asson-Batres MA, Ahmad O, Smith WB (2003) Expression of the cellular retinoic acid binding proteins, type II and type I, in mature rat olfactory epithelium. Cell Tissue Res 312:9–19
3. Astrom A, Tavakkol A, Pettersson U, Cromie M, Elder JT, Voorhees JJ (1991) Molecular cloning of two human cellular retinoic acid binding proteins (CRABP). J Biol Chem 266:17662–17666
4. Astrom A, Pettersson U, Chambon P, Voorhees JJ (1994) Retinoic acid induction of human cellular retinoic acid binding protein II gene transcription is mediated by retinoic acid receptor-retinoid X receptor heterodimers bound to one far upstream retinoic acid responsive element with 5 base pair spacing. J Biol Chem 269:22334–22339
5. Bailey JS, Siu C-H (1988) Purification and partial characterization of novel binding protein for retinoic acid from neonatal rat. J Biol Chem 263:9326–9332
6. Basher MM, Chytil F (1975) Cellular retinol-binding protein. Biochim Biophys Acta 411:87–96
7. Berry DC, Noy N (2009) All-trans-retinoic acid represses obesity and insulin resistance by activating both peroxisome proliferation-activated receptor beta/delta and retinoic acid receptor. Mol Cell Biol 29:3286–3296
8. Berry DC, Soltanian H, Noy N (2010) Repression of cellular retinoic acid-binding protein II during adipocyte differentiation. J Biol Chem 285:15324–15332
9. Berry DC, DeSantis D, Soltanian H, Croniger CM, Noy N (2012) Retinoic acid upregulates preadipocyte genes to block adipogenesis and suppress diet-induced obesity. Diabetes 61:1112–1121
10. Bi J, Hu X, Zhou FC, Wei L-N (2001) Upregulation of cellular retinoic acid binding protein I expression by ethanol. Dev Growth Differ 43:553–561
11. Boylan JF, Gudas LF (1991) Overexpression of the cellular retinoic acid binding protein I (CRABP-I) results in a reduction in differentiation specific gene expression in F9 teratocarcinoma cells. J Cell Biol 112:965–979
12. Budhu AS, Noy N (2002) Direct channeling of retinoic acid between cellular retinoic acid-binding protein II and retinoic acid receptor sensitizes mammary carcinoma cells to retinoic acid-induced growth arrest. Mol Cell Biol 22:2632–2641
13. Chang L, Wei L-N (1997) Characterization of a negative response-DNA element in the upstream region of the cellular retinoic acid-binding protein-I gene of the mouse. J Biol Chem 272:10144–10150
14. Chuang Y-S, Huang WH, Park SW, Persaud SD, Hung CH, Ho PC, Wei L-N (2011) Promyelocytic leukemia protein in retinoic acid-induced chromatin remodeling of *Oct4* gene promoter. Stem Cells 29:660–669
15. de Bruijn DR, Oerlemans F, Hendriks W, Baats E, Baats WE, Ploemacher R, Wieringa B, van Geurts Kessel A (1994) Normal development, growth and reproduction in cellular retinoic acid binding protein-I (CRABPI) null mutant mice. Differentiation 58:141–148
16. Delva L, Bastie JN, Rochette-Egly C, Kraiba R, Balitrand N, Despouy G, Chambon P, Chomienne C (1999) Physical and functional interactions between cellular retinoic acid binding protein II and the retinoic acid-dependent nuclear complex. Mol Cell Biol 19:7158–7167

17. Dong D, Ruuska SE, Levinthal DJ, Noy C (1999) Distinct roles for cellular retinoic acid-binding proteins I and II in regulating signaling by retinoic acid. J Biol Chem 274:23695–23698
18. Donovan M, Olofsson B, Gustafson AL, Decker L, Eriksson U (1995) The cellular retinoic acid binding proteins. J Steroid Biochem Mol Biol 55:479–485
19. Durand B, Saunders M, Leroy P, Leid M, Chambon P (1992) All-trans and 9-cis retinoic acid induction of CRABPII transcription is mediated by RAR-RXR heterodimers bound to DR1 and DR2 repeated motifs. Cell 71:73–85
20. Fogh K, Voorhees JJ, Astrom A (1993) Expression, purification and binding properties of human cellular retinoic acid binding protein type I and II. Arch Biochem Biophys 300:751–755
21. Gaub MP, Lutz Y, Ghyselinck NB, Scheuer I, Pfister V, Chambon P, Rochette-Egly C (1998) Nuclear detection of cellular retinoic acid binding proteins I and II with new antibodies. J Histochem Cytochem 46:1103–1111
22. Giguere V, Lyn S, Yip P, Siu C, Amin S (1990) Molecular cloning of cDNA encoding a second cellular retinoic acid binding protein. Proc Natl Acad Sci U S A 87:6233–6237
23. Gorry P, Lufkin T, Dierich A, Rochette-Egly C, Decimo D, Dolle P, Mark M, Durand B, Chambon P (1994) The cellular retinoic acid binding protein I is dispensable. Proc Natl Acad Sci U S A 91:9032–9036
24. Gupta P, Huq MDM, Ha SG, Park SW, Khan AA, Tsai N-P, Wei L-N (2008) Retinoic acid-stimulated sequential phosphorylation, PML recruitment and SUMOylation of nuclear receptor TR2 to suppress Oct4. Proc Natl Acad Sci U S A 105:11424–11429
25. Kleywegt GJ, Bergfors T, Senn H, Motte PL, Gsell B, Shudo K, Jones TA (1994) Crystal structures of cellular retinoic acid binding proteins I and II in complex with all-trans-retinoic acid and a synthetic retinoid. Structure 2:1241–1258
26. Lampron C, Rochette-Egly C, Gorry P, Dolle P, Mark M, Lufkin T, Lemur M, Chambon P (1994) Mice deficient in cellular retinoic acid binding protein II or in both CRABPI and CRABPII are essentially normal. Development 121:539–548
27. Li E, Norris AW (1996) Structure/function of cytoplasmic vitamin A binding proteins. Annu Rev Nutr 16:205–234
28. Li XH, Ong DE (2003) Cellular retinoic acid binding protein II gene expression is directly induced by estrogen but not retinoic acid in rat uterus. J Biol Chem 278:35819–35825
29. Majumbar A, Petrescu AD, Xiong Y, Noy N (2011) Nuclear translocation of cellular retinoic acid binding protein II is regulated by retinoic acid-controlled Sumoylation. J Biol Chem 286:42749–42757
30. McPherson LA, Woodfield GW, Weigel RJ (2007) AP2 transcription factors regulate expression of CRABPII in hormone responsive breast carcinoma. J Surg Res 138:71–78
31. Napoli JL (1996) Retinoic acid biosynthesis and metabolism. FASEB J 10:993–1001
32. Norris AW, Cheng L, Giguere V, Rosenberger M, Li E (1994) Measurement of sub-nanomolar retinoic acid binding affinities for cellular retinoic acid bind protein by fluorometric titration. Biochim Biophys Acta 1037:192–199
33. Norris AW, Rong D, d'Avignon DA, Rosenberger M, Tasaki K, Li E (1995) Nuclear magnetic resonance studies demonstrate differences in the interaction of retinoic acid with two highly homologous cellular retinoic acid binding proteins. Biochemistry 34:15564–15573
34. Noy N (2000) Retinoid binding proteins: mediators of retinoid action. Biochem J 348:481–495
35. Ong DE, Chytil F (1975) Retinoic acid-binding protein in rat tissue: partial purification and comparison to rat tissue retinol-binding protein. J Biol Chem 250:6113–6117
36. Ong DE, Newcomer ME, Chytil F (1994) Cellular retinoid-binding proteins. In: Sport MB, Roberts AB, Goodman DS (eds) The retinoids: biology, chemistry and medicine, vol 2. Raven, New York, pp 283–317

37. Park SW, Li G, Lin YP, Faroqui MJ, Ge K, Roeder RG, Wei L-N (2005) Thyroid hormone-induced juxtaposition of regulatory elements/factors and chromatin remodeling of *Crabp1* dependent on MED1/TRAP220. Mol Cell 19:643–653
38. Park SW, Hu X, Gupta P, Lin YP, Ha SG, Wei L-N (2007) Sumoylation of Tr2 orphan receptor involves PML and fine-tunes Oct4 expression in stem cells. Nat Struct Mol Biol 14:68–75
39. Park SW, Huang WH, Persaud SD, Wei L-N (2009) RIP140 in thyroid hormone-repression and chromatin remodeling of Crabp1 gene during adipocyte differentiation. Nucleic Acid Res 37:7085–7094
40. Parmar MB, Lee JJA, Wright JM (2013) Duplicated crabp1 and crabp2 genes in medaka (Oryzlas latipes): gene structrure, phylogenetic relationship and tissue specific distribution of transcripts. Comp Biochem Physiol B Biochem Mol Biol 165:10–18
41. Perez-Castro AV, Tran VT, Nguyen-Huu MC (1993) Defective lens fiber differentiation and pancreatic tumorigenesis caused by ectopic expression of the cellular retinoic acid-binding protein I. Development 119:363–375
42. Persaud SD, Lin Y-W, Wu C-Y, Kagechika H, Wei LN (2013) Cellular retinoic acid binding protein I mediates rapid non-canonical activation of ERK1/2 by all-trans retinoic acid. Cell Signal 25:19–25
43. Plum LA, Clagett-Dame M (1995) 9-cis-retinoic acid selectively activates the cellular retinoic acid binding protein-II gene in human neuroblastoma cells. Arch Biochem Biophy 319:457–463
44. Ruberte E, Friederich V, Morriss-Kay G, Chambon P (1992) Differential distribution patterns of CRABPI and CRABPII transcripts during mouse embryogenesis. Development 115:973–987. FEBS
45. Ruff SJ, Ong DE (2000) Cellular retinoic acid binding protein is associated with mitochondria. FEBS Lett 487:282–286
46. Sani BP, Hill DL (1974) Retinoic acid: a binding protein in metatarsal skin. Biochem Biophys Res Commun 61:1276–1282
47. Sanquer S, Gilchrest BA (1994) Characterization of human cellular retinoic acid-binding proteins I and II: ligand binding properties and distribution in skin. Arch Biochem Biophys 311:86–94
48. Schug TT, Berry DC, Shaw NS, Travis SN, Noy N (2007) Opposing effects of retinoic acid on cell growth results from alternate activation of two different nuclear receptors. Cell 129:723–733
49. Sessler RJ, Noy N (2005) A ligand-activated nuclear localization signal in cellular retinoic acid binding protein-II. Mol Cell 18:343–353
50. Shubeita HE, Sambrook JF, McCormick AM (1987) Molecular cloning and analysis of functional cDNA and genomic clones encoding bovine cellular retinoic acid-binding protein. Proc Natl Acad Sci U S A 84:5645–5649
51. Sundelin J, Das S, Eriksson U, Rask L, Peterson P (1985) The primary structure of bovine cellular retinoic acid binding protein. J Biol Chem 260:6494–6499
52. Thomas JR, Bratt JM, Banaszak LJ (1995) Crystal structures of cellular retinoic binding protein I show increased access to the binding cavity due to formation of an intermolecular b-sheet. J Mol Biol 252:433–446
53. Vahlquist A, Andersson E, Coble BI, Rollman O, Torma H (1996) Increased concentrations of 3,4-didehydroretinol and retinoic acid binding protein (CRABPII) in human squamous cell carcinoma and keratoacanthoma but not in basal cell carcinoma of the skin. J Invest Dermatol 106:1070–1074
54. Vo HP, Crowe DL (1998) Transcriptional regulation of retinoic acid responsive genes by cellular retinoic acid binding protein II modulates RA mediated tumor cell proliferation and invasion. Anticancer Res 18:217–224
55. Vreeland AC, Yu S, Levi L, de Barros RD, Noy N (2014) Transcript stabilization by the RNA-binding protein HuR is regulated by cellular retinoic acid binding protein 2. Mol Cell Biol 34:2135–2146

56. Wei L-N (2012) Chromatin remodeling and epigenetic regulation of the CrabpI gene in adipocyte differentiation. BBAMCB 1821:206–212
57. Wei L-N, Chang L (1996) Promoter and upstream regulatory activities of the gene for retinoic acid binding protein I from the mouse. J Biol Chem 271:5073–5078
58. Wei L-N, Lee C-H (1994) Demethylation in the 5'-flanking region of mouse CRABP-I gene is associated with its high level of expression in mouse embryos and facilitates its induction by retinoic acid in P19 embryonal carcinoma cells. Dev Dyn 201:1–10
59. Wei L-N, Mertz JR, Goodman DS, Nguyen-Huu MC (1987) Cellular retinoic acid- and cellular retinol-binding proteins: cDNA cloning, chromosomal assignment and tissue specific expression. Mol Endocrinol 1:526–534
60. Wei L-N, Blaner WS, Goodman DS, Nguyen-Huu MC (1989) Regulation of the cellular retinoid-binding proteins and their mRNAs during P19 embryonal carcinoma cell differentiation induced by retinoic acid. Mol Endocrinol 3:454–463
61. Wei L-N, Tsao J-L, Chu Y-S, Jeannotte L, Nguyen-Huu MC (1990) Molecular cloning and transcriptional mapping of the mouse cellular retinoic acid- binding protein gene. DNA Cell Biol 9:471–478
62. Wei L-N, Chen JC, Tsao J-L, Chu Y-S, Nguyen-Huu MC (1991) A 3 kb sequence of the mouse cellular retinoic acid-binding protein gene upstream region mediates spatial and temporal LacZ expression in transgenic mouse embryos. Development 112:847–854
63. Wei L-N, Lee C-H, Chang SL, Chu YS (1991) Pathogenesis in transgenic mice expressing bovine cellular retinoic acid-binding protein. Dev Growth Differ 34:479–488
64. Wei L-N, Lee C-H, Chang L (1995) Retinoic acid induction of mouse cellular retinoic acid-binding protein-I gene expression is enhanced by sphinganine. Mol Cell Endocrinol 111:207–211
65. Wei L-N, Chang L, Lee C-H (1997) Studies of over-expressing cellular retinoic acid binding protein-I in cultured cells and transgenic mice. Trangenics 2:201–209
66. Wei L-N, Lee C-H, Filipcik P, Chang L (1997) Regulation of the mouse cellular retinoic acid-binding protein I gene by thyroid hormone and retinoids in transgenic mouse embryos and P19 cells. J Endocrinol 155:35–46
67. Wei L-N, Chang L, Hu X (1999) Studies of the type I cellular retinoic acid binding protein mutants and their biological activities. Mol Cell Biochem 200:69–76
68. Wolf G (2009) Cellular Retinoic acid-binding protein II: coactivator of the transactivation by the retinoic acid receptor complex RAR.RXR. Nutr Rev 58:151–153
69. Yuan J, Tang Z, Yang S, Li K (2013) CRABP2 promotes myoblast differentiation and is modulated by the transcription factors MyoD and Sp1 in C2C12 cells. PLoS One 8:e55479

Chapter 7
Non-classical Transcriptional Activity of Retinoic Acid

Noa Noy

Abstract It has long been established that the transcriptional activity of retinoic acid (RA) is mediated by members of the nuclear receptor family of ligand-activated transcription factors termed RA receptors (RARs). More recent observations have established that RA also activates an additional nuclear receptor, PPARβ/δ. Partitioning RA between RARs and PPARβ/δ is governed by different intracellular lipid-binding proteins: cellular RA binding protein 2 (CRABP2) delivers RA to nuclear RARs and a fatty acid binding protein (FABP5) delivers the hormone from the cytosol to nuclear PPARβ/δ. Consequently, RA signals through RARs in CRABP2-expressing cells, but activates PPARβ/δ in cells that express a high level of FABP5. RA elicits different and sometimes opposing responses in cells that express different FABP5/CRABP2 ratios because PPARβ/δ and RARs regulate the expression of distinct sets of genes. An overview of the observations that led to the discovery of this non-classical activity of RA are presented here, along with a discussion of evidence demonstrating the involvement of the dual transcriptional activities of RA in regulating energy homeostasis, insulin responses, and adipocyte and neuron differentiation.

Keywords Retinoic acid • PPAR • CRABP2 • FABP5 • RAR • Transcription • Cancer • Adipogenesis • Neuronal differentiation • Intracellular lipid binding proteins • Nuclear receptors

Abbreviations

ANGLP4	angiopoietin-like protein 4
C/EBP	CCAAT enhancer-binding protein
CRABP	cellular retinoic-binding protein

N. Noy (✉)
Department of Cellular and Molecular Medicine, Lerner Research Institute, Cleveland Clinic Foundation, Cleveland, OH 44195, USA
e-mail: noyn@ccf.org

M.A. Asson-Batres, C. Rochette-Egly (eds.), *The Biochemistry of Retinoid Signaling II*, Subcellular Biochemistry 81, DOI 10.1007/978-94-024-0945-1_7

CRBP	cellular retinol-binding protein
ERK	extracellular signal-regulated kinase
FABP	fatty acid binding protein
ILK	integrin-linked kinase
iLBP	intracellular lipid-binding protein
PDPK1	3-Phosphoinositide Dependent Protein Kinase
PLIN2	perilipin 2
PPAR	peroxisome proliferator activated receptor
Pref-1	preadipocyte factor 1
PTEN	phosphatase and tensin homolog
RA	retinoic acid
RAR	RA receptor
VEGF	vascular endothelial growth factor

Introduction

Forty-eight genes encoding different nuclear hormone receptors have been identified in the human genome (see http://www.nursa.org/nursa/molecules/index.jsf). Of these, it has long been established that retinoic acid receptors (RARα, RARβ, and RARγ, also referred to as NR1B1, NR1B2, and NR1B3) mediate the transcriptional activities of the vitamin A metabolite all-trans-retinoic acid (RA) [1]. However, various reports over the last decade have suggested that RA displays biological activities that are independent of its ability to activate RAR. A rationale for such additional activities has been provided by the discovery that RA can function as an agonist for a different nuclear receptor, namely PPARβ/δ (NR1C2) [2, 3]. RARs and PPARβ/δ regulate the expression of distinct sets of genes, and, consequently, RA elicits different cellular responses depending on whether RARs or PPARβ/δ are activated. The broad distribution of RARs and PPARβ/δ in vertebrate tissues, often in the same cells, suggests that mechanisms must exist to partition RA between these receptors. Indeed, it has been shown that two members of the family of intracellular lipid-binding proteins are responsible for selectively providing RA to RAR and to PPARβ/δ [4–6]. Specifically, cellular RA-binding protein 2 (CRABP2) delivers RA to RAR, while its homolog fatty acid-binding protein 5 (FABP5) shuttles RA (and other ligands) to PPARβ/δ. Both CRABP2 and FABP5 reside in the cytosol in the absence of ligand, but upon binding RA, they move to the nucleus where they form a complex with either RAR or PPARβ/δ. Within this complex, RA is directly 'channeled' from the binding protein to the cognate nuclear receptor. CRABP2 and FABP5 facilitate the ligation and markedly enhance the transcriptional activity of their cognate receptors. Hence, the biological activities of RA are largely governed by the relative expression levels of CRABP2 and FABP5.

RA mainly activates RARs in cells that express CRABP2, even if this protein is expressed at low levels. In contrast, RA only efficiently activates PPARβ/δ when the FABP5/CRABP2 ratio is very high. This is because the binding affinities of

CRABP2 and RAR for RA are in the sub-nanomolar (nM) range, whereas the affinities of FABP5 and PPARβ/δ for RA are two orders of magnitude weaker. Thus, RA activates RAR in some cells, PPARβ/δ in others, and, under some circumstances, both. Recent observations show that a precise balance between CRABP2/RAR and FABP5/PPARβ/δ signaling paths is critical for RA-mediated regulation of biological processes, including adipocyte and neuronal differentiation, sugar and lipid homeostasis, and insulin responses. Perturbations in the balance between the two pathways have been shown to interfere with cognitive function, promote the development of some cancers, and contribute to high fat feeding-induced obesity.

History and Development of the Field

PPARβ/δ

It was established in the late 1980s that RA can regulate gene transcription by activating RARα, RARβ, and RARγ, but subsequent reports suggested that some biological activities of RA may be mediated by a different mechanism. While it was known that RA is essential for skin maintenance [7], the effect of RA on this tissue was still present even when RARs were inactivated [8]. Clinical use of RA in therapy of some cancers originated from the observations that RA-induced activation of RAR often results in inhibition of cell proliferation [9, 10], but it was discovered that some cancers were RA-resistant or, paradoxically, responded to RA with enhanced proliferation (e.g. [11, 12]). The mechanism by which RA exerts RAR-independent activities remained unknown until it was reported in 2003 that, in addition to activating RARs, the hormone also activates a different nuclear receptor, namely a subtype of the peroxisome proliferator activated receptors (PPARs) termed PPARβ/δ [2].

Like RARs and other members of the superfamily of nuclear hormone receptors, PPARs interact with the retinoid X receptor (RXR) to form heterodimers. Upon binding their cognate ligands, the activated heterodimer complexes bind to PPAR response elements (PPRE) in regulatory regions of specific genes to induce target gene transcription [13, 14]. PPARβ/δ is ubiquitously expressed, but has particularly high expression in brain, adipose tissue, skeletal muscle, and skin [15]. It has been reported that this receptor is involved in neuronal development [16], inflammation [17, 18], keratinocyte differentiation, and epidermal barrier recovery [19, 20].

Activation of PPARβ/δ during skin injury protects against apoptosis, and thus, enhances skin repair [17, 18]. The anti-apoptotic activities are mediated in part by the ability of PPARβ/δ to directly upregulate the expression of the kinase PDPK1, thereby activating the survival factor Akt1 and protecting keratinocytes from pro-apoptotic inflammatory cytokines [17, 18, 21]. PPARβ/δ is the primary PPAR isotype in skeletal muscle [22, 23] where it participates in fiber formation and maintenance [24]. Like other PPARs, PPARβ/δ is also involved in regulating lipid and sugar metabolism and insulin responses. Hence, activation of PPARβ/δ in adipose tissue and muscle results in upregulation of multiple proteins that enhance lipid oxidation, mitochondrial respiration, and sensitization of cells to insulin signaling. An end result of PPARβ/δ signaling is that lipid stores are depleted, circulating lipid levels are decreased, and high fat-fed animals are protected against insulin resistance [25].

The involvement of PPARβ/δ in cancer cell biology has been somewhat controversial, but accumulating evidence shows that this receptor potently promotes cancer cell proliferation and survival and displays pronounced oncogenic activities [18, 21, 26–36]. It has been suggested that these activities are mediated in part by decreasing the expression of the tumor suppressor PTEN [18] and by direct induction of the survival factors ILK and PDPK1 [18, 21], the angiogenic factor VEGFa [26, 37], the pro-oncogenic adipokine ANGPTL4 [30, 38], and the oncogene Src [36].

PPARβ/δ Ligands

The exact nature of endogenous, physiologically-relevant ligands that activate PPARβ/δ is incompletely defined. The ligand-binding pocket of the receptor is considerably larger than those of other nuclear receptors, displaying a total volume of ~1300 Å, [39–42]. Ligands can access the pocket through a solvent-exposed channel that has an entrance area of approximately 100 Å and an ability to open wider due to the flexibility of surrounding helices [41]. Consistent with its large ligand-binding pocket, PPARβ/δ can associate with a broad spectrum of ligands, including long chain fatty acids (LCFA), leukotrienes, and prostaglandins [41–44]. However, affinities for these ligands are highly variable and not all compounds that bind PPARβ/δ function as activators. For example, the unsaturated LCFA, linoleic acid, binds with high affinity and efficiently activates PPARβ/δ, while the saturated LCFA, palmitate, binds with low affinity and does not activate the receptor at physiologically relevant concentrations [45].

It was a surprise to discover that RA potently and selectively activates PPARβ/δ, but not PPARα or PPARγ [2]. The affinity of PPARβ/δ for RA is ~15 nM [2]. This discovery raised the possibility that 'non-classical' activities of RA could be mediated by this receptor. Indeed, it was shown that, in keratinocytes, RA induces the expression of well-established PPARβ/δ targets including perilipin 2 (PLIN2) [46], angiopoietin-like protein 4 (ANGLP4) [47], and PDPK1 [21]. Expression of these genes was induced by RA and by a selective synthetic PPARβ/δ ligand, but not by an RAR-selective agonist [31]. Subsequent studies confirmed that PPARβ/δ mediates certain transcriptional activities of RA in different tissues and that these activities are distinct from those mediated by RARs. It became apparent that detailed knowledge of how RA signaling partitions between these two classes of nuclear receptors was a critically-important new area of investigation in the field of retinoid biology.

Intracellular Lipid-Binding Proteins

Activation of RARs and PPARβ/δ, which are normally present in the nucleus in an inactivated state, requires mobilization of their ligands to the nucleus. How this is achieved is not a trivial matter since the hydrophobic nature of nuclear receptor ligands limits their ability to freely move across the aqueous milieu of the cell. Soluble proteins that belong to the family of intracellular lipid binding proteins (iLBPs) provide one mechanism for transporting lipophilic ligands. Members of the iLBP family are small (~15 KDa)proteins that include fatty acid binding proteins (FABPs), which associate with LCFAs and various other lipophilic compounds, and proteins that specifically bind retinoids (cellular retinol-binding proteins, CRBP 1–4, and cellular retinoic acid (RA) binding proteins, CRABP 1, 2) [48–50]. It has long been thought that iLBPs were involved in solubilizing and protecting hydrophobic and labile ligands in the aqueous milieu of cells. However, it is becoming increasingly clear that these proteins also regulate the biological activities of their ligands. The individual functions of many iLBPs are unknown, but it has been established that several of them interact with specific nuclear hormone receptors. Specifically, CRABP2, FABP1, FABP4, and FABP5 transport ligands from the cytosol to the nuclear receptors RAR, PPARα, PPARγ, and PPARβ/δ in the nucleus [4, 6, 51–54]. Upon binding to a cognate receptor, the ligand is "channeled" from the binding protein to the receptor. CRABP2 and FABP5 selectively transfer RA to RAR and PPARβ/δ, respectively (Fig. 7.1).

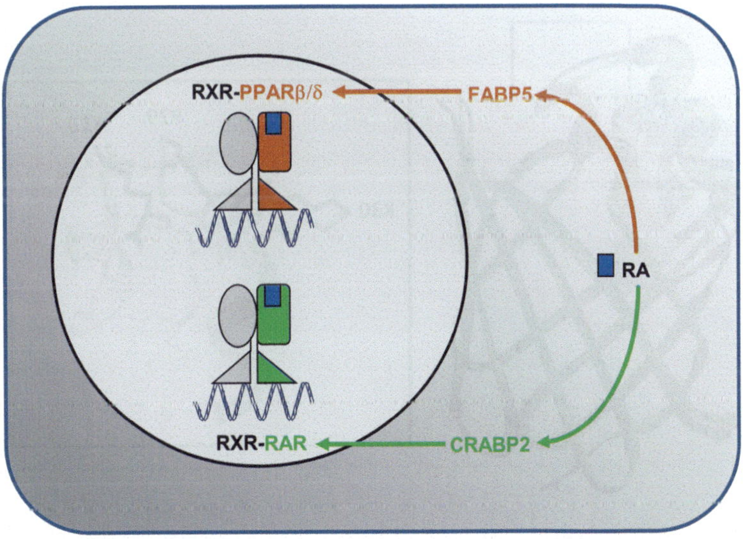

Fig. 7.1 The dual transcriptional activity of RA. RA regulates gene expression by activating two types of nuclear receptors, the classical RA receptors RARs and the alternative receptor PPARβ/δ. The partitioning of RA between the two signaling paths is governed by CRABP2, which shuttles RA from the cytosol to nuclear RAR, and by FABP5, which selectively delivers it to PPARβ/δ

CRABP2 tightly and selectively binds RA and mobilizes it to the nucleus. In contrast, FABPs bind a large array of compounds, including LCFAs, fatty acid metabolites, and various synthetic drugs and probes [4, 6, 31, 50, 53–56]. Although FABPs bind these compounds with similar affinity, only particular ligands trigger their nuclear import. Hence, nuclear translocation of FABP5 is induced by ligands that activate PPARβ/δ, such as RA and some unsaturated LCFAs, while FABP4 is mobilized to the nucleus upon binding of PPARγ agonists [6, 31, 52, 57, 58]. These observations raise the important question of how 'activating' ligands selectively trigger the nuclear mobilization of these proteins.

Classical nuclear localization signal (NLS) sequence motifs are comprised of a series of basic residues in the form K(K/R)X(K/R) [59–63]. These motifs are recognized by adapter proteins known as α-importins which mediate nuclear translocation of NLS-containing cargo to the nucleus [59]. The iLBPs do not possess such a signal in their primary sequence. Instead, their folded, 3-dimensional, structure contains a patch of basic residues that forms a functional NLS. The iLBPs display a highly conserved 3-dimensional structure comprised of two orthogonal β-sheets folded into a β-barrel that forms the ligand-binding pocket. The entrance to the pocket is covered by a helix-loop-helix lid [49, 50, 64] (Fig. 7.2, left). In CRABP2, the NLS patch has been mapped to residues K20, R29, and K30 at the helix-loop-helix region of the protein. While not consecutive in the primary sequence, these residues assemble to form a ligand-controlled NLS in the folded protein [5] (Fig. 7.2, right). A similar "3-dimensional NLS" comprised of three basic residues in the helix-loop-helix region is present in both FABP4 and FABP5 [57, 58].

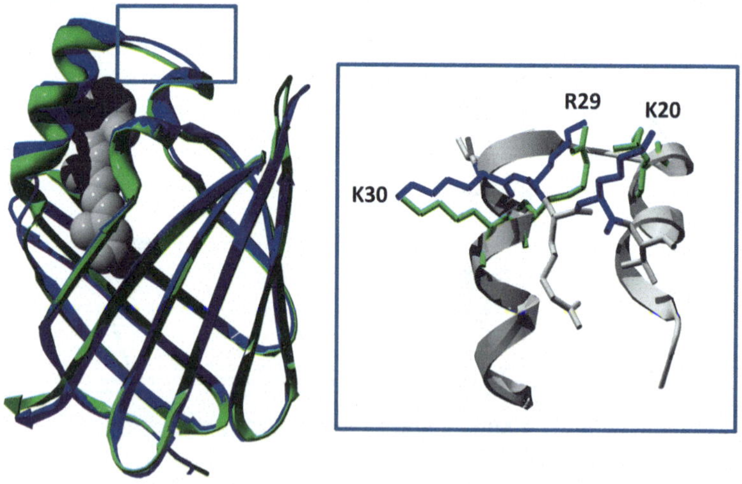

Fig. 7.2 The nuclear localization signal of CRABP2. *Left*: Superposition of the of apo-CRABP2 (*blue*) [148] and holo-CRABP2 (*green*) [149]. Bound RA is shown in gray. *Right*: Superposition of residues K20, R29 and K30 of holo-CRABP2 (*green*) with the classical NLS of the SV40 NLS peptide (*blue*) [150]

The mechanism by which ligands activate the NLS in CRABP2, FABP4 and FABP5 is different for each protein. In CRABP2, the side chains of the NLS residues are directed towards the protein in the absence of ligand, but they shift their orientation upon binding RA, rendering them accessible to the surface and placing them in close alignment with a classical NLS ([5]. Apo-FABP4 forms dimers that shield the NLS residues (K21, R30, R31) from the surrounding solution. The NLS becomes exposed when an activating ligand binds to FABP4 and induces dissociation of the dimers [57]. While a dimer rearrangement underlies the nuclear import of FABP4, nuclear import of FABP5 is driven by a ligand-induced intra-molecular allosteric communication between the ligand-sensing region and the residues that comprise the NLS [58].

Current State of the Field

The Dual Transcriptional Activity of RA in Adipose Tissue Biology

Adipogenesis, the process by which mature adipose cells are formed from preadipocytes, entails a timed progression of alterations in gene expression, culminating in terminal differentiation. Cultured preadipocytes can be induced to differentiate by treating them with insulin, isobutylmethylxanthine (IBMX), a phosphodiesterase inhibitor which elevates cellular cAMP levels, and a glucocorticoid receptor agonist [65, 66]. This differentiation mixture triggers an initial growth arrest, followed by one or two cell divisions known as clonal expansion. Cells then exit the cell cycle and proceed to terminal differentiation, driven by the transcription factors, PPARγ and CCAAT/enhancer binding proteins (C/EBPs). Mature adipocytes are characterized by their ability to accumulate lipids and by the appearance of various adipocyte markers [67, 68]. The individual components of the differentiation mixture facilitate different aspects of adipogenesis. Glucocorticoids activate glucocorticoid receptor (GR), which represses the expression of the negative regulator of adipocyte differentiation Pref-1 [69] and, concomitantly, induce the expression of C/EBPδ, in turn upregulating PPARγ [70, 71]. Elevation of cAMP levels by IBMX results in activation of protein kinase A (PKA) and, in turn, the transcription factor cAMP response element binding protein (CREB). CREB upregulates the expression of multiple adipogenic genes, including C/EBPβ which promotes cell cycle arrest and clonal expansion [67, 72–74]. CREB also induces the expression of the cAMP response element modulator gene (CREM) which functions as a cAMP-responsive negative regulator of gene expression [75–77]. The downstream effectors that mediate the involvement of insulin in adipocyte differentiation are poorly understood but it has been suggested that they include the IGF-1 receptor [67].

RA is a potent inhibitor of adipogenesis [78–83], and this activity is mediated by the CRABP2/RAR pathway. Through this path, RA induces the expression of several adipocyte differentiation inhibitors. Some of these are components of a signal-

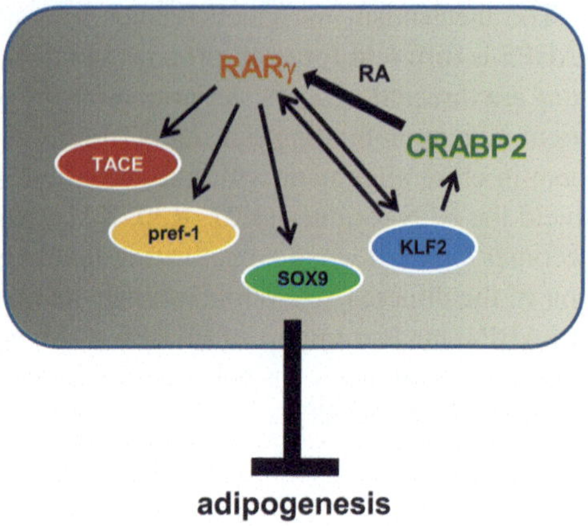

Fig. 7.3 Suppression of adipocyte differentiation by RA. In preadipocytes, RA signals through the CRABP2/ RARγ path to induce expression of the adipogenesis inhibitors Pref-1, TACE, Sox9, and KLF2. KLF2, RARγ and CRABP2 are placed in a positive feedback loop that further potentiates inhibition of adipocyte differentiation by RA

ing cascade initiated by Pref-1. Pref-1 is a plasma membrane protein that is cleaved by extracellular TNF-alpha-converting enzyme (TACE, also termed ADAM17) to produce an extracellular soluble protein that triggers ERK signaling. In turn, ERK upregulates expression of the transcription factor SOX9 which blocks adipocyte differentiation by suppressing the expression of the adipogenic factors C/EBPβ and C/EBPδ [84, 85]. The genes that encode Pref-1, TACE and SOX9 all contain RAR response elements in their promoters and they are directly targeted and upregulated by RARγ [86] (Fig. 7.3).

Another well-known adipogenesis inhibitor that is regulated by RAR is the transcription factor Kruppel-like factor 2 (KLF2). KLF2 inhibits adipocyte differentiation by suppressing the expression of PPARγ, the master regulator of adipogenesis, as well as the adipogenic factors C/EBPα and SREBP1c [87, 88]. Recent work demonstrated that RARγ directly activates KLF2 in preadipocytes [86]. Interestingly, the promoters of both RARγ and CRABP2 contain KLF2 response elements that positively regulate their expression so that when KLF2 is upregulated, it also results in the induction of CRABP2 and RARγ expression. Thus, the KLF2 and CRABP2/RAR pathways are part of a positive feedback loop that promotes RA signaling to inhibit adipogenesis [86] (Fig. 7.3).

Adipogenesis cannot proceed as long as the RA-activated CRABP2/RAR axis is active. Indeed, down-regulating CRABP2 and RARs, upregulating FABP5, and shifting RA signaling towards the FABP5/PPARβ/δ pathway [81, 89], is necessary to allow the formation of mature adipocytes. CRABP2 expression is suppressed upon induction of differentiation by the classical adipogenic hormonal mixture containing a glucocorticoid receptor (GR) ligand, insulin, and IBMX [90]. IBMX activates PKA, leading to activation of the transcription factor CREM, which in turn, binds to a cognate response element in the CRABP2 promoter to directly suppress CRABP2 expression. CRABP2 is also directly targeted by GR which represses its expression during adipogenesis. As a consequence, CRABP2 levels are markedly lower in mature adipocytes than in preadipose cells. CRABP2 expression in mature adipocytes is maintained at a low level by direct repression by the adipogenic factor C/EBPα [86].

The marked increase in the FABP5/CRABP2 ratio that accompanies adipogenesis allows RA to function through both RAR and PPARβ/δ in mature adipocytes. In these cells, PPARβ/δ plays key roles in regulating energy balance. It has been reported that activation of this receptor facilitates lipid catabolism and thus protects animals from excess weight gain and, consequently, improves insulin sensitivity [24, 25, 91–94]. By activating PPARβ/δ, RA induces the expression of PPARβ/δ target genes involved in "lipid burning", including genes that encode lipases, mitochondrial uncoupling proteins, and enzymes that mediate fatty acid oxidation [89]. RA also induces the expression of PPARβ/δ target genes involved in insulin responses. These include the kinase PDPK-1 and the insulin-responsive glucose transporter GLUT4. Accordingly, treatment of mice with RA raises body temperature, decreases body weight, and reduces blood levels of triglycerides and insulin in obese mice [89], results that are similar to those produced by treatment with synthetic selective PPARβ/δ ligands [25, 91, 92, 95]), Importantly, RA was shown to be more effective than a synthetic PPARβ/δ-selective ligand in inducing weight loss and in reducing insulin and triglyceride levels [89].

The protective activities of RA against adiposity and insulin resistance stem from its ability to activate both RAR and PPARβ/δ (Fig. 7.4). Uncoupling protein 1 (UCP1), which mediates energy dissipation, and hormone sensitive lipase (HSL), which catalyzes lipid hydrolysis, are activated by both receptors [96–98]. RA regulates adipose tissue biology and protects animals form accumulation of excess weight and development of insulin resistance by two distinct activities: in preadipocytes, it activates CRABP2 and RAR to inhibit formation of fat cells, while in

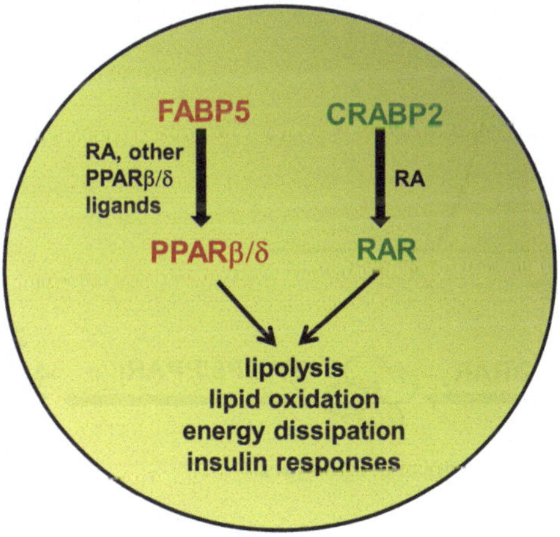

Fig. 7.4 RA prevents adipocyte hypertrophy by activating both of its receptors in adipocytes. Adipocyte differentiation is accompanied by downregulation of CRABP2 and RAR and upregulation of FABP5 and PPARβ/δ, enabling RA to activate both receptors in mature adipocytes. Both pathways contribute to the ability of RA to induce the expression of genes that enhance energy expenditure and that promote insulin responses

mature adipocytes, it activates both the CRABP2/RAR and the FABP5/PPARβ/δ paths to promote lipid oxidation and insulin responsiveness.

The Dual Transcriptional Activity of RA in Neuronal Differentiation

RA plays important roles in central nervous system development [99] and has been used to study neurogenesis in cultured embryonic stem cells such as P19 embryonal carcinoma cells (e. g. [100, 101]). The process entails two major steps. In the first, RA-treated stem cells differentiate into neuronal progenitors, typified by expression of specific markers such as nestin. Subsequently, these progenitor cells differentiate into mature neurons and glial cells. It was recently reported that RA-induced neurogenesis critically relies on temporal shifts in RA signaling, allowing the CRABP2/RAR and the FABP5/PPARβ/δ paths to be differentially employed across the differentiation process ([102], Fig. 7.5). Specifically, it was shown that induction of the differentiation of P19 cells into neuronal progenitors is driven by CRABP2 and RAR, but the final stage of neurogenesis, entailing generation of mature neurons, relies on transcriptional signaling by FABP5 and PPARβ/δ. Moreover, while the early stage of neuronal differentiation requires activation of RAR, the FABP5/

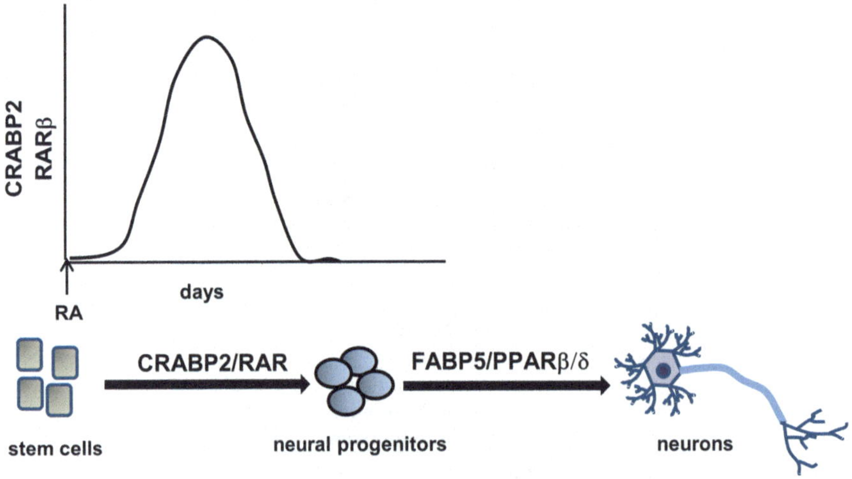

Fig. 7.5 Induction of neuronal differentiation by RA. RA promotes differentiation of stem cells into neural progenitor cells by activating the CRABP2/RAR pathway. The alternative FABP5/PPARβ/δ path can interfere with this step. The inhibition is avoided by a RA-induced dramatic increase in the levels of CRABP2 and RARβ in early neurogenesis, shiftin RA signaling to RAR. The increase is transient and the levels of CRABP2 and RARβ subsequently decrease, enabling RA to activate PPARβ/δ in late stage of neuronal differentiation. At this stage, the FABP5/PPARβ/δ pathway promotes the differentiation of progenitor neural cells into mature neurons

PPARβ/δ path suppresses the commitment of stem cells to the neuronal lineage. The inhibitory activity of PPARβ/δ at this stage is mediated by its target genes Ajuba and SIRT1, which, in turn, suppress the transcriptional activity of RAR [103, 104]. To enable differentiation at this stage, RA signaling is shifted towards RAR by a dramatic upregulation of RARβ and CRABP2 which rapidly occurs following RA treatment. The increase in the CRABP2/FABP5 ratio and in expression of RARβ peaks 3–4 days following induction of differentiation and subsequently decreases, enabling RA to activate PPARβ/δ and thus, to support generation of mature neurons (Fig. 7.5). This activity is mediated at least in part by induction of the kinase PDPK1, a direct PPARβ/δ target gene involved in neuronal differentiation [105, 106].

In agreement with the conclusion that FABP5/PPARβ/δ is necessary for differentiation of neuronal progenitors to mature neurons, hippocampi of FABP-null mice contain excess neuronal progenitor cells but fewer mature neurons than their WT counterparts [102]. Using reporter mice in which luciferase expression is driven by a PPAR response element, it was demonstrated that ablation of FABP5 almost completely abolishes PPARβ/δ activities in the brain [107]. Expression of PPARβ/δ target genes in these FABP null mice, including cognition-associated genes, was found to be markedly reduced. Moreover, a behavioral phenotyping screen revealed that the mice display pronounced impairment in hippocampal-based learning and memory [107]. Hence, the FABP5/PPARβ/δ pathway plays key roles in regulating cognitive function.

The Dual Transcriptional Activity of RA in Cancer

RA displays distinct anticarcinogenic activities and is currently used in or is being tested as a therapeutic agent in several human cancers, most notably acute promyelocytic leukemia [9, 10, 108]. The growth-inhibitory activities of the hormone are mediated primarily by RARs which, when activated, can induce expression of genes that promote cell cycle arrest, apoptosis, and differentiation [109–119]. In contrast, other cells are RA-resistant or, paradoxically, respond to RA treatment by enhanced proliferation and survival. For example, RA functions as an anti-apoptotic agent in neurons [120–124], keratinocytes [125], capillary endothelial cells [126], hematopoietic cells [127, 128], cardiomyocytes [129], retinal progenitor cells [130], and lung epithelial cells [131]. RA has also been shown to induce hyperproliferation of basal keratinocytes [132, 133] and to enhance mammary tumor growth in mouse model of breast cancer [12, 31]. The ability of RA to enhance proliferation and suppress apoptosis is unlikely to be mediated by RAR, whose target genes are often involved in inhibition of cell growth. Instead, the pro-proliferative actions of RA can be traced to activation of an alternative transcriptional pathway, one that is mediated by PPARβ/δ. Indeed, direct targets for PPARβ/δ include pro-oncogenic genes that promote proliferation, survival, and angiogenesis [18, 21, 26–36].

Alternate activation of its two receptors allows RA to exert opposing effects on cell growth. RA inhibits growth in cells where there is high expression of CRABP2 which leads to activation of RAR, but promotes survival and proliferation in cells where there is high expression of FABP5 that results in activation of PPARβ/δ [31, 32]. Dysregulation of the FABP5/CRABP2 ratio, often originating from marked upregulation of FABP5, occurs in many human cancers, including glioblastoma, astrocytic gliomas, oral and head-and-neck squamous cell carcinomas, and pancreatic, bladder, prostate, breast, skin, and esophageal cancers ([34, 134–145] and the Cancer Genome Atlas, (https://tcga-data.nci.nih.gov/tcga/). Accordingly, expression of FABP5 is high in the triple-negative MDA-MB231 mammary carcinoma cells and in the aggressive PC3M prostate cancer cells, both of which are profoundly RA resistant. It has been demonstrated that a cancer-associated switch in RA signaling drives tumor development in the RA-resistant MMTV-*neu* mouse model of breast cancer. In this model, similarly to a significant fraction of human breast tumors, the epidermal growth factor receptor (EGFR) HER2/ErbB2/neu is over-expressed in mammary tissue, resulting in mammary tumor development in 100 % of female mice [146]. Tumor development in these mice is accompanied both by upregulation of FABP5 and by downregulation of CRABP2, resulting in an aberrantly high FABP5/CRABP2 ratio [32]. Due to the high intra-tumor FABP5/CRABP2 expression ratio, treatment of MMTV-*neu* mice with RA leads to activation of PPARβ/δ but not RAR, and promotes tumor development [32]. Moreover, reducing the FABP5/CRABP2 in these mice, either by ablation of FABP5 [34] or by mammary over-expression of CRABP2, shifts RA signaling from PPARβ/δ to RAR and markedly inhibits tumorigenesis [12, 32].

How expression of FABP5 and CRABP2 is regulated and how molecular events alter their levels during tumorigenesis are not understood. One clue was provided by the observation that expression of FABP5 in mammary carcinomas is directly induced by EGFR functioning through activation of NFkB [35]. This observation suggests that this binding protein plays a critical role in mediating the ability of EGFR to promote tumorigenesis. In support of this notion, it was reported that reducing the level of FABP5 in mammary carcinoma cells completely abolishes EGFR-induced proliferation of mammary carcinoma cells [35].

It may be worth noting that the role of PPARβ/δ in cancer has been controversial, with some groups suggesting that the receptor exerts pro-carcinogenic activities and others proposing that it displays anti-proliferative functions. The debate emanated from conflicting reports on tumorigenesis-associated alterations in the expression of PPARβ/δ in various cancers and from discrepancies in reported effects of PPARβ/δ agonists on the growth of different carcinomas [147]. The discovery that transcriptional activation by PPARβ/δ critically relies on expression of FABP5 suggests a different focus. Indeed, expression of FABP5 appears to be a better predictor than expression of the receptor itself for involvement of PPARβ/δ in cancer cell biology as well as for the ability of agonists to activate the receptor. Taken together, available information demonstrates that, in some cancers, CRABP2 functions as a tumor suppressor while FABP5 is highly pro-oncogenic. These binding proteins, and especially FABP5, may thus comprise novel targets for therapy of RA-resistant cancers.

Relevance

Obesity stems from excess lipid accumulation leading to adipocyte hypertrophy, and from generation of new adipocytes through adipogenesis. The observations that RA both inhibits adipocyte differentiation and promotes lipid oxidation to deplete lipid stores provide a rationale for the long-noted, but poorly understood function of vitamin A in energy balance regulation. Available data suggest that, due to its ability to activate both RAR and PPARβ/δ, RA is a uniquely effective agent in overcoming obesity and its associated pathologies. In support of this notion, it was demonstrated that treatment with RA protects mice from high fat feeding-induced weight gain and hepatic steatosis, and markedly improves glucose tolerance.

RA and other PPARβ/δ ligands can promote proliferation and survival in cancer cells where, due to a high level of FABP5, PPARβ/δ is efficiently activated. The pro-oncogenic activities of FABP5 are further emphasized by the observation that this protein is markedly upregulated in multiple types of human cancers. Hence FABP5 is a promising therapeutic drug target for some cancers. Inhibitors of this protein are expected to suppress tumorigenesis by blocking the activation of PPARβ/δ and by diverting RA to RAR, thereby inducing the expression of anti-proliferative RAR target genes. Such inhibitors are expected to comprise a novel class of anticarcinogenic agents.

The observations that FABP5 plays a key role in promoting hippocampal-associated learning and memory raises the possibility that upregulation of this protein in the brain may comprise a new strategy for treatment of some cognitive disorders.

Future Directions

The discovery that RA regulates gene transcription not only through classical RAR type nuclear receptors but also by activating an alternative PPARβ/δ receptor markedly broadened our understanding of the biological activities of vitamin A. The spectrum of genes regulated by RA through the CRABP2/RAR and the FABP5/PPARβ/δ paths and the biological activities controlled by these paths in different tissues remains to be elucidated.

The mechanisms that control the expression of CRABP2 and FABP5 and that determine the activation status of the RA receptors in different cells are poorly understood. Similarly, the molecular basis for the upregulation of FABP5 that accompanies the development of many human cancers remains to be clarified.

The broad ligand-binding characteristic of PPARβ/δ has long hampered identification of physiologically-relevant ligands for the receptor. The observation that activity of PPARβ/δ is critically supported by FABP5 provides an additional criterion for identifying *bona fide* endogenous PPARβ/δ ligands. Such ligands are expected to not only activate the receptor, but also to trigger nuclear localization of FABP5.

While CRABP2 specifically binds RA, FABP5 can associate with many ligands with a similar affinity. Some of these ligands induce FABP5 to move to the nucleus and function as PPARβ/δ agonists. However, other compounds that bind to FABP5 do not activate nuclear import and do not serve as PPARβ/δ ligands. The functional significance of the association of FABP5 with "non-activating" ligands remains to be explored.

Acknowledgments Work in the author's laboratory was supported by NIH grants RO1 DK060684, RO1 DK088669, and R01 CA166955

References

1. Germain P, Chambon P, Eichele G, Evans RM, Lazar MA, Leid M, De Lera AR, Lotan R, Mangelsdorf DJ, Gronemeyer H (2006) International union of pharmacology. Lx. Retinoic acid receptors. Pharmacol Rev 58:712–725
2. Shaw N, Elholm M, Noy N (2003) Retinoic acid is a high affinity selective ligand for the peroxisome proliferator-activated receptor beta/delta. J Biol Chem 278:41589–41592
3. Berry DC, Noy N (2007) Is pparbeta/delta a retinoid receptor? PPAR Res 2007:73256
4. Dong D, Ruuska SE, Levinthal DJ, Noy N (1999) Distinct roles for cellular retinoic acid-binding proteins i and ii in regulating signaling by retinoic acid. J Biol Chem 274:23695–23698
5. Sessler RJ, Noy N (2005) A ligand-activated nuclear localization signal in cellular retinoic acid binding protein-ii. Mol Cell 18:343–353
6. Tan NS, Shaw NS, Vinckenbosch N, Liu P, Yasmin R, Desvergne B, Wahli W, Noy N (2002) Selective cooperation between fatty acid binding proteins and peroxisome proliferator-activated receptors in regulating transcription. Mol Cell Biol 22:5114–5127
7. Wolbach SB, Howe PR (1978) Nutrition classics. J Exp Med 42:753–777. Tissue changes following deprivation of fat-soluble a vitamin. S. Burt wolbach and percy r. Howe. Nutrition reviews 1978;36:16–19
8. Chapellier B, Mark M, Messaddeq N, Calleja C, Warot X, Brocard J, Gerard C, Li M, Metzger D, Ghyselinck NB, Chambon P (2002) Physiological and retinoid-induced proliferations of epidermis basal keratinocytes are differently controlled. EMBO J 21:3402–3413
9. Soprano DR, Qin P, Soprano KJ (2004) Retinoic acid receptors and cancers. Annu Rev Nutr 24:201–221
10. Altucci L, Gronemeyer H (2001) The promise of retinoids to fight against cancer. Nat Rev Cancer 1:181–193
11. Verma AK, Conrad EA, Boutwell RK (1982) Differential effects of retinoic acid and 7,8-benzoflavone on the induction of mouse skin tumors by the complete carcinogenesis process and by the initiation-promotion regimen. Cancer Res 42:3519–3525
12. Manor D, Shmidt EN, Budhu A, Flesken-Nikitin A, Zgola M, Page R, Nikitin AY, Noy N (2003) Mammary carcinoma suppression by cellular retinoic acid binding protein-ii. Cancer Res 63:4426–4433
13. Mangelsdorf DJ, Thummel C, Beato M, Herrlich P, Schutz G, Umesono K, Blumberg B, Kastner P, Mark M, Chambon P, Evans RM (1995) The nuclear receptor superfamily: the second decade. Cell 83:835–839
14. Bocher V, Pineda-Torra I, Fruchart JC, Staels B (2002) Ppars: Transcription factors controlling lipid and lipoprotein metabolism. Ann N Y Acad Sci 967:7–18
15. Peters JM, Lee SS, Li W, Ward JM, Gavrilova O, Everett C, Reitman ML, Hudson LD, Gonzalez FJ (2000) Growth, adipose, brain, and skin alterations resulting from targeted dis-

ruption of the mouse peroxisome proliferator-activated receptor beta(delta). Mol Cell Biol 20:5119–5128

16. Cimini A, Bernardo A, Cifone MG, Di Marzio L, Di Loreto S (2003) Tnfalpha downregulates ppardelta expression in oligodendrocyte progenitor cells: implications for demyelinating diseases. Glia 41:3–14
17. Tan NS, Michalik L, Desvergne B, Wahli W (2004) Peroxisome proliferator-activated receptor-beta as a target for wound healing drugs. Expert Opin Ther Targets 8:39–48
18. Tan NS, Michalik L, Noy N, Yasmin R, Pacot C, Heim M, Fluhmann B, Desvergne B, Wahli W (2001) Critical roles of ppar beta/delta in keratinocyte response to inflammation. Genes Dev 15:3263–3277
19. Michalik L, Wahli W (2007) Peroxisome proliferator-activated receptors (ppars) in skin health, repair and disease. Biochim Biophys Acta 1771:991–998
20. Icre G, Wahli W, Michalik L (2006) Functions of the peroxisome proliferator-activated receptor (ppar) alpha and beta in skin homeostasis, epithelial repair, and morphogenesis. J Investig dermatol 126(Suppl):30–35
21. Di-Poi N, Tan NS, Michalik L, Wahli W, Desvergne B (2002) Antiapoptotic role of pparbeta in keratinocytes via transcriptional control of the akt1 signaling pathway. Mol Cell 10:721–733
22. Braissant O, Foufelle F, Scotto C, Dauca M, Wahli W (1996) Differential expression of peroxisome proliferator-activated receptors (ppars): tissue distribution of ppar-alpha, -beta, and -gamma in the adult rat. Endocrinology 137:354–366
23. Lemberger T, Braissant O, Juge-Aubry C, Keller H, Saladin R, Staels B, Auwerx J, Burger AG, Meier CA, Wahli W (1996) Ppar tissue distribution and interactions with other hormone-signaling pathways. Ann N Y Acad Sci 804:231–251
24. Wang YX, Zhang CL, Yu RT, Cho HK, Nelson MC, Bayuga-Ocampo CR, Ham J, Kang H, Evans RM (2004) Regulation of muscle fiber type and running endurance by ppardelta. PLoS Biol 2:e294
25. Wang YX, Lee CH, Tiep S, Yu RT, Ham J, Kang H, Evans RM (2003) Peroxisome-proliferator-activated receptor delta activates fat metabolism to prevent obesity. Cell 113:159–170
26. Wang D, Wang H, Guo Y, Ning W, Katkuri S, Wahli W, Desvergne B, Dey SK, DuBois RN (2006) Crosstalk between peroxisome proliferator-activated receptor delta and vegf stimulates cancer progression. Proc Natl Acad Sci U S A 103:19069–19074
27. Di-Poi N, Michalik L, Tan NS, Desvergne B, Wahli W (2003) The anti-apoptotic role of pparbeta contributes to efficient skin wound healing. J Steroid Biochem Mol Biol 85:257–265
28. Wagner KD, Benchetrit M, Bianchini L, Michiels JF, Wagner N (2011) Peroxisome proliferator-activated receptor beta/delta (pparbeta/delta) is highly expressed in liposarcoma and promotes migration and proliferation. J Pathol 224:575–588
29. Tan NS, Icre G, Montagner A, Bordier-ten-Heggeler B, Wahli W, Michalik L (2007) The nuclear hormone receptor peroxisome proliferator-activated receptor beta/delta potentiates cell chemotactism, polarization, and migration. Mol Cell Biol 27:7161–7175
30. Adhikary T, Brandt DT, Kaddatz K, Stockert J, Naruhn S, Meissner W, Finkernagel F, Obert J, Lieber S, Scharfe M, Jarek M, Toth PM, Scheer F, Diederich WE, Reinartz S, Grosse R, Muller-Brusselbach S, Muller R (2013) Inverse pparbeta/delta agonists suppress oncogenic signaling to the angptl4 gene and inhibit cancer cell invasion. Oncogene 32:5241–5252
31. Schug TT, Berry DC, Shaw NS, Travis SN, Noy N (2007) Opposing effects of retinoic acid on cell growth result from alternate activation of two different nuclear receptors. Cell 129:723–733
32. Schug TT, Berry DC, Toshkov IA, Cheng L, Nikitin AY, Noy N (2008) Overcoming retinoic acid-resistance of mammary carcinomas by diverting retinoic acid from pparbeta/delta to rar. Proc Natl Acad Sci U S A 105:7546–7551
33. Morgan E, Kannan-Thulasiraman P, Noy N (2010) Involvement of fatty acid binding protein 5 and pparbeta/delta in prostate cancer cell growth. PPAR Res 2010:9

34. Levi L, Lobo G, Doud MK, von Lintig J, Seachrist D, Tochtrop GP, Noy N (2013) Genetic ablation of the fatty acid binding protein fabp5 suppresses her2-induced mammary tumorigenesis. Cancer Res 73:4770–4780
35. Kannan-Thulasiraman P, Seachrist DD, Mahabeleshwar GH, Jain MK, Noy N (2010) Fatty acid-binding protein 5 and pparbeta/delta are critical mediators of epidermal growth factor receptor-induced carcinoma cell growth. J Biol Chem 285:19106–19115
36. Montagner A, Delgado MB, Tallichet-Blanc C, Chan JS, Sng MK, Mottaz H, Degueurce G, Lippi Y, Moret C, Baruchet M, Antsiferova M, Werner S, Hohl D, Saati TA, Farmer PJ, Tan NS, Michalik L, Wahli W (2014) Src is activated by the nuclear receptor peroxisome proliferator-activated receptor beta/delta in ultraviolet radiation-induced skin cancer. EMBO Mol Med 6:80–98
37. Fauconnet S, Lascombe I, Chabannes E, Adessi GL, Desvergne B, Wahli W, Bittard H (2002) Differential regulation of vascular endothelial growth factor expression by peroxisome proliferator-activated receptors in bladder cancer cells. J Biol Chem 277:23534–23543
38. Tan MJ, Teo Z, Sng MK, Zhu P, Tan NS (2012) Emerging roles of angiopoietin-like 4 in human cancer. Mol Can Res: MCR 10:677–688
39. Nolte RT, Wisely GB, Westin S, Cobb JE, Lambert MH, Kurokawa R, Rosenfeld MG, Willson TM, Glass CK, Milburn MV (1998) Ligand binding and co-activator assembly of the peroxisome proliferator-activated receptor-gamma. Nature 395:137–143
40. Xu HE, Lambert MH, Montana VG, Plunket KD, Moore LB, Collins JL, Oplinger JA, Kliewer SA, Gampe RT Jr, McKee DD, Moore JT, Willson TM (2001) Structural determinants of ligand binding selectivity between the peroxisome proliferator-activated receptors. Proc Natl Acad Sci U S A 98:13919–13924
41. Xu HE, Lambert MH, Montana VG, Parks DJ, Blanchard SG, Brown PJ, Sternbach DD, Lehmann JM, Wisely GB, Willson TM, Kliewer SA, Milburn MV (1999) Molecular recognition of fatty acids by peroxisome proliferator- activated receptors. Mol Cell 3:397–403
42. Fyffe SA, Alphey MS, Buetow L, Smith TK, Ferguson MA, Sorensen MD, Bjorkling F, Hunter WN (2006) Recombinant human ppar-beta/delta ligand-binding domain is locked in an activated conformation by endogenous fatty acids. J Mol Biol 356:1005–1013
43. Kersten S, Desvergne B, Wahli W (2000) Roles of ppars in health and disease. Nature 405:421–424
44. Forman BM, Chen J, Evans RM (1997) Hypolipidemic drugs, polyunsaturated fatty acids, and eicosanoids are ligands for peroxisome proliferator-activated receptors alpha and delta. Proc Natl Acad Sci U S A 94:4312–4317
45. Desvergne B, Wahli W (1999) Peroxisome proliferator-activated receptors: nuclear control of metabolism. Endocr Rev 20:649–688
46. Schmuth M, Haqq CM, Cairns WJ, Holder JC, Dorsam S, Chang S, Lau P, Fowler AJ, Chuang G, Moser AH, Brown BE, Mao-Qiang M, Uchida Y, Schoonjans K, Auwerx J, Chambon P, Willson TM, Elias PM, Feingold KR (2004) Peroxisome proliferator-activated receptor (ppar)-beta/delta stimulates differentiation and lipid accumulation in keratinocytes. J Investig Dermatol 122:971–983
47. Kersten S, Mandard S, Tan NS, Escher P, Metzger D, Chambon P, Gonzalez FJ, Desvergne B, Wahli W (2000) Characterization of the fasting-induced adipose factor fiaf, a novel peroxisome proliferator-activated receptor target gene. J Biol Chem 275:28488–28493
48. Noy N (2000) Retinoid-binding proteins: mediators of retinoid action. Biochem J 348(Pt 3):481–495
49. Storch J, Corsico B (2008) The emerging functions and mechanisms of mammalian fatty acid-binding proteins. Annu Rev Nutr 28:73–95
50. Smathers RL, Petersen DR (2011) The human fatty acid-binding protein family: evolutionary divergences and functions. Hum Genomics 5:170–191
51. Hostetler HA, McIntosh AL, Atshaves BP, Storey SM, Payne HR, Kier AB, Schroeder F (2009) L-fabp directly interacts with pparalpha in cultured primary hepatocytes. J Lipid Res 50:1663–1675

52. Ayers SD, Nedrow KL, Gillilan RE, Noy N (2007) Continuous nucleocytoplasmic shuttling underlies transcriptional activation of ppargamma by fabp4. Biochemistry 46:6744–6752
53. Budhu A, Gillilan R, Noy N (2001) Localization of the rar interaction domain of cellular retinoic acid binding protein-ii. J Mol Biol 305:939–949
54. Budhu AS, Noy N (2002) Direct channeling of retinoic acid between cellular retinoic acid-binding protein ii and retinoic acid receptor sensitizes mammary carcinoma cells to retinoic acid-induced growth arrest. Mol Cell Biol 22:2632–2641
55. Marcelino AM, Smock RG, Gierasch LM (2006) Evolutionary coupling of structural and functional sequence information in the intracellular lipid-binding protein family. Proteins 63:373–384
56. Coe NR, Simpson MA, Bernlohr DA (1999) Targeted disruption of the adipocyte lipid-binding protein (ap2 protein) gene impairs fat cell lipolysis and increases cellular fatty acid levels. J Lipid Res 40:967–972
57. Gillilan RE, Ayers SD, Noy N (2007) Structural basis for activation of fatty acid-binding protein 4. J Mol Biol 372:1246–1260
58. Armstrong EH, Goswami D, Griffin PR, Noy N, Ortlund EA (2014) Structural basis for ligand regulation of the fatty acid binding protein 5, peroxisome proliferator activated receptor beta/delta (fabp5-pparbeta/delta) signaling pathway. J Biol Chem 289:14941–14954
59. Chook YM, Blobel G (2001) Karyopherins and nuclear import. Curr Opin Struct Biol 11:703–715
60. Kalderon D, Roberts BL, Richardson WD, Smith AE (1984) A short amino acid sequence able to specify nuclear location. Cell 39:499–509
61. Robbins J, Dilworth SM, Laskey RA, Dingwall C (1991) Two interdependent basic domains in nucleoplasmin nuclear targeting sequence: identification of a class of bipartite nuclear targeting sequence. Cell 64:615–623
62. Hodel MR, Corbett AH, Hodel AE (2001) Dissection of a nuclear localization signal. J Biol Chem 276:1317–1325
63. Dingwall C, Laskey RA (1991) Nuclear targeting sequences--a consensus? Trends Biochem Sci 16:478–481
64. Chmurzynska A (2006) The multigene family of fatty acid-binding proteins (fabps): function, structure and polymorphism. J Appl Genet 47:39–48
65. Green H, Meuth M (1974) An established pre-adipose cell line and its differentiation in culture. Cell 3:127–133
66. Green H, Kehinde O (1975) An established preadipose cell line and its differentiation in culture Ii. Factors affecting the adipose conversion. Cell 5:19–27
67. Rosen ED, Spiegelman BM (2000) Molecular regulation of adipogenesis. Annu Rev Cell Dev Biol 16:145–171
68. Rosen ED, Walkey CJ, Puigserver P, Spiegelman BM (2000) Transcriptional regulation of adipogenesis. Genes Dev 14:1293–1307
69. Smas CM, Chen L, Zhao L, Latasa MJ, Sul HS (1999) Transcriptional repression of pref-1 by glucocorticoids promotes 3 t3-l1 adipocyte differentiation. J Biol Chem 274:12632–12641
70. Shi XM, Blair HC, Yang X, McDonald JM, Cao X (2000) Tandem repeat of c/ebp binding sites mediates ppargamma2 gene transcription in glucocorticoid-induced adipocyte differentiation. J Cell Biochem 76:518–527
71. Pantoja C, Huff JT, Yamamoto KR (2008) Glucocorticoid signaling defines a novel commitment state during adipogenesis in vitro. Mol Biol Cell 19:4032–4041
72. Zhang JW, Klemm DJ, Vinson C, Lane MD (2004) Role of creb in transcriptional regulation of ccaat/enhancer-binding protein beta gene during adipogenesis. J Biol Chem 279:4471–4478
73. Petersen RK, Madsen L, Pedersen LM, Hallenborg P, Hagland H, Viste K, Doskeland SO, Kristiansen K (2008) Cyclic amp (camp)-mediated stimulation of adipocyte differentiation requires the synergistic action of epac- and camp-dependent protein kinase-dependent processes. Mol Cell Biol 28:3804–3816

74. Reusch JE, Colton LA, Klemm DJ (2000) Creb activation induces adipogenesis in 3 t3-l1 cells. Mol Cell Biol 20:1008–1020
75. Foulkes NS, Borrelli E, Sassone-Corsi P (1991) Crem gene: Use of alternative DNA-binding domains generates multiple antagonists of camp-induced transcription. Cell 64:739–749
76. Don J, Stelzer G (2002) The expanding family of creb/crem transcription factors that are involved with spermatogenesis. Mol Cell Endocrinol 187:115–124
77. Sassone-Corsi P (1995) Transcription factors responsive to camp. Annu Rev Cell Dev Biol 11:355–377
78. Murray T, Russell TR (1980) Inhibition of adipose conversion in 3 t3-l2 cells by retinoic acid. J Supramol Struct 14:255–266
79. Sato M, Hiragun A, Mitsui H (1980) Preadipocytes possess cellular retinoid binding proteins and their differentiation is inhibited by retinoids. Biochem Biophys Res Commun 95:1839–1845
80. Kuri-Harcuch W (1982) Differentiation of 3 t3-f442a cells into adipocytes is inhibited by retinoic acid. Differentiation 23:164–169
81. Schwarz EJ, Reginato MJ, Shao D, Krakow SL, Lazar MA (1997) Retinoic acid blocks adipogenesis by inhibiting c/ebpbeta-mediated transcription. Mol Cell Biol 17:1552–1561
82. Xue JC, Schwarz EJ, Chawla A, Lazar MA (1996) Distinct stages in adipogenesis revealed by retinoid inhibition of differentiation after induction of ppargamma. Mol Cell Biol 16:1567–1575
83. Marchildon F, St-Louis C, Akter R, Roodman V, Wiper-Bergeron NL (2010) Transcription factor smad3 is required for the inhibition of adipogenesis by retinoic acid. J Biol Chem 285:13274–13284
84. Sul HS (2009) Minireview: Pref-1: Role in adipogenesis and mesenchymal cell fate. Mol Endocrinol 23:1717–1725
85. Wang Y, Zhao L, Smas C, Sul HS (2010) Pref-1 interacts with fibronectin to inhibit adipocyte differentiation. Mol Cell Biol 30:3480–3492
86. Berry DC, DeSantis D, Soltanian H, Croniger CM, Noy N (2012) Retinoic acid upregulates preadipocyte genes to block adipogenesis and suppress diet-induced obesity. Diabetes 61:1112–1121
87. Banerjee SS, Feinberg MW, Watanabe M, Gray S, Haspel RL, Denkinger DJ, Kawahara R, Hauner H, Jain MK (2003) The kruppel-like factor klf2 inhibits peroxisome proliferator-activated receptor-gamma expression and adipogenesis. J Biol Chem 278:2581–2584
88. Wu J, Srinivasan SV, Neumann JC, Lingrel JB (2005) The klf2 transcription factor does not affect the formation of preadipocytes but inhibits their differentiation into adipocytes. Biochemistry 44:11098–11105
89. Berry DC, Noy N (2009) All-trans-retinoic acid represses obesity and insulin resistance by activating both peroxisome proliferation-activated receptor beta/delta and retinoic acid receptor. Mol Cell Biol 29:3286–3296
90. Rangwala SM, Lazar MA (2000) Transcriptional control of adipogenesis. Annu Rev Nutr 20:535–559
91. Barish GD, Narkar VA, Evans RM (2006) Ppar delta: a dagger in the heart of the metabolic syndrome. J Clin Invest 116:590–597
92. Lee CH, Olson P, Hevener A, Mehl I, Chong LW, Olefsky JM, Gonzalez FJ, Ham J, Kang H, Peters JM, Evans RM (2006) Ppardelta regulates glucose metabolism and insulin sensitivity. Proc Natl Acad Sci U S A 103:3444–3449
93. Burkart EM, Sambandam N, Han X, Gross RW, Courtois M, Gierasch CM, Shoghi K, Welch MJ, Kelly DP (2007) Nuclear receptors pparbeta/delta and pparalpha direct distinct metabolic regulatory programs in the mouse heart. J Clin Invest 117:3930–3939
94. Narkar VA, Downes M, Yu RT, Embler E, Wang YX, Banayo E, Mihaylova MM, Nelson MC, Zou Y, Juguilon H, Kang H, Shaw RJ, Evans RM (2008) Ampk and ppardelta agonists are exercise mimetics. Cell 134:405–415
95. Evans RM, Barish GD, Wang YX (2004) Ppars and the complex journey to obesity. Nat Med 10:355–361

96. Gervois P, Torra IP, Fruchart JC, Staels B (2000) Regulation of lipid and lipoprotein metabolism by ppar activators. Clin Chem Lab Med 38:3–11
97. Rabelo R, Reyes C, Schifman A, Silva JE (1996) A complex retinoic acid response element in the uncoupling protein gene defines a novel role for retinoids in thermogenesis. Endocrinology 137:3488–3496
98. Rottman JN, Widom RL, Nadal-Ginard B, Mahdavi V, Karathanasis SK (1991) A retinoic acid-responsive element in the apolipoprotein ai gene distinguishes between two different retinoic acid response pathways. Mol Cell Biol 11:3814–3820
99. Maden M (2007) Retinoic acid in the development, regeneration and maintenance of the nervous system. Nat Rev Neurosci 8:755–765
100. Jones-Villeneuve EM, McBurney MW, Rogers KA, Kalnins VI (1982) Retinoic acid induces embryonal carcinoma cells to differentiate into neurons and glial cells. J Cell Biol 94:253–262
101. McBurney MW, Jones-Villeneuve EM, Edwards MK, Anderson PJ (1982) Control of muscle and neuronal differentiation in a cultured embryonal carcinoma cell line. Nature 299:165–167
102. Yu S, Levi L, Siegel R, Noy N (2012) Retinoic acid induces neurogenesis by activating both retinoic acid receptors (rars) and peroxisome proliferator-activated receptor beta/delta (ppar-beta/delta). J Biol Chem 287:42195–42205
103. Kang MR, Lee SW, Um E, Kang HT, Hwang ES, Kim EJ, Um SJ (2010) Reciprocal roles of sirt1 and skip in the regulation of rar activity: Implication in the retinoic acid-induced neuronal differentiation of p19 cells. Nucleic Acids Res 38:822–831
104. Goyal RK, Lin P, Kanungo J, Payne AS, Muslin AJ, Longmore GD (1999) Ajuba, a novel lim protein, interacts with grb2, augments mitogen-activated protein kinase activity in fibroblasts, and promotes meiotic maturation of xenopus oocytes in a grb2- and ras-dependent manner. Mol Cell Biol 19:4379–4389
105. Oishi K, Watatani K, Itoh Y, Okano H, Guillemot F, Nakajima K, Gotoh Y (2009) Selective induction of neocortical gabaergic neurons by the pdk1-akt pathway through activation of mash1. Proc Natl Acad Sci U S A 106:13064–13069
106. Fishwick KJ, Li RA, Halley P, Deng P, Storey KG (2010) Initiation of neuronal differentiation requires pi3-kinase/tor signalling in the vertebrate neural tube. Dev Biol 338:215–225
107. Yu S, Levi L, Casadesus G, Kunos G, Noy N (2014) Fatty acid-binding protein 5 (fabp5) regulates cognitive function both by decreasing anandamide levels and by activating the nuclear receptor peroxisome proliferator-activated receptor beta/delta (pparbeta/delta) in the brain. J Biol Chem 289:12748–12758
108. Dos Santos GA, Kats L, Pandolfi PP (2013) Synergy against pml-rara: targeting transcription, proteolysis, differentiation, and self-renewal in acute promyelocytic leukemia. J Exp Med 210:2793–2802
109. Strickland S, Mahdavi V (1978) The induction of differentiation in teratocarcinoma stem cells by retinoic acid. Cell 15:393–403
110. Altucci L, Rossin A, Raffelsberger W, Reitmair A, Chomienne C, Gronemeyer H (2001) Retinoic acid-induced apoptosis in leukemia cells is mediated by paracrine action of tumor-selective death ligand trail. Nat Med 7:680–686
111. Battle TE, Roberson MS, Zhang T, Varvayanis S, Yen A (2001) Retinoic acid-induced blr1 expression requires raralpha, rxr, and mapk activation and uses erk2 but not jnk/sapk to accelerate cell differentiation. Eur J Cell Biol 80:59–67
112. Breitman TR, Collins SJ, Keene BR (1981) Terminal differentiation of human promyelocytic leukemic cells in primary culture in response to retinoic acid. Blood 57:1000–1004
113. Elstner E, Muller C, Koshizuka K, Williamson EA, Park D, Asou H, Shintaku P, Said JW, Heber D, Koeffler HP (1998) Ligands for peroxisome proliferator-activated receptorgamma and retinoic acid receptor inhibit growth and induce apoptosis of human breast cancer cells in vitro and in bnx mice. Proc Natl Acad Sci U S A 95:8806–8811

114. Mangiarotti R, Danova M, Alberici R, Pellicciari C (1998) All-trans retinoic acid (atra)-induced apoptosis is preceded by g1 arrest in human mcf-7 breast cancer cells. Br J Cancer 77:186–191
115. Toma S, Isnardi L, Riccardi L, Bollag W (1998) Induction of apoptosis in mcf-7 breast carcinoma cell line by rar and rxr selective retinoids. Anticancer Res 18:935–942
116. Donato LJ, Noy N (2005) Suppression of mammary carcinoma growth by retinoic acid: pro-apoptotic genes are targets for retinoic acid receptor and cellular retinoic acid-binding protein ii signaling. Cancer Res 65:8193–8199
117. Donato LJ, Suh JH, Noy N (2007) Suppression of mammary carcinoma cell growth by retinoic acid: the cell cycle control gene btg2 is a direct target for retinoic acid receptor signaling. Cancer Res 67:609–615
118. Niizuma H, Nakamura Y, Ozaki T, Nakanishi H, Ohira M, Isogai E, Kageyama H, Imaizumi M, Nakagawara A (2006) Bcl-2 is a key regulator for the retinoic acid-induced apoptotic cell death in neuroblastoma. Oncogene 25:5046–5055
119. Mrass P, Rendl M, Mildner M, Gruber F, Lengauer B, Ballaun C, Eckhart L, Tschachler E (2004) Retinoic acid increases the expression of p53 and proapoptotic caspases and sensitizes keratinocytes to apoptosis: a possible explanation for tumor preventive action of retinoids. Cancer Res 64:6542–6548
120. Henion PD, Weston JA (1994) Retinoic acid selectively promotes the survival and proliferation of neurogenic precursors in cultured neural crest cell populations. Dev Biol 161:243–250
121. Rodriguez-Tebar A, Rohrer H (1991) Retinoic acid induces ngf-dependent survival response and high-affinity ngf receptors in immature chick sympathetic neurons. Development 112:813–820
122. Plum LA, Parada LF, Tsoulfas P, Clagett-Dame M (2001) Retinoic acid combined with neurotrophin-3 enhances the survival and neurite outgrowth of embryonic sympathetic neurons. Exp Biol Med (Maywood) 226:766–775
123. Jacobs S, Lie DC, DeCicco KL, Shi Y, DeLuca LM, Gage FH, Evans RM (2006) Retinoic acid is required early during adult neurogenesis in the dentate gyrus. Proc Natl Acad Sci U S A 103:3902–3907
124. Hornberg M, Gussing F, Berghard A, Bohm S (2009) Retinoic acid selectively inhibits death of basal vomeronasal neurons during late stage of neural circuit formation. J Neurochem 110:1263–1275
125. Sorg O, Tran C, Carraux P, Grand D, Hugin A, Didierjean L, Saurat JH (2005) Spectral properties of topical retinoids prevent DNA damage and apoptosis after acute uv-b exposure in hairless mice. Photochem Photobiol 81:830–836
126. Vorotnikova E, Tries M, Braunhut S (2004) Retinoids and timp1 prevent radiation-induced apoptosis of capillary endothelial cells. Radiat Res 161:174–184
127. Oritani K, Kaisho T, Nakajima K, Hirano T (1992) Retinoic acid inhibits interleukin-6-induced macrophage differentiation and apoptosis in a murine hematopoietic cell line, y6. Blood 80:2298–2305
128. Iwata M, Mukai M, Nakai Y, Iseki R (1992) Retinoic acids inhibit activation-induced apoptosis in t cell hybridomas and thymocytes. J Immunol 149:3302–3308
129. Choudhary R, Baker KM, Pan J (2008) All-trans retinoic acid prevents angiotensin ii- and mechanical stretch-induced reactive oxygen species generation and cardiomyocyte apoptosis. J Cell Physiol 215:172–181
130. Kholodenko R, Kholodenko I, Sorokin V, Tolmazova A, Sazonova O, Buzdin A (2007) Anti-apoptotic effect of retinoic acid on retinal progenitor cells mediated by a protein kinase a-dependent mechanism. Cell Res 17:151–162
131. Besnard V, Nabeyrat E, Henrion-Caude A, Chadelat K, Perin L, Le Bouc Y, Clement A (2002) Protective role of retinoic acid from antiproliferative action of tnf-alpha on lung epithelial cells. Am J Physiol Lung Cell Mol Physiol 282:L863–L871
132. Zouboulis CC (2001) Retinoids--which dermatological indications will benefit in the near future? Skin Pharmacol Appl Ski Physiol 14:303–315

133. Kang S, Duell EA, Fisher GJ, Datta SC, Wang ZQ, Reddy AP, Tavakkol A, Yi JY, Griffiths CE, Elder JT et al (1995) Application of retinol to human skin in vivo induces epidermal hyperplasia and cellular retinoid binding proteins characteristic of retinoic acid but without measurable retinoic acid levels or irritation. J Investig Dermatol 105:549–556
134. Gupta S, Pramanik D, Mukherjee R, Campbell NR, Elumalai S, Dewilde RF, Hong SM, Goggins M, De Jesus-Acosta A, Laheru D, Maitra A (2011) Molecular determinants of retinoic acid sensitivity in pancreatic cancer. Clin Cancer Res 18:280–289
135. Chen R, Feng C, Xu Y (2011) Cyclin-dependent kinase-associated protein cks2 is associated with bladder cancer progression. Journal Int Med Res 39:533–540
136. Liu RZ, Graham K, Glubrecht DD, Germain DR, Mackey JR, Godbout R (2011) Association of fabp5 expression with poor survival in triple-negative breast cancer: implication for retinoic acid therapy. Am J Pathol 178:997–1008
137. Barbus S, Tews B, Karra D, Hahn M, Radlwimmer B, Delhomme N, Hartmann C, Felsberg J, Krex D, Schackert G, Martinez R, Reifenberger G, Lichter P (2011) Differential retinoic acid signaling in tumors of long- and short-term glioblastoma survivors. J Natl Cancer Inst 103:598–606
138. Fang LY, Wong TY, Chiang WF, Chen YL (2010) Fatty-acid-binding protein 5 promotes cell proliferation and invasion in oral squamous cell carcinoma. J Oral Pathol Med 39:342–348
139. Forootan SS, Bao ZZ, Forootan FS, Kamalian L, Zhang Y, Bee A, Foster CS, Ke Y (2010) Atelocollagen-delivered sirna targeting the fabp5 gene as an experimental therapy for prostate cancer in mouse xenografts. Int J Oncol 36:69–76
140. Han J, Kioi M, Chu WS, Kasperbauer JL, Strome SE, Puri RK (2009) Identification of potential therapeutic targets in human head & neck squamous cell carcinoma. Head Neck Oncol 1:27
141. Ogawa R, Ishiguro H, Kuwabara Y, Kimura M, Mitsui A, Mori Y, Mori R, Tomoda K, Katada T, Harada K, Fujii Y (2008) Identification of candidate genes involved in the radiosensitivity of esophageal cancer cells by microarray analysis. Dis Esophagus 21:288–297
142. Vanaja DK, Ballman KV, Morlan BW, Cheville JC, Neumann RM, Lieber MM, Tindall DJ, Young CY (2006) Pdlim4 repression by hypermethylation as a potential biomarker for prostate cancer. Clin Cancer Res 12:1128–1136
143. Morgan EA, Forootan SS, Adamson J, Foster CS, Fujii H, Igarashi M, Beesley C, Smith PH, Ke Y (2008) Expression of cutaneous fatty acid-binding protein (c-fabp) in prostate cancer: potential prognostic marker and target for tumourigenicity-suppression. Int J Oncol 32:767–775
144. Adamson J, Morgan EA, Beesley C, Mei Y, Foster CS, Fujii H, Rudland PS, Smith PH, Ke Y (2003) High-level expression of cutaneous fatty acid-binding protein in prostatic carcinomas and its effect on tumorigenicity. Oncogene 22:2739–2749
145. Richardson AL, Wang ZC, De Nicolo A, Lu X, Brown M, Miron A, Liao X, Iglehart JD, Livingston DM, Ganesan S (2006) X chromosomal abnormalities in basal-like human breast cancer. Cancer Cell 9:121–132
146. Muller WJ, Sinn E, Pattengale PK, Wallace R, Leder P (1988) Single-step induction of mammary adenocarcinoma in transgenic mice bearing the activated c-neu oncogene. Cell 54:105–115
147. Peters JM, Shah YM, Gonzalez FJ (2012) The role of peroxisome proliferator-activated receptors in carcinogenesis and chemoprevention. Nat Rev Cancer 12:181–195
148. Chen X, Tordova M, Gilliland GL, Wang LC, Li Y, Yan HG, Ji XH (1998) Crystal structure of apo-cellular retinoic acid-binding protein type ii (r111m) suggests a mechanism of ligand entry. J Mol Biol 278:641–653
149. Kleywegt GJ, Bergfors T, Senn H, Le Motte P, Gsell B, Shudo K, Jones TA (1994) Crystal structures of cellular retinoic acid binding proteins i and ii in complex with all-trans-retinoic acid and a synthetic retinoid. Structure 2:1241–1258
150. Conti E, Uy M, Leighton L, Blobel G, Kuriyan J (1998) Crystallographic analysis of the recognition of a nuclear localization signal by the nuclear import factor karyopherin alpha. Cell 94:193–204

Chapter 8
Vitamin A as PKC Co-factor and Regulator of Mitochondrial Energetics

Ulrich Hammerling

Abstract For the past century, vitamin A has been considered to serve as a precursor for retinoids that facilitate vision or as a precursor for retinoic acid (RA), a signaling molecule that modulates gene expression. However, vitamin A circulates in plasma at levels that far exceed the amount needed for vision or the synthesis of nanomolar levels of RA, and this suggests that vitamin A alcohol (i.e. retinol) may possess additional biological activity. We have pursued this question for the last 20 years, and in this chapter, we unfold the story of our quest and the data that support a novel and distinct role for vitamin A (alcohol) action. Our current model supports direct binding of vitamin A to the activation domains of serine/threonine kinases, such as protein kinase C (PKC) and Raf isoforms, where it is involved in redox activation of these proteins. Redox activation of PKCs was first described by the founders of the PKC field, but several hurdles needed to be overcome before a detailed understanding of the biochemistry could be provided. Two discoveries moved the field forward. First, was the discovery that the PKCδ isoform was activated by cytochrome c, a protein with oxidoreduction activity in mitochondria. Second, was the revelation that both PKCδ and cytochrome c are tethered to p66Shc, an adapter protein that brings the PKC zinc-finger substrate into close proximity with its oxidizing partner. Detailed characterization of the PKCδ signalosome complex was made possible by the work of many investigators. Our contribution was determining that vitamin A is a vital co-factor required to support an unprecedented redox-activation mechanism. This unique function of vitamin A is the first example of a general system that connects the one-electron redox chemistry of a heme protein (cytochrome c) with the two-electron chemistry of a classical phosphoprotein (PKCδ). Furthermore, contributions to the regulation of mitochondrial energetics attest to biological significance of vitamin A alcohol action.

Keywords PKC • Mitochondria • Redox • Retinol • p66shc • Cytochrome • Vitamin A • Kinase

---

U. Hammerling (✉)
Member Emeritus, Immunology Program, Sloan-Kettering Institute for Cancer Research, 10065 New York, NY, USA
e-mail: uhammerling@gmail.com

M.A. Asson-Batres, C. Rochette-Egly (eds.), *The Biochemistry of Retinoid Signaling II*, Subcellular Biochemistry 81, DOI 10.1007/978-94-024-0945-1_8

Abbreviations

AR	anhydro-retinol
DAG	diacyl-glycerol
DHR	13,14-dihydroxy-retinol
ETC	electron transfer chain
Holo-RBP	vitamin A - retinol binding protein complex
14HRR	14-hydroxy-retro-retinol
Hsp33	bacterial heat shock protein 33
IMS	mitochondrial intermembrane space
JAK	Janus kinase
LHC	light harvesting complex
MEFs	mouse embryo fibroblasts
PDHC	pyruvate dehydrogenase complex
PDHC	E1 pyruvate dehydrogenase complex regulatory subunit E1
PDK1-4	pyruvate dehydrogenase kinase isoforms 1–4
PDP1,2	pyruvate dehydrogenase phosphatase isoforms 1 & 2
PKA	protein kinase A
PKCα	protein kinase C alpha
PKCδ	protein kinase C delta
PKCε	protein kinase C epsilon
PPARγ	peroxisome proliferator-activated receptor γ
PSII	photosynthetic system II
P66Shc	SRC homology domain containing transforming protein 1 p66 isoform
RA	retinoic acid
Raf	Raf kinase
RAR	retinoic acid receptor
RBP4	retinol binding protein 4
ROS	reactive oxygen species
RXR	retinoid acid X receptor
sAC	soluble adenylate cyclase
SOCS3	suppressor of cytokine signaling
STAT	signal transducer and activator of transcription
STRA6	stimulated by retinoic acid 6

Introduction

By all rights, we should have abandoned the search for a new human B cell growth factor. Not only was the retinol binding protein (RBP) we had isolated an ordinary serum protein that originated from the fetal calf serum used to grow the B cells, but the growth promoting property was clearly mediated by vitamin A (aka, retinol), the cargo of holo-RBP [10, 11]. The experts cautioned us that our finding might be trivial and most likely attributed to the actions of *all-trans* retinoic acid, the predictable product of vitamin A, which was well known to regulate multiple aspects of lymphocyte physiology, not least cell viability. However, when Jochen Buck, the post-doctoral fellow who spearheaded the effort, diligently tested retinoic acid, he found no growth-promoting capacity associated with this retinoid. Over the same wide dose range where *all-trans*-retinol provided a distinct survival advantage, retinoic acid proved completely ineffective [11].

Still, everybody knew that vitamin A was necessary for normal development and growth of animals, and was it not discovered 80 years earlier for its ability to promote the growth of chicks [74, 87]. And were we as scientists not guilty for committing the sin of which Albert *Einstein* professed little patience "of tak[ing] a board of wood, look[ing] for its thinnest part, and drilling *a great number of* holes *where* drilling *is easy*"? Indeed, drilling would have been comparatively easy, had we not chanced upon a knot so hard that it took us 20 years to begin to understand the biological relevance of what we had glimpsed through that original knothole, namely that vitamin A possesses a physiology quite distinct from that of the reigning retinoic acid paradigm of gene regulation [51]. And who could have anticipated that the holo-RBP complex we had "re-discovered" would re-emerge 15 years later as a major risk factor of metabolic disease [107]. Hence unbeknownst to us, we had been working all these years on one of the fundamental pathways that control energy homeostasis. To make gains towards understanding the unique physiology of vitamin A, we had to dig deeper, as well as to circumnavigate obstacles presented by perceived dogmas and entrenched beliefs. Or, to quote Albert Einstein again: "You can never solve a problem on the level on which it was created".

History

Early Observations Revealing a Unique Biological Function of Vitamin A

Our route to discovery of a broader biological role for vitamin A beyond its role as precursor for retinoic acid met with a succession of challenges that required us to overcome entrenched scientific convictions and yes, even dogmas. From the beginning, we ran up against the notion that vitamin A was solely the reservoir fueling vision and gene regulation. What drove our interests in defining a contrasting mechanism of action for vitamin A was our observation that the survival of cultured cells was strictly dependent upon physiological concentrations of retinol equivalent to those present in plasma [11, 22, 27]. Moreover, retinoic acid was incapable of substituting for retinol over a wide dose range, and, instead, actually accelerated a decline in cell viability [12]. Therefore, our findings were at odds with the opinion held by many researchers that vitamin A solely served as a storage form and

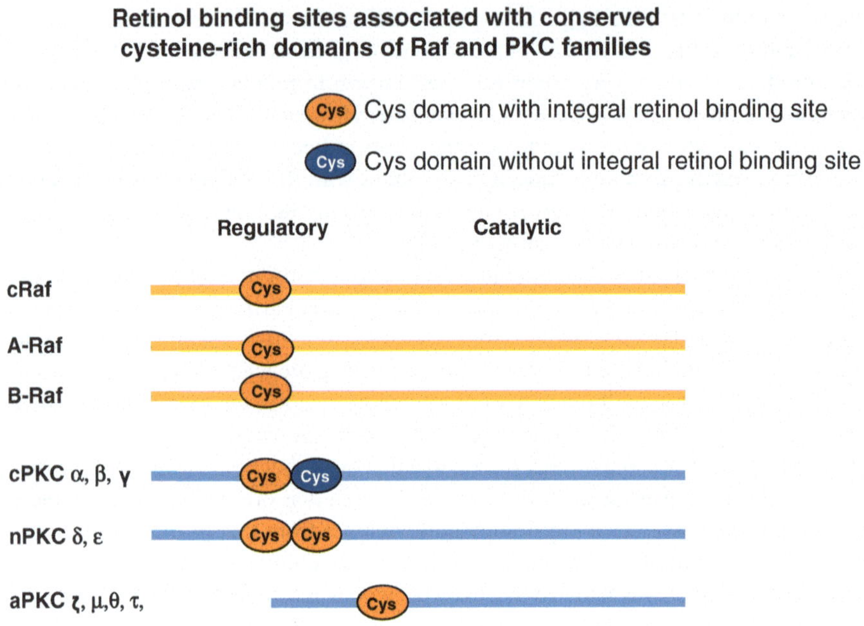

Fig. 8.1 Retinol binding sites are associated with the cysteine rich domains of the Raf and PKC family members. The classic α,β,γ PKC isoforms contain one binding site in the C1a domain, but none in the C1b domain. The novel δ, ε PKC isoforms harbor one binding site in each of the C1a and C1b domains. The binding affinities are in the tens of nanomoles [48]. The cysteine rich domains are synonymous with the activation domains

precursor for the most prominent bioactive, metabolites of vitamin A [8, 34], including retinaldehyde (active in vision [104]) and retinoic acid (active in gene transcription [30, 37, 90].

Until very recently, no coherent theory was formulated that could explain a novel, cell survival promoting activity of vitamin A. Nonetheless, older animal feeding experiments documented multiple examples demonstrating that not all biological effects of vitamin A deficiency could be corrected by retinoic acid supplementation [19, 99, 106]. In addition, the large surplus of vitamin A circulating in plasma that far exceeds the amount needed for vision or for the synthesis of nanomolar levels of retinoic acid, along with observations that many cells possess the biochemical machinery to take up vitamin A, convert it to retinyl esters for storage, and retrieve free vitamin A by ester-hydrolases for conversion to various metabolites, strongly suggested to us that vitamin A alcohol (i.e. retinol) possesses intrinsic biological activity beyond its function as a precursor for other bioactive retinoids. In light of these findings, it seemed imperative to investigate whether vitamin A had other physiological functions.

A breakthrough came when, in the course of a fishing expedition, we defined a group of target molecules that bound vitamin A with moderately high affinities in the tenths of nM (Fig. 8.1). These belong to the super-family of serine/threonine kinases that includes several protein kinase C (PKC) and Raf isoforms. A survey revealed that the vitamin A binding sites on PKCs were conserved from drosophila to man [48], convincing us that they were not random hydrophobic sites. And if they were real, these findings argued that retinol targets are evolutionarily conserved across all phyla in contrast with retinoic acid binding protein targets which appear at the invertebrate/vertebrate boundary. Most intriguingly, the retinol binding sites of kinases were invariably localized to their activation domains [54, 61, 62], suggesting they were relevant to activation mechanisms [49].

Major obstacles arose when we discovered that many other natural retinoids, including retinoic acid (RA), anhydroretinol (AR) and 14-hydroxy-retro-retinol (14-HRR), also bound to PKCs with affinities similar to retinol [57]. How was this lack of specificity compatible with regulation of any kind when selectivity is a key element in cellular control? And why would retinol have similarly high affinity for 11 PKC and 3 Raf isoforms, presumably attaching itself to all expressed isoforms simultaneously? Next to being constantly available, this lack of target selectivity was contrary to any notion that retinol served as a kinase activator. At best, we surmised, retinol was a co-factor that needed to occupy specific sites in the activation domains as prerequisite for kinase activation. This hypothesis, too, seemed untenable when we found that neither PKC nor Raf kinases were dependent on retinol for activation by classic second messengers such as diacyl-glycerol (DAG) or phorbol ester [45], a pharmaco-mimetic of DAG. Similarly, cRaf activation by the Ras pathway did not require retinol.

Nonetheless, we were reminded that PKCs and cRaf could also be activated by oxidative stress [66]. We guessed, and subsequently demonstrated, that this alternate activation mode required vitamin A as co-factor. In a key experiment, we observed that cRaf was converted into the active enzyme by UV-induced oxidative stress when retinol was present, but only to a much lesser degree when absent [49]. In later work we showed that abolishing the retinol binding sites by genetic mutation also blocked the redox activation pathway of PKC and cRaf, while preserving activation by classic second messengers [50, 57].

Development of the Field

PKCs have high-affinity binding sites for vitamin A that are located within their activation domains [48]. We have observed that these sites need to be occupied by retinol as a prerequisite for activation by the redox stress pathway [57]. So far, retinol-dependent redox activation of PKC has only been studied in detail in mitochondria [1, 2], but since defined retinol binding sites are associated with the activation domains of all PKC isoforms and their cousins, the Raf family of kinases [48, 50], redox activation may well be more general. If so, our findings may presage the beginnings of a broader role of vitamin A in redox biology relevant to virtually all vertebrate cells. They would also change how we regard vitamin A physiology beyond its accredited role as precursor of retinoic acid.

Redox activation of PKCs was first described by the founders of the PKC field [63, 66, 83] but several hurdles needed to be overcome before a detailed understanding of the biochemistry could be provided. The first hurdle was the widely held notion that reactive oxygen species (ROS), acting like a second messenger, reacted with a redox sensitive site of PKC, zinc-fingers [35]. The second one was that the time was not ripe for accepting the concept that PKC zinc-fingers are reversible hinges, similar to their bacterial orthologs [55, 58].

Two discoveries solved the problem: first, the discovery that the PKCδ isoform was activated by cytochrome c, a protein with oxidoreduction activity in mitochondria; second, the revelation that both PKCδ and cytochrome c are tethered to p66Shc, an adapter protein that brings the PKC zinc-finger substrate into close proximity with its oxidizing partner [2]. Detailed characterization of the PKCδ signalosome complex was made possible by the work of many investigators. Our contribution was determining that vitamin A is a vital co-factor required to support an unprecedented redox-activation mechanism.

Flexible Zinc-Finger Domains: Orthologs of Bacterial Hsp33?

The activation domains of PKC and Raf serine/ threonine kinases harbor two intertwined zinc-coordination centers [47, 80, 108]. In simplified terms each center consists of a Zn^{2+} ion surrounded by 3 cysteine and one histidine residue, arranged into a perfect tetrahedron. The symmetry of the 4 thiolate/histidinyl anions surrounding the Zn^{2+} cation makes for a super-strong ionic bond. This device is widely used for the stabilization of the tertiary structure of cytoplasmic and nuclear proteins. However, its Achilles heel is that Zn-coordination is only stable in the reducing environment inside cells, but is lost when oxidizing conditions arise. Bacteria adopted this paradigm to regulate the activity of the Hsp33 chaperone [55, 56, 58]. This enzyme is inactive in resting bacteria when reducing conditions prevail, but is activated by oxidative conditions associated with high metabolic rates during bacterial growth, coincident with high demands for repair of misfolded proteins. The pivotal change from inactive to active foldase is the dissolution of the zinc-coordination center, caused by the oxidation of crucial cysteines of Hsp33 [70]. This event sets off a local conformation change that is followed by allosteric large-scale rearrangement of the protein molecule to yield the active enzyme.

Mammalian PKC zinc-fingers appeared to us to function as orthologs of bacterial ones serving, like these, as flexible hinges. As such, the activation of mammalian PKCs and Raf by the alternative redox-stress pathway implied oxidation of redox sensitive moieties – cysteines being the most likely ones. We tested for cysteine oxidation, but for technical reasons, we did not succeed in pinpointing such modifications. However, we and others did observe that PKCα from redox-stress activated cells possessed significantly less high-affinity bound zinc ions than inactive PKCα isolated from resting cells, suggesting that a proportion of zinc-coordination centers had disassembled during conversion to active enzyme [64, 67].

We also observed that stimulation of cells with phorbol ester led within minutes to a similar decline in high-affinity bound Zn^{2+} of PKCα [67], once again suggesting a structural change. By use of a chimeric Hsp33 reporter protein, where the endogenous zinc-finger was replaced by the PKCε C1B zinc-finger and which remained sensitive to both redox and phorbol signals, we demonstrated that Zn^{2+} coordination was lost upon contact with the activating agents. Concomitant with the release of Zn^{2+} ions, phorbol ester caused the cysteine residues visible in the NMR HSQC spectrum to change from well dispersed to unstructured [109]. Taken together these findings suggested that exposure to either oxidative conditions or to lipid second messenger caused loss of tertiary structure. We surmised that conformation change induced in this manner in the zinc-finger domain represented the initiating event that caused partial unfolding. The accompanying exposure of hydrophobic recognition surfaces could enable PKC to become anchored at select membrane sites [4, 82] where it would receive further allosteric signals and chemical modifications for full activation.

Our interpretation of the PKC activation sequence strikes a discordant note with previous models of PKC activation since there are no discernible conformation changes between the published structures of the PKCδ C1b zinc-finger co-crystal with phorbol ester and the unligated crystal [108]. If we consider PKC zinc-fingers as reversible hinges, that is: folds inherently unstable to phorbol ligands, why was it possible to produce a crystal of the zinc-finger domain complexed with phorbol ester? One likely explanation is that in order to construct their crystal, the authors used phorbol-13-acetate (P13A) as ligand, a weak agonist of PKCδ and, in our hands, one that does not disperse zinc-coordination [109]. The admirable structure the P13A-ligated zinc-finger portrays is of course correct, but the message the image projects that zinc-finger domains are static, needs revision.

Mitochondrially Located PKCs Are Redox-Activated in Situ

With the conceptual advance that redox activation of PKCs is theoretically possible, we looked for a system to test our model. We found that mitochondrially localized PKCδ and PKCε were good candidates. The key observations leading us to mitochondria were raised in studies involving two retinol antagonists, AR, a natural retinoid [13, 38], and fenretinide (alias 4-hydroxyphenyl retinamide, 4-HPR) a synthetic retinoid. These compounds are known to cause necrotic cell death in numerous cell lines [14, 40, 85, 86]. Initially we thought that cells died because AR and 4-HPR displaced retinol from the PKC binding sites, but were incapable of co-activating PKC. The first premise was true as we showed by binding competition studies [57].

However, the second was not: paradoxically, these retinoids are activation cofactors that are as potent as retinol for both the δ and ε PKC isoforms [33]. Why their action results in cell death remains unclear. One reason might be that the PKCε isoform predominates when cells are exposed to AR or fenretinide. PKCε suppresses mitochondrial respiration, and hence it is conceivable that under these circumstances energy reserves become depleted to a degree that adversely impacts cell survival [15].

Our findings concerning energy depletion in AR or fenretinide-supplemented cells were at odds with reports in the literature that extolled the exact opposite, namely that PKCδ signaling diminished, whereas PKCε augmented, respiration [16]. Our interpretation also differed from the model of others in one fundamental aspect, namely that in order to preserve substrate specificity, we postulated that PKCs were to be activated *in situ*. This concept was based on the micro-domain theory of Buck and Levin who had demonstrated that mitochondrial PKA was associated with its own biochemical machinery for the generation of the second messenger, cyclic AMP, by the dedicated adenylate cyclase, sAC [3]. Alluding to this model, we thought that under no circumstances was *activated* PKC to be imported into mitochondria, lest it react en route with other substrates, of which there are many. Cells could not tolerate such haphazard signaling.

Focusing on PKCδ, we confirmed the findings of others that this isoform translocates to the mitochondrial intermembrane space (IMS) where it forms a complex with the adapter protein, p66Shc [2]. Our observations were supported by findings that p66Shc regulates mitochondrial metabolism with profound influences on redox stress and aging [25, 81]. Independently, a group of Italian scientists reported that p66Shc played a role in redox regulation [76]. These investigators convincingly demonstrated that presence of p66Shc exacerbated mitochondrial oxygen stress, whereas genomic inactivation of p66Shc conferred resistance. Mice carrying an inactive p66Shc gene were protected from chronic damage by oxygen radicals and displayed an astonishing life-span extension [76]. Hence, they named p66Shc the "longevity" gene and assumed, on the basis of in vitro experiments, that p66Shc possessed intrinsic oxido-reductase activity, which was deemed responsible for the production of damaging ROS [77]. Cumulative oxygen damage in normal mice is a hallmark of aging as theorized by Harman [43], so it was no surprise that protection from oxidative stress in p66Shc knockout mice was associated with profound life-span extension.

P66Shc was found to bind cytochrome c, potentially providing a convenient electron donor for ROS synthesis [31]. However, do oxygen radicals really emanate from p66Shc in intact cells, and if so, what would be the physiological relevance? We found it difficult to accept that p66Shc functions as an oxido-reductase in the absence of a demonstration that transfer of electrons from reduced cytochrome c to oxygen actually occurs in living cells. Such a demonstration is still outstanding. Moreover, a rigorous biochemical analysis of enzyme parameters, such as Michaelis-Menten kinetics, is needed to substantiate the idea. Therefore, the characterization of p66Shc as an oxido-reductase ought to be viewed with caution. Further complicating the issue is the absence of a definable reaction center that one would expect in an oxido-reductase. Iron, copper, and manganese ions do not seem to be associated with p66Shc. In one report, it was necessary to artificially supplement the *in vitro* preparation of p66Shc with copper ions in order to demonstrate ROS production [28, 29]. In this circumstance, the Fenton reaction might well have confounded the redox biochemistry.

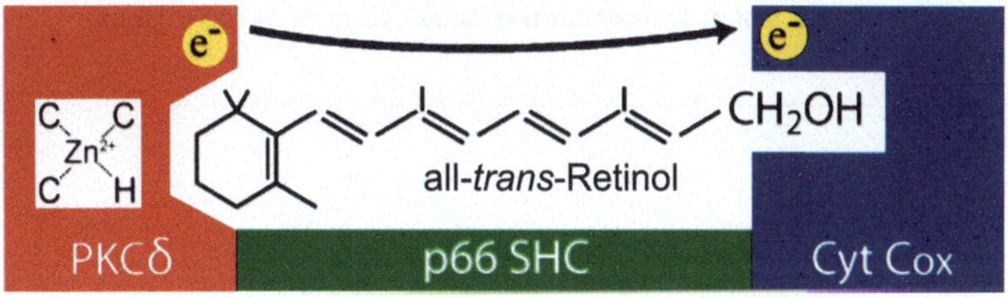

Fig. 8.2 Schematic of the PKCδ signalosome. The presence of the tetrameric PKCδ complex in the intermembrane space of mitochondria, comprising besides PKCδ the adapter protein, p66Shc, cytochrome c and retinol, was vetted by biochemical pull-down experiments [2, 52, 76]. An exhaustive genetic analysis indicated that all four components are required for *in situ* activation of PKCδ. Mutational studies of protein-protein interactions sites, as well as of the PKCδ–retinol binding site yielded a crude map of the signalosome

We postulate that p66Shc, within its classic definition as adapter protein, serves as a platform to promote the assembly of PKCδ in close apposition to cytochrome c. Biochemical experiments and genetic dissection confirm the existence of a heterotrimeric PKCδ/p66Shc/cytochrome c protein complex in the IMS [2]. Vitamin A is the fourth component of this complex (Fig. 8.2) [51]. We hypothesize that this PKCδ signal complex regulates oxidative phosphorylation and, in that capacity, may well be linked to the all-too-real ROS release caused by slippage within an overloaded electron transfer chain. Oxygen stress is a normal occurrence of hard-working mitochondria, a price to pay for living in an oxygen world. Reduction of the workload by down-regulating oxidative phosphorylation, as might be achieved by disabling p66Shc, would be accompanied by reduced oxygen stress, a fitting explanation for the observed reduced metabolic stress of p66Shc null mice, but also for their observed reduced metabolic fitness [32].

Forward Targets of PKC Signaling in Mitochondria

From our work, we developed a model that can account for the following features: operates semi-autonomously in mitochondria; is centered on the PKCδ signalosome composed of PKCδ, the signal adapter, p66Shc, cytochrome c and vitamin A; depends on the concerted action of all four components to achieve PKCδ activation *in situ*; targets the pyruvate dehydrogenase complex; functions as a sensor in real-time to adjust fuel flux with the workload of the electron transfer chain; integrates both stimulatory and inhibitory signals in yin-yang fashion emanating from PKCδ and PKCε, respectively. This model is schematically represented in Fig. 8.3.

Evidence gained from different approaches indicates that PKCδ and PKCε indirectly regulate the fuel flux in mitochondria via the pyruvate dehydrogenase complex (PDHC) [1]. Because of its central role in catalyzing the final step in glycolysis, this enzyme complex is regulated at multiple levels [4, 44, 46, 65, 88, 103], responding

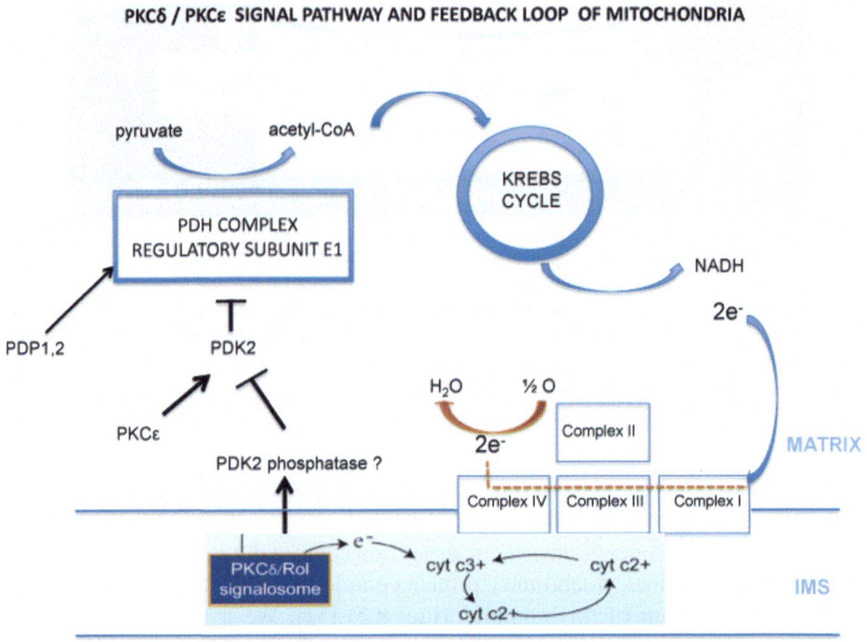

Fig. 8.3 Schematic representation of the PKCδ/ε signal pathway of mitochondria. The PKCδ signalosome resides in the intermembrane space of mitochondria where it functions as redox sensor. The 4 membrane-bound complexes of the mitochondrial respiratory chain are shown. In its oxidized state cytochrome c oxidizes the activation domain of PKCδ using vitamin A (Rol) as redox coupler. Cytochrome c^{2+} is generated and reoxidized by complex IV, thus connecting the PKCδ signaling module to the ETC. While the primary substrate of PKCδ remains unknown, the active enzyme causes the dephosphorylation of PDK2, necessitating the participation of an intermediate "PDK2-phosphatase". Active PDK2 normally phosphorylates the E1 regulatory subunit of the PDHC resulting in diminished acetyl-CoA production. However, when PDK2 is inactivated by dephosphorylation the inhibitory influence is removed, allowing the PDP phosphatase to maintain the PDHC in the activated state. Increased acetyl-CoA production promotes higher Krebs cycle turnover, increasing the production of reducing equivalents. The high-potential electron of NADH enters the ETC, stepwise driving the three proton pumps that generate the proton motive force. The electron is finally transferred to oxygen. As proposed, the feedback loop coordinates the fuel production with energy demands. PKCε works in opposition to PKCδ. Its upstream activator is unknown but it likely activates PDK2. The Yin yang action of the two protein kinase isoforms is thought to prevent fuel production from exceeding the capacity of the ETC

to signals emanating from within mitochondria and without. Best understood among the direct internal regulators are two phosphatases, PDP1 and PDP2 [53, 60, 97], which dephosphorylate the PDHC regulatory E1 subunit and thereby activate PDHC. Conversely, four PDH-kinases, PDK1 – 4 [39, 101], inhibit PDHC by phosphorylation of PDHC E1 [9]. The PDKs are themselves regulated by phosphorylation and by de-phosphorylation (causing PDK activation or inhibition, respectively). While the PDK-kinases and phosphatases are largely unknown, previous studies pointed to PKCδ as the PDK2-kinase, but the evidence for this rested solely on potentially misleading in vitro experiments [16, 17]. PKCδ and PDK2 reside in different compartments, the IMS and the matrix, respectively. Thus PKCδ, separated from its putative target by the inner membrane, cannot directly access the putative PDK2 substrate unless migrating across the membrane into the matrix, a process that is possible but cumbersome, given that PKCδ exists as a hetero-trimeric complex. The drawback of the experiments identifying PDK2 as the direct substrate of PKCδ was that the results were not confirmed in intact cells [17]. PDK2 may well become phosphorylated by PKCδ in the test tube, but this result might be meaningless in the structural context of mitochondria.

To identify the presumptive targets of PKCs, we analyzed the phosphorylation patterns of all four PDKs by two-dimensional gel electrophoresis. We observed that the isoelectric points of immunoprecipitated PDK1 and 2, but not those of PDK3 and 4, isolated from resting cells were on average more basic than those of activated cells. High isolectric points of PDK1 and 2 (signifying de-phosphorylation) correlated with the presence of active PKCδ (as indicted by the auto-phosphorylation at Thr 505), whereas low pIs correlated with inactive PKCδ. We concluded therefore that both PDK1 and 2 were potential indirect targets of PKCδ. However, genetic inactivation of PKCδ preserved the PDK1 phosphorylation pattern characteristic of resting versus activated cells, whereas highly phosphorylated species of PDK2 predominated in resting or active cells. This result suggested that PDK2, but not PDK1, was in the PKCδ signal path [1]. Moreover, the fact that active PKCδ caused PDK2 to become dephosphorylated suggested the involvement of an intermediary PDK2-phosphatase, the nature of which remains unknown. That PDK2 was the likely downstream target of PKCδ was also supported by genetic experiments. Both siRNA knockdown and genomic inactivation of PDK2 abolished the regulatory control of PKCδ over the PDHC (Hoyos and Hammerling, unpublished results).

PKCδ operates in mitochondria in partnership with PKCε [33]. This PKC isoform requires retinol as co-factor, just like PKCδ. However, the ε isoform is unlikely to be activated by cytochrome c since it resides in the matrix where cytochrome c is not available. The upstream activator of PKCε is still unknown, whereas preliminary evidence suggests that its downstream target may be PDK2. The correlations between PDK2 and PDHC E1 phosphorylation patterns and the PKCε activation status were precisely the converse of those elicited by PKCδ [33]. The activation of PKCε in PKCδ null cells resulted in the accumulation of phosphorylated PDK2 species, signifying activation of this kinase. Simultaneously, the E1 subunit of PDHC was hyper-phosphorylated indicating down-regulation of PDHC. Taken together our results suggest that PKCε may represent the long-sought PDK2-kinase.

It was long suspected that PKCδ and ε oppose each other [42, 72, 73]. As mentioned before, an earlier model was based on the erroneous assumption that PKCδ activated PDK2 and thereby down-regulated PDHC function [17]. In that model PKCε logically featured as an activator of PDHC. However, our findings that PKCε down-regulates PDHC challenge this view [33]. We showed that PKCε knock-out cells display increased respiratory capacity, whereas PKCδ knock-out cells have reduced oxygen consumption rates compared to wild type cells. Furthermore, cells transfected with a conditional PKCε gene suffered a drastic decline in oxygen consumption upon transcriptional activation [33].

These findings support a coherent signal system operating in yin-yang fashion upstream of the PDHC and entrusted with the important task of matching fuel flux with the metabolic demands of mitochondria (Fig. 8.3). PKCδ is activated by oxidized cytochrome c [2] and thus the PKCδ/ε signal system may well be tied to the activity of complexes III and IV of the mitochondrial respiratory chain, serving as a sensor for the workload of the electron transfer chain (ETC) and as gate-keeper for the control of fuel production from glycolytic sources. How PKCε fits in and, in particular, what lays upstream of this kinase is still obscure. Nevertheless, the yin-yang relation between PKCδ and PKCε highlights the importance of securing on the one hand, an adequate fuel supply via PKCδ, but on the other, of preventing, via PKCε, oxidative phosphorylation from exceeding the capacity of the ETC. The risk of ROS production by chronically overworked ETC seems best tempered by a negative feedback signal.

Current State of the Field

Definition of Cytoplasmic Receptors of Vitamin A in Members of the PKC and Raf Kinase Families

We identified cRaf kinase and PKCα as kinases that share common cysteine-rich subdomains designated C1a or C1b, respectively [62]. We predicted that retinoid binding sites are associated with these subdomains. Fluorimetric binding assays [18] performed with recombinant versions of the cysteine-rich domains of cRaf and PKCα confirmed the presence of predicted high-affinity binding sites in both [48]. The nominal binding affinities were approximately 20 nM. When analyses were extended to other members of the cRaf and PKC families, it was found that they also harbor retinoid binding sites, with affinities ranging from 10 to 50 nM [57]. Conventional PKC isoforms with two tandem cysteine-rich domains only bind retinoid in the C1a subdomain [50], whereas the novel family members, PKCδ and ε, possess binding sites in both subdomains. An atypical isoform, PKCζ, and the zinc-fingers of A and B Raf isoforms possess a single binding site (Fig. 8.1).

Protein kinase C homologs are found in eukaryotic species from drosophila to man. Their cysteine rich domains contain retinoid binding sites that are conserved over a vast evolutionary distance and, incongruously, even extend to yeast PKC, although this organism is not known to use vitamin A [57].

Definition of Contact Amino Acids Mediating Retinol Binding

While the cysteine-rich domains display considerable amino acid sequence diversity, their structural homology is nearly perfect. These domains are organized into intertwined dual zinc-finger folds, the overall structure of which is unchanged in all members of this serine/threonine kinase family [47, 80, 108]. The question arose whether the binding sites would be conserved at the same relative location within these folds. Efforts to co-crystallize cRaf C1or PKCδ C1b domains with retinol failed, but a mutagenesis study defined the presumptive contact amino acids that mediate binding of retinol. They comprise a trio: Phe-7, Phe-8 and Trp-22 (counting from the first conserved histidine of the zinc finger) [50]. The Phe duo is believed to contact the β-ionone head group of retinoids, as deduced from the fact that within a group of retinoids, retinoic acid, anhydroretinol, 14-hydroxy-retro-retinol all displayed binding affinities similar to retinol [57]. The chemical structure of the head group is conserved in all of these metabolites, whereas the tail ends contain variable hydrophilic substitutions. These results agree with computational docking trials that consistently place the head group into the hydrophobic binding groove delineated by Phe-7 and 8. These anchor sites, or close homologs thereof, were present at the same locale in all zinc-finger domains that bound retinoids, but were absent from those that failed to bind retinol. To further assess the nature of the binding sites, the Phe-7/Phe-8/Trp trio was engineered into the non-binding PKCα C1b domain. The reconfigured αC1b domain now bound retinol at high affinity. Conversely, mutating the presumptive anchoring amino acids of the PKCα C1a domain abolished vitamin A binding [50].

The conserved contact Trp-22 amino acid deserves special attention, as it may be the key for understanding the physiologic function of retinol. We showed that UV excitation of Trp resulted in fluorescence energy transfer to retinol, quenching the fluorescence emission by Trp at λ_{max} = 330 nm, and exciting specific fluorescence emission of retinol at λ_{max} = 466nm [50]. Thus, retinol is electronically coupled with the zinc-finger peptide. At λ_{max} = 325 nm the absorption peak of retinol overlaps closely with the peak of Trp emission, a condition for energy transfer [48]. The position of Trp-22 in the binding pocket relative to the bound retinol seems equally important for VanDerWaal contacts.

Biochemistry of the PKCδ Signalosome of Mitochondria

PKCδ is widely expressed in animal tissues where it participates in numerous signal pathways, including differentiation, cell cycle regulation, survival, endocytosis, and secretion [96, 100]. A number of excellent reviews on the biology and biochemistry of PKCδ are available [98]. We focus here on the unorthodox function of PKCδ in mitochondria.

Although devoid of overt import pre-sequences, PKCδ localizes to mitochondria. Kowalczyk et al traced PKCδ to mitochondria by immuno-electron-microscopy [69] and Acin-Perez et al narrowed the location to the IMS [1]. They used the protease protection assay to show that PKCδ resisted proteolytic degradation in intact cells, but was extinguished after permeabilizing the outer membrane with digitonin. Most convincing, however, was the demonstration that the PKCδ heterotrimeric complex found in mitochondria includes cytochrome c [2]. The location of the latter in the IMS is not in question.

The second partner of PKCδ is the signal adapter protein, p66Shc. This is a member of the Shc superfamily comprising Shc A, B, C and RalP [23]. ShcA encodes three isoforms that are designated according to their molecular weights, the largest one being p66Shc [93]. As is typical for signal adapters, p66Shc contains a number of recognition domains that enable integration into multiple signal complexes of diverse function. These comprise an N-terminal phosphotyrosine-binding domain (PTB), a collagen homology domain (CH1), a C-terminal Src-homology domain (SH2) and a cytochrome c binding domain (CB) [75, 89]. Although lacking a canonical mitochondrial import sequence p66Shc, like PKCδ, translocates to the IMS. This process is not well understood, but involves the mandatory phosphorylation by PKCβ, followed by allosteric modification via the prolyl-isomerase, PIM1. The bulk of p66Shc localizes to the IMS, and to a minor degree also to the matrix [91].

As first described in 2007 by Hu et al. p66Shc exists in several cell compartments as a hetero-dimeric complex with PKCδ [52]. Morita et al confirmed the propensity of p66Shc to bind PKCδ in the cytosol [79], and Acin-Perez et al defined a similar complex in the IMS, as mentioned. Reciprocal immunoprecipitation experiments showed that PKCδ immunoprecipitates contained p66Shc and, vice versa, that p66Shc pulled down PKCδ. The interaction was mediated by the phospho-Y332 motif on PKCδ that was recognized by the p66Shc SH2 domain. Phosphorylation of Y332 was a prerequisite for binding, since the mutation of Y332 to F abolished hetero-dimer formation *in vitro* and *in vivo* [2, 33].

Because cell disruption can produce artificial protein aggregation, it was imperative to confirm the p66Shc/PKCδ linkage in intact mitochondria by genetic engineering. Our group analyzed the importance of p66Shc and PKCδ in intact cells, using oxygen consumption and ATP production as functional readouts, as well as the phosphorylation of PDHC E1 as biochemical readout. As detailed above, PKCδ generates, when activated, a forward signal that activates PDHC, increasing acetyl-CoA production (Fig. 8.3). Increased PDHC activity is associated with decreased E1 phosphorylation. With more fuel entering the Krebs cycle the level of reducing equivalents rises, oxygen consumption accelerates and ATP synthesis increases. We showed that genetic inactivation of PKCδ prevented the dephosphorylation of PDHC E1 and furthermore compromised oxidative phosphorylation of isolated liver mitochondria and mouse embryo fibroblasts alike, diminishing both the respiratory capacity and ATP synthesis [1]. In genetic complementation experiments, re-introduction of the full-length PKCδ gene restored normal respiration and ATP production. Likewise, liver mitochondria, or mouse embryonic fibroblasts (MEFs) carrying an inactive p66Shc gene, were incapable of responding to PKCδ-mediated signals. Re-introduction of a cDNA gene encoding full-length p66Shc repaired the defect. Interestingly, stimulation of p66Shc$^{-/-}$ MEFs with phorbol ester up-regulated their respiration, indicating that the blockade imposed by the absence of functional p66Shc could be bypassed [2]. The downstream signal path evidently was intact and responsive to pharmacologically activated PKCδ. These observations suggested that p66Shc was needed for the redox activation of PKCδ *in situ*, but not for signal propagation. To further examine the importance of the PKCδ/p66Shc dimer, PKCδ$^{-/-}$ MEFs were reconstituted with a PKCδ transgene carrying the Y332F mutation. Consistent with the inability to be phosphorylated and recognized by the Shc SH2 domain the mutated PKCδ did not restore respiration to normal levels. However, phorbol ester rescued the pathway [2], at least temporarily. These results imply that heterodimer formation is a prerequisite for PKCδ activation in mitochondria.

Role of Cytochrome c as Redox Activator of PKCδ

In the course of their pioneering work, Pelicci and coworkers described the formation of heterodimers between p66Shc and cytochrome c [31, 76]. Their genetic dissection revealed that the p66Shc N terminal CH2- PTB domain was responsible for cytochrome c binding. Two glutamate residues E132, E133 and tryptophan W134 are highly conserved in eukaryotes from yeast to man, implying an essential function. Mutation of these residues disrupted the p66Shc/ cytochrome c heterodimer indicating that E132 and E133 formed one half of the cytochrome c recognition site, and W134 the other half. Genetic mutation of these sites also attenuated the production of ROS.

Taking the cue that cytochrome c partners with p66Shc, we wondered whether the placement of cytochrome c into close apposition to PKCδ on the p66Shc platform was important for signaling. By expressing the E132Q, E133Q cytochrome c non-binding p66Shc mutant in p66Shc$^{-/-}$ MEFs, close contact was indeed shown to be important, since the mutation of the p66Shc/PKC interface disrupted PKCδ signaling. As expected, stimulation by phorbol ester bypassed this block [2].

The question of whether oxidized or reduced cytochrome c was required to energize PKCδ signaling was solved by the use of mitoplasts. The respiration rates of mitochondria isolated from cytochrome c knockdown MEFs are below basal levels compared to WT MEFs. Restoration of normal cytochrome c levels and near-normal rates of respiration is feasible by perforating the outer membrane with digitonin and adding high concentrations of cytochrome c protein into the medium. This technique allowed us to test the efficacy of oxidized, compared to reduced, cytochrome c. Acin-Perez et al demonstrated that, as expected, oxidized cytochrome c, but not the reduced form, restored PKCδ-mediated regulation of the PDHC, normalizing respiration. Mitoplasts derived from PKCδ knockout cells failed to upregulate respiration in response to cytochrome c^{3+} [2]. These results support a redox-based model of PKCδ activation in which the activation domain is oxidized by cytochrome c, leading to kinase activation.

Role of Retinol in the Redox Activation of the PKCδ Signalosome

It has been proposed that freely diffusible reactive oxygen species (ROS), notably H_2O_2, serve as second messengers for the activation of PKC [35, 66], but this concept seems inherently flawed. For one thing ROS, spreading through a cell like a plume, would encounter large concentrations of glutathione, and those few ROS escaping neutralization could only be haphazardly directed to their specific protein targets. Should these by chance reach PKCδ, ROS would be prone to randomly modify any susceptible amino acids, like cysteines or methionines, of which there are many, causing more harm than good. To circumvent such difficulties, we proposed that redox reactions are catalyzed by oxido-reductases. Like kinases or phosphatases, an oxido-reductase can be targeted to specific amino acids (such as cysteines of the PKC activation domain) thanks to complementary recognition sites embedded in enzyme and substrate protein. Our evidence shows that this concept is in large part realized in the mitochondrial PKCδ complex. The target to be oxidized would be the C1b zinc-finger domain of PKCδ, where one or more cysteines (yet to be specified) donate electrons, i.e. they become oxidized. Electrons are accepted by oxidized cytochrome c and transferred to cytochrome c oxidase (COX) for further disposal [51]. Only one of the pair of electrons transferred to cytochrome c^{3+}, can be accommodated by the Fe^{3+} core of the heme, leaving the other temporarily stranded, and creating a potentially harmful radical. The recruitment of a second cytochrome c^{3+} would complete the oxidation step and resolve the danger. It is conceivable that the PKCδ complex is physically integrated into the cytochrome C oxido-reductase supercomplex, for greater electron transfer efficiency.

As is well known, electron transfer from one protein to another is a slow process. Electronically coupling PKCδ to cytochrome c would speed up the process. This is where retinol comes into play. We propose that retinol operates as an electron transfer agent, much like ubiquinol does in the electron transfer chain, but without the need for physical movement. With its extended linear system of conjugated double bonds, retinol is an eminently suitable "molecular wire" [26] that could channel electrons from one protein to another and momentarily stabilize unpaired electrons. The location of retinol binding sites within the PKC activation domains together with the potential electronic coupling between the Trp-22 residue of C1b domain and retinol [50] hardly appears accidental.

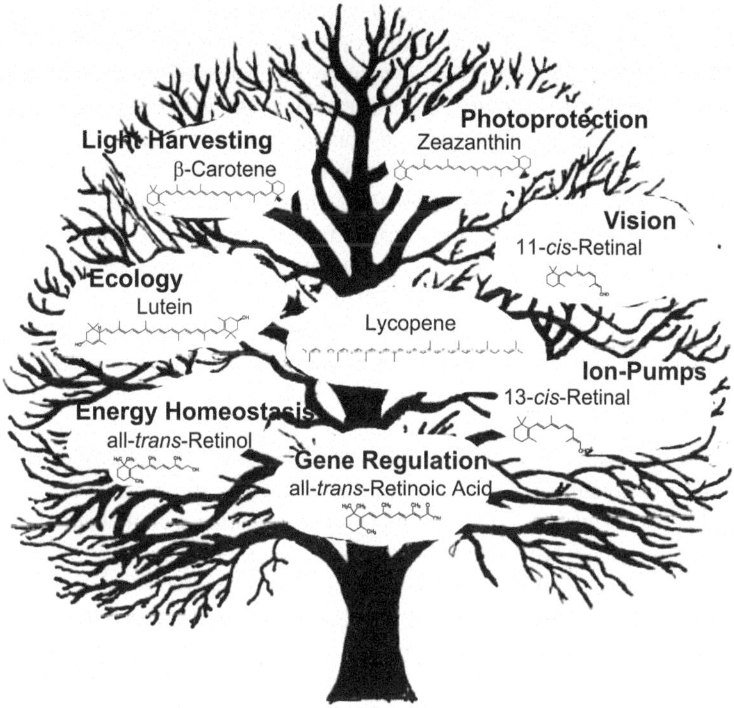

Fig. 8.4 The conjugated double bond system underlies most biological functions of retinoids and carotenoids. In its function as essential cofactor in the regulation of mitochondrial energy homeostasis retinol joins a family of retinoids and carotenoids that are active in vision, photosynthesis, ecology, and ion pumping, all critically dependent on the π electron system of conjugated double bonds. Retinoic acid, functioning as transcriptional transactivator, is the exception

Working Model and Unresolved Problems of the Electron-Transfer Capacity of Retinol

Natural retinoids, as well as their carotenoid parents, derive their utility from their unique electronic structure, the core feature of which is the linear system of conjugated double bonds [41]. The delocalization of π electrons of carotenes and retinoids underlies much of the function of these versatile molecules, be it light absorption (in photosynthesis, vision, and biopigment systems), photoprotection (by deflection of excess electrons) or electron transfer within photosystem II (Fig. 8.4). Electron transfer also defines the retinoid cofactors of the PKCδ signalosome within mitochondria.

We proposed that retinoids facilitate the oxido-reduction process by transfer of electrons from the PKCδ zinc-finger to oxidized cytochrome c, as well as in the reverse direction. Switching PKCδ off in a timely manner is as important as turning it on, since prolonged PKCδ activity causes cytotoxicity. As a case in point, expression of the constitutively active PKCδ kinase domain in mitochondria produces redox stress that may lead to apoptosis [71]. But even though this problem is recognized, no mechanism has come to light of how PKCδ might be inactivated.

In bacteria, reversibly switching Hsp33 from an active to an inactive state is accomplished by alternately oxidizing or reducing the zinc-finger, depending on the redox state of the microenvironment. On/off cycling would be a desirable feature for PKCδ as well, but would be tied to the redox status of cytochrome c. Whether or not cytochrome c exists at any given moment in the oxidized or reduced form is determined by the metabolic workload of the ETC. In the PKCδ signalosome (Fig. 8.2) the zinc-finger "cysteine/cystine" redox couple is electronically linked to the cytochrome c $^{3+}$/ c $^{2+}$ redox couple. By responding to changes in polarity of the cytochrome c redox potential, the zinc-finger could theoretically toggle between open (active) and closed (inactive) states. In this manner, the kinase could become rapidly activated when needed, but switched off when no longer required. The cysteine pair comprising the cysteine/cystine redox couple has not been identified.

The fact that oxidation of cysteine thiols requires the movement of two electrons, whereas the prosthetic heme of cytochrome c can accept only one electron at a time presents a conundrum. By necessity the unpaired electron left behind needs to be stabilized until the lead electron has moved on to cytochrome C oxidase. How this feat is accomplished is unclear. A similar situation arises during the flow of electrons from ubiquinol to cytochrome c. In this case, the conflict is resolved by the formation of a transient semiquinone radical intermediate. The question is whether during the transfer of electrons from PKCδ to cytochrome c retinol acts as a semiquinone-like stabilizer. Retinol possesses a hydroxyl group that in deprotonated form could donate one electron to conjugated cytochrome c^{3+}. Forming a powerful oxidizer, the retinol-radical carrying the transiently unpaired electron could then draw an electron from a cysteine sulfhydryl. This would restore retinol to its original structure but create a new radical in the zinc-finger peptide. With the deployment of a second cytochrome c^{3+} molecule the electron transfer cycle could repeat itself leading to the safe disposal of the zinc-finger unpaired electron, without chemically changing the retinol catalyst. While this scenario is speculative, it is noteworthy that retinol and ubiquinol are chemically, and perhaps evolutionarily, related due to their being synthesized from isoprene subunits.

PKCε localizes to the matrix where it regulates PDK2 in opposition to PKCδ. The PKCε, like the δ isoform, harbors high affinity retinol binding sites in the activation domains. Much evidence indicates that retinol binding is required for PKCε activation, suggesting a redox mechanism. However, the responsible upstream oxidant is unknown.

Other Retinoids as Cofactors for PKCδ Activation in Mitochondria

When vitamin A was initially identified as a co-factor for PKC activation, the question arose whether vitamin A itself, that is, the alcohol form, or one of its metabolites was the active form. This issue was solved by the demonstration that isolated mitochondria responded within seconds to retinol by upregulating the respiratory rate [1]. The absence of the cytoplasmic biochemical machinery and the immediacy of the response make it likely that retinol itself has electron transfer capacity.

But this does not disprove the possibility that retinol metabolites also serve as genuine co-activators of PKCs. At present we know of four major natural metabolites: Retinoic acid (RA), 14HRR [12], 13,14-dihydroxyretinol (DHR) [21], and anhydroretinol (AR) [13]. 14HRR and DHR were identified in the nineties, but it is unresolved whether they act as substitute ligands for retinol in similar redox pathways, whether they have their own special targets, or whether they can perform both functions. While their physiological relevance remains uncertain, the fact that 14HRR, DHR and AR are evolutionarily conserved from insect to man [38] suggests they likely meet a biological need. However, they have not yet been linked to specific *in vivo* systems. The enzymes involved in the hydroxylation of retinol are unknown, denying researchers the genetic tools that might reveal their biological significance. The biochemistry of AR received a boost by the discovery of a specific retinol-dehydratase in insects [38], but the mammalian homolog has not been defined.

Retinol, 14HRR, DHR, and RA share similar PKC binding affinities, but differences in the bioavailability of these retinoids may dictate that retinol is the more physiologically relevant effector of PKC activity. PKC receptor affinities are in the range of 10–100 nM. At the nominal concentration of 2 μM in the circulation, retinol would readily saturate the PKC receptors, but RA, 14HRR and DHR would not since their concentrations are below the nanomolar range in plasma. Therefore, unless high local concentrations are generated in microdomains, retinol would generally outcompete all three of these retinoids *in vivo*. Further, even when generated in specialized tissues, restrictions may be in place to prevent diffusion into the surrounding areas. For instance, the field of influence of RA is tightly walled-off by the Cyp26 family of retinoid-degrading enzymes [95, 102, 105].

Anhydroretinol as Contragonist of Retinol

AR fits the definition of a contragonist. This term was coined by British neuroscientist David Nutt to define agents that displace natural ligands from their receptors but, far from being biologically inert, set events in motion that countermand the original message [84].

Despite representing a natural retinoid of unknown biological relevance, AR has been a useful tool for mechanistic studies of retinoid action in cell culture systems. AR induces programmed cell death in a variety of cell lines of different origins and pedigrees. Retinol rescues cells from AR-mediated damage, but only during the first four to 6 h post exposure to AR. This time course is closely paralleled by changes in NADPH and ATP levels, which remain constant during an initial lag phase but decline rapidly thereafter, contributing to cell death [15]. The drastic ATP depletion preceding cell death suggests that mitochondria are the site where AR acts in a specific, energy-disrupting manner. AR-mediated effects are dramatically dependent on PKCε as indicated by the demonstration that AR-induced cell death is significantly delayed when PKCε is genetically inactivated, but accelerated when PKCε is over-expressed as a transgene [33]. Of note, over-expression of a mutant PKCε protein that lacks the retinol-binding site, fails to accelerate cell death. This finding ties AR directly to redox-activation of PKCε.

Interestingly, the time window during which cells are rescued by retinol is entirely eliminated in PKCε over-expressing cells, ie. the process of energy depletion leading to necrosis starts immediately after AR stimulation. In wild-type cells, the phase of rapid energy decline correlates with significant import of PKCε from the cytosol, implying AR-mediated signaling. These observations are consistent with the assumption that AR, contrary to its decades-old sobriquet of the "inert retinoid", is in fact an active co-factor of mitochondrial PKC and potent regulator of oxidative phosphorylation [33].

The Pathology Induced by AR May Explain How Retinol Works

It is often said that in order to understand normal physiology one must study what goes wrong in pathology. Therefore, if we knew why AR causes such dramatic imbalance in energy homeostasis we would also learn how retinol works. On the one hand, AR functions as PKC activation co-factor at similar concentrations as retinol, but on the other, the activation kinetics of AR are different from those of retinol; they are biphasic for AR, but monophasic for retinol [15]. The paradoxical behavior of AR can be partially explained by the recruitment of PKCε from the cytosol, causing inhibition of the PDHC, as discussed, although it begs the question of why AR, but not retinol, signals the influx of PKCε. Further, what types of AR-selective signals initiate the import of extra PKCε into the mitochondrial matrix is still unclear, but they are likely the same as those causing ROS release, if ROS itself is not the responsible signal.

Chen et al observed that AR, but not retinol, induced intracellular oxidative stress in lymphocytes in a time and dose-dependent manner, the severity of which positively correlated with cell death by necrosis. Interestingly, α-tocopherol, a potent scavenger of ROS, prevented AR-induced cell death [14]. Chiu et al obtained similar results and pointed out that retinol rescued cells, as long as it was given during the first 2 h after AR stimulation [15]. This early window in time roughly correlated with the positive-stimulatory phase of AR described above. Korichneva et al studied AR-specific effects in MEFs and cardiomyocytes and confirmed the time course of ROS production, showing a concomitant mitochondrial membrane hyperpolarization, another indicator of early AR-induced stress [68].

To reconcile these partially conflicting findings, especially the ability of AR to co-activate PKCδ, yet cause harmful redox stress, we propose that AR upsets the dynamism of PKCδ. The working model postulates that in order to maintain energy homeostasis PKCδ alternates between oxidized (active) and reduced (inactive) states, as dictated by the metabolic rates of the ETC, and that retinol is instrumental in catalyzing the transitions in both directions. To account for its paradoxical behavior we propose that AR functions as a forward co-activator of PKCδ, but is deficient in switching the kinase off. By prolonging its active phase PKCδ would in effect cause an overload of the ETC, resulting in the release of damaging amounts of ROS from complex III. The activation of the inhibitory PKCε system, including the influx of extra PKCε protein, may be part of the cell's strategy to mitigate mitochondrial overload [33].

Similar observations were raised with the synthetic retinoid, fenretinide. This retinoid was previously shown by others to kill a variety of cancer cells [24, 40, 86]. The authors suggested that compromised function of complex III contributed to programmed cell death but did not offer further proof. In our opinion the action of fenretinide was consistent with the disruption of the PKC signal network upstream of the PDHC. This would indeed impact the ETC including possible disturbances of Complex III. Fenretinide bound both PKCδ and ε isoforms and induced necrotic cell death by a mechanism not dissimilar from that caused by AR. Like AR, fenretinide engaged during the first 60 min the stimulatory PKCδ pathway and thereafter activated the inhibitory PKCε pathway. Also like AR, the experimental inactivation of the PKCε gene attenuated fenretinide-induced cell death, whereas over-expression of PKCε accelerated cell death, as ATP production plummeted below sustainable levels [33]. That AR and fenretinide elicited the same sequence of events with an eventually negative outcome re-affirmed the importance of opposing functions of the two PKC isoforms for controlling energy homeostasis, but did not explain why these two retinoids would initially serve as seemingly normal, positively acting co-factors, yet in the long run cause a energy crisis so severe as to commit cells towards programmed cell death. These observations also raise the question of what sets these two retinoids apart from retinol, which does not cause overt cell damage. The answer must lie in differences of their chemical and/or electronic structures and in their inability to control the orderly stepwise electron transfer. Indeed, in their presence, the PKC zinc-finger might disassemble and cause kinase activation, but due to the extended length of their conjugated double bond systems they may be incapable of catalyzing the reverse reaction making it difficult to restore the zinc-coordinated fold and to silence the kinase.

Relevance

Retinol Regulates Normal Energy Homeostasis, Whereas Imbalance in Retinol Homeostasis Causes Aberrant PKCδ Signaling Which Contributes to Metabolic Disease

As described above, vitamin A functions as an indispensible cofactor of a regulatory circuit that controls glycolysis in mitochondria [1]. Therefore, the news that over-expression of the holo-RBP complex correlated with obesity, and might even be responsible for progression to type 2 diabetes, struck us like lightening [36, 107]. Had we not argued and amassed much supportive evidence that vitamin A deficiency damaged oxidative phosphorylation [15], whereas over-supply of vitamin A drove glycolysis to potentially unsustainable levels [1]? Was not chronic reliance on glucose as fuel at the expense of fatty acids a hallmark of metabolic disease? These possibilities were exciting and remain so to date.

Recent research on the relationship between holo-RBP and metabolic syndrome has proven to be increasingly complex. The transporter controls vitamin A uptake in a variety of cells, but also has hormone-like functions that regulate adiposity and insulin responses [5–7]. Moreover, RBP4 has been found to produce inflammation in adipose tissues, which contributes to insulin resistance [78]. Apart from cytokine functions, vitamin A is also involved in short-term regulation of the PDHC of mitochondria [1]. How these different RBP4 and vitamin signal pathways may interact with each other to maintain energy homeostasis remains to be seen.

Since the PDHC controls the final step of glycolysis in mitochondria it should however not come as a surprise that imbalances of the upstream signals regulating the PDHC have pathological consequences. The fault-line of oxidative phosphorylation lies in the risk that when fuel production becomes disconnected from the workload of the ETC excessive production of harmful ROS might occur. Since the focal point of PKC signaling is the PDK2 and its downstream PDHC target, imbalances in that network will affect PDHC function with two predictable outcomes: (1) Signals that tip the balance towards decreased PDK2 activity (by activating the PKCδ axis, for example) will boost the PDHC output of glycolytic fuel, thereby increasing the risks of ETC-overload, promoting ROS production and generally contributing to metabolic disease; and (2) the converse is true for signals that increase PDK2 activity (via PKCε activation) which dampen the PHC and thus generally have beneficial consequences, but also carry the risk of subpar metabolic fitness.

We proposed that PKCδ activation in mitochondria depends on the formation of a specific signal complex, comprising p66Shc, cytochrome c and vitamin A. If so, manipulations of the components of this complex that lead to either over- or under-performance of PKCδ should result in disease phenotypes that resemble the PKCδ overexpression or knockout-phenotypes, respectively. Indeed, supra-normal holo-RBP4 levels, have been found to predispose mice to obesity and type 2 diabetes [107]. The proposed mechanism of pathogenesis was attributed to cytokine-like action of the RBP protein. Accordingly, stimulation of the STRA6 surface receptor by holo-RBP4 led to the activation of a JAK/STAT signal cascade that upregulated the transcription of genes that inhibit insulin signaling (SOCS3) or enhance adiposity (PPARγ) [5]. Recently, a second RBP4 pathway leading to insulin resistance was described. This pathway involved the activation of antigen-presenting cells in adipose tissue, causing a local inflammation (Moraes-Vieira et al. [78]).

We offer a complementary explanation that is based on the accumulation of vitamin A in tissues, especially in muscle, that is driven by the overabundance of the retinol transporter [94]. In our view, excess vitamin A is likely to chronically increase PKCδ activity causing mitochondrial stress. The reason why increased abundance of the cofactor should increase the function of the effector, PKCδ, is not intuitive. However, the occupancy of PKCδ binding sites by retinol must conform to mass action law. The retinol binding affinity for PKCδ (50 nM) is in near-equilibrium with that of RBP4 (20 nM). Therefore, the proportions of retinol-bound versus free PKCδ species will be in equilibrium with the holoRBP concentration in plasma, and increasing the extracellular holoRBP concentration would augment the pool of retinol-primed PKCδ in mitochondria, leading to a stronger PKCδ signal to PDHC. This in turn will result in enhanced glucose utilization potentially increasing the risk of metabolic stress.

PKCε opposes PKCδ in mitochondria. Our genetic and biochemical evidence from cell culture experiments indicated that the activation of PKCε results in the activation of PDK2, the opposite outcome of PKCδ activation. Activated PDK2 in turn inhibited the PDHC. It would be of interest to explore to what extent an active PKCε pathway contributes to the amelioration of the metabolic syndrome. In fact, mice expressing a constitutively active PKCε transgene in the normoxic heart displayed profound changes in glucose and energy metabolism, exemplified by lower levels of glucose, lactate, glutamin and creatine, but higher levels of choline, glutamate, and adenosine nucleotides all indicative of reduced energetics. On the other hand, a strategy to compensate for diminished glycolysis in these hearts resulted in increased levels of PKCδ in mitochondria [73]. The biochemical logic of shifting during times of need the balance away from the inhibitory action of PKCε towards enhanced PDHC output by strengthening PKCδ signaling is compelling. Since the fundamentals of mitochondria are broadly conserved among animal cells, the balancing function of PKCε vis-à-vis PKCδ likely goes beyond cardiac cells and has general physiological significance.

Stimulation of cells *in vitro* with fenretinide results in the recruitment of extra PKCε to mitochondria, causing the balance to shift from the stimulatory PKCδ towards the suppressive PKCε action [33]. Because of strong antitumor activity *in vitro*, fenretinide was previously used in clinical trials for the chemoprevention of secondary breast cancer [20]. While the hoped-for risk reduction was indeed observed, the primary effect of fenretinide was not attributable to the purported tumoricidal activity, but was attributed to a profound reduction in adiposity. Hence lowering the cancer risk associated with adiposity was the reason for the observed drop in cancer incidence [59]. These observations are consistent with the fenretinide-mediated suppression of glycolysis [33]. The compensatory switch to fat metabolism promoted a lean body mass. Similar to these results, chronic fenretinide treatment of mice prevented, or even reversed, high-fat diet-induced obesity, insulin resistance and hepatosteatosis [92].

The Future

The discovery of the PKCδ signalosome marks the beginning of a novel paradigm of vitamin A function that is distinct from the well-established retinoic acid paradigm. While conceptual advances steadily accrue, a long list of unresolved questions remains, including resolution of the biochemistry, kinetics, the details of one-electron redox chemistry that control PKC activity, the thermodynamics of PKC activation, and the molecular structure of the PKCδ signalosome. One priority will be to understand the mechanics and control of reversibility of PKCδ action, since monitoring the ETC in real-time is unlikely to work unless PKCδ has the dynamic ability to switch between active and inactive states.

Many of these endeavors will require new technologies. By comparison, it seems technically feasible to fill gaps in the PKCδ signal pathway, such as the identification of the "missing" PDK2 phosphatase, the nature of the primary substrate of PKCδ and the definition of the phosphorylation sites on PDK2. From the thermodynamics point of view, the identification of a chaperone participating in the refolding of PKCδ would be a major step forward. The PKCε isoform provides a important counterweight to PKCδ, yet the upstream activator was not identified. Like PKCδ, PKCε requires vitamin A as cofactor, suggesting a redox dependent activation mechanism. Since PKCε resides in the matrix, cytochrome c is excluded, the search for an equivalent oxidoreductase mediating PKCε activation might focus on a different cytochrome.

The unique function of vitamin A as an electron carrier is the first example of a general system that connects the one-electron redox chemistry of a heme protein (cytochrome c) with the two-electron chemistry of a classical phosphoprotein (PKCδ). Whether this paradigm extends to other members of the PKC and Raf families, or even beyond, seems a worthy line of inquiry. All of these kinases contain vitamin A binding sites in their zinc-finger activation domains, and some of these kinases (eg C-Raf) are activated by redox stress. It is conceivable that the redox network underpinning phosphoprotein signaling systems is extensive. The non-genomic physiology of vitamin A may be considerably broader than hitherto suspected.

Acknowledgments I wish to express my appreciation to all friends and colleagues who contributed wisdom, advice, encouragement to carry on, and also caution, happily ignored, not to overreach: William Blaner, Jochen Buck, Lorraine Gudas, Marilyn Gunner, Beatrice Hoyos, Lonny Levin, Filippo Mancia, Noa Noy, James Pendergrast, Loredana Quadro, Elena Shabrova, Lawrence Shapiro. The order of names is not hierarchical as all contributions are valuable. However I would like to acknowledge specially the inner circle of my research group: BH, ES, LQ, JB and LL. Thanks are also due to Sloan-Kettering Institute for Cancer Research for the generous support over the many years I spent at that illustrious institution.

References

1. Acin-Perez R, Hoyos B, Zhao F, Vinogradov V, Fischman DA, HArris RA, Leitges M, Wongsiriroy N, Blaner WS, MAnfredi G, Hammerling U (2009) Control of oxidative phosphorylation by vitamin A illuminates a fundamental role in mitochondrial energy homoeostasis. FASEB J 24:627–36
2. Acin-Perez R, Hoyos B, Gong J, Vinogradov V, Fischman DA, Leitges M, Borhan B, Starkov A, Manfredi G, Hammerling U (2010) Regulation of intermediary metabolism by the PKCdelta signalosome in mitochondria. FASEB J 24:5033–5042
3. Acin-Perez R, Gatti DL, Bai Y, Manfredi G (2011) Protein phosphorylation and prevention of cytochrome oxidase inhibition by ATP: coupled mechanisms of energy metabolism regulation. Cell Metab 13:712–719
4. Adams DR, Ron D, Kiely PA (2011) RACK1, A multifaceted scaffolding protein: structure and function. Cell Commun Signal 9:22
5. Berry DC, Noy N (2012) Signaling by vitamin A and retinol-binding protein in regulation of insulin responses and lipid homeostasis. Biochim Biophys Acta 1821:168–176
6. Berry DC, O'Byrne SM, Vreeland AC, Blaner WS, Noy N (2012) Cross talk between signaling and vitamin A transport by the retinol-binding protein receptor STRA6. Mol Cell Biol 32:3164–3175
7. Berry DC, Jacobs H, Marwarha G, Gely-Pernot A, O'Byrne SM, DeSantis D, Klopfenstein M, Feret B, Dennefeld C, Blaner WS, Croniger CM, Mark M, Noy N, Ghyselinck NB (2013) The STRA6 receptor is essential for retinol-binding protein-induced insulin resistance but not for maintaining vitamin A homeostasis in tissues other than the eye. J Biol Chem 288:24528–24539
8. Blaner WS, Olson JA (eds) (1994) Retinol and retinoic acid metabolism. Ravenn Press, New York, pp 229–255
9. Bowker-Kinley MM, Davis WI, Wu P, Harris RA, Popov KM (1998) Evidence for existence of tissue-specific regulation of the mammalian pyruvate dehydrogenase complex. Biochem J 329(Pt 1):191–196
10. Buck J, Hammerling U, Hoffmann MK, Levi E, Welte K (1987) Purification and biochemical characterization of a human autocrine growth factor produced by Epstein-Barr virus-transformed B cells. J Immunol 138:2923–2928
11. Buck J, Ritter G, Dannecker L, Katta V, Cohen SL, Chait BT, Hammerling U (1990) Retinol is essential for growth of activated human B cells. J Exp Med 171:1613–1624
12. Buck J, Derguini F, Levi E, Nakanishi K, Hammerling U (1991) Intracellular signaling by 14-hydroxy-4,14-retro-retinol. Science 254:1654–1656
13. Buck J, Grun F, Kimura S, Noy N, Derguini F, Hammerling U (1993) Anhydroretinol: a naturally occurring inhibitor of lymphocyte physiology. J Exp Med 178:675–680
14. Chen Y, Buck J, Derguini F (1999) Anhydroretinol induces oxidative stress and cell death. Cancer Res 59:3985–3990
15. Chiu H-J, Fischman DA, Hammerling U (2008) Vitamin A-depletion causes oxidative stress, mitochondrial dysfunction and PARP-1-dependent energy deprivation. FASEB J 22:3738–3787
16. Churchill EN, Szweda LI (2005) Translocation of deltaPKC to mitochondria during cardiac reperfusion enhances superoxide anion production and induces loss in mitochondrial function. Arch Biochem Biophys 439:194–199
17. Churchill EN, Murreil CL, Chen CH, Mochley-Rosen D, Szweda LI (2005) Reperfusion-induced translocation of delta-PKC to cardiac mitochondria prevents pyruvate dehydrogenase reactivation. Circ Res 97:78–85
18. Cogan U, Kopelman M, Mokady S, Shinitzky M (1976) Binding affinities of retinol and related compounds to retinol binding proteins. Eur J Biochem 65:71–78
19. De Luca LM (1991) Retinoids and their receptors in differentiation, embryogenesis, and neoplasia. FASEB J 5:2924–2933

20. Decensi A, Zanardi S, Argusti A, Bonanni B, Costa A, Veronesi U (2007) Fenretinide and risk reduction of second breast cancer. Nat Clin Pract Oncol 4:64–65
21. Derguini F, Nakanishi K, Hammerling U, Buck J (1994) Intracellular signaling activity of synthetic (14R)-, (14S)-, and (14RS)-14-hydroxy-4,14-retro-retinol. Biochemistry 33:623–628
22. Eppinger TM, Buck J, Hammerling U (1993) Growth control or terminal differentiation: endogenous production and differential activities of vitamin A metabolites in HL-60 cells. J Exp Med 178:1995–2005
23. Fagiani E, Giardina G, Luzi L, Cesaroni M, Quarto M, Capra M, Germano G, Bono M, Capillo M, Pelicci P, Lanfrancone L (2007) RaLP, a new member of the Src homology and collagen family, regulates cell migration and tumor growth of metastatic melanomas. Cancer Res 67:3064–3073
24. Ferrari N, Morini M, Pfeffer U, Minghelli S, Noonan DM, Albini A (2003) Inhibition of Kaposi's sarcoma in vivo by fenretinide. Clin Cancer Res 9:6020–6029
25. Finkel T, Holbrook NJ (2000) Oxidants, oxidative stress and the biology of ageing. Nature 408:239–247
26. Frank HA, Brudvig GW (2004) Redox functions of carotenoids in photosynthesis. Biochemistry 43:8607–8615
27. Garbe A, Buck J, Hammerling U (1992) Retinoids are important cofactors in T cell activation. J Exp Med 176:109–117
28. Gertz M, Steegborn C (2010) The Lifespan-regulator p66Shc in mitochondria: redox enzyme or redox sensor? Antioxid Redox Signal 13:1417–1428
29. Gertz M, Fischer F, Wolters D, Steegborn C (2008) Activation of the lifespan regulator p66Shc through reversible disulfide bond formation. Proc Natl Acad Sci U S A 105:5705–5709
30. Giguere V, Ong ES, Segui P, Evans RM (1987) Identification of a receptor for the morphogen retinoic acid. Nature 330:624–629
31. Giorgio M, Migliaccio E, Orsini F, Paolucci D, Moroni M, Contursi C, Pellicia G, Luzi L, Minucci S, Marcaccio M, Pinton P, Rizzuto R, Bernardi P, Paolucci F, Pelicci PG (2005) Electron transfer between cytochrome c and p66Shc generates reactive oxygen species that trigger mitochondrial apoptosis. Cell 122:221–233
32. Giorgio M, Berry A, Berniakovich I, Poletaeva I, Trinei M, Stendardo M, Hagopian K, Ramsey JJ, Cortopassi G, Migliaccio E, Notzli S, Amrein I, Lipp HP, Cirulli F, Pelicci PG (2012) The p66Shc knocked out mice are short lived under natural condition. Aging Cell 11:162–168
33. Gong J, Hoyos B, Acin-Perez R, Vinogradov V, Shabrova E, Zhao F, Leitges M, Fischman D, Manfredi G, Hammerling U (2012) Two protein kinase C isoforms, delta and epsilon, regulate energy homeostasis in mitochondria by transmitting opposing signals to the pyruvate dehydrogenase complex. FASEB J 26:3537–3549
34. Goodman D (1984) Vitamin A and retinoids in health and disease. New Engl J Med 310:1023–1031
35. Gopalakrishna R, Gundimeda U, Schiffman JE, McNeill TH (2008) A direct redox regulation of protein kinase C isoenzymes mediates oxidant-induced neuritogenesis in PC12 cells. J Biol Chem 283:14430–14444
36. Graham TE, Yang Q, Bluher M, Hammarstedt A, Ciaraldi TP, Henry RR, Wason CJ, Oberbach A, Jansson PA, Smith U, Kahn BB (2006) Retinol-binding protein 4 and insulin resistance in lean, obese, and diabetic subjects. N Engl J Med 354:2552–2563
37. Green S, Chambon P (1988) Nuclear receptors enhance our understanding of transcription regulation. Trends Genet 4:309–314
38. Grun F, Noy N, Hammerling U, Buck J (1996) Purification, cloning, and bacterial expression of retinol dehydratase from Spodoptera frugiperda. J Biol Chem 271:16135–16138
39. Gudi R, Bowker-Kinkley MM, Kedishvili NY, Zhao Y, Popov KM (1995) Diversity of the pyruvate dehydrogenase kinase gene family in humans. J Biol Chem 270:28989–28994

40. Hail NJ, Kim HJ, Lotan R (2006) Mechanisms of fenretinide-induced apoptosis. Apoptosis 11:1677–1694
41. Hammerling U (2013) The centennial of vitamin A: a century of research in retinoids and carotenoids. FASEB J 27:3887–3890
42. Hanlon PR, Fu P, Wright GL, Steenbergen C, Arcasoy MO, Murphy E (2005) Mechanisms of erythropoietin-mediated cardioprotection during ischemia-reperfusion injury: role of protein kinase C and phosphatidylinositol 3-kinase signaling. FASEB J 19:1323–1325
43. Harman D (2006) Free radical theory of aging: an update: increasing the functional life span. Ann N Y Acad Sci 1067:10–21
44. Harris RA, Huang B, Wu P (2001) Control of pyruvate dehydrogenase kinase gene expression. Adv Enzym Regul 41:269–288
45. Hecker E (1967) Phorbol esters from croton oil. Chemical nature and biological activities. Naturwissenschaften 54:282–284
46. Holness MJ, Sugden MC (2003) Regulation of pyruvate dehydrogenase complex activity by reversible phosphorylation. Biochem Soc Trans 31:1143–1151
47. Hommel U, Zurini M, Luyten M (1994) Solution structure of a cysteine rich domain of rat protein kinase C. Nat Struct Biol 1:383–388
48. Hoyos B, Imam A, Chua R, Swenson C, Tong G-X, Levi E, Noy N, Hammerling U (2000) The cysteine-rich regions of the regulatory domains of Raf and protein kinase C as retinoid receptors. J Exp Med 192:835–845
49. Hoyos B, Imam A, Korichneva I, Levi E, Chua R, Hammerling U (2002) Activation of c-Raf kinase by ultraviolet light. J Biol Chem 277:23949–23947
50. Hoyos B, Jiang S, Hammerling U (2005) Location and functional significance of retinol binding sites on the serine/threonine kinase, cRaf. J Biol Chem 280:6872–6878
51. Hoyos B, Acin-Perez R, Fischman DA, Manfredi G, Hammerling U (2012) Hiding in plain sight: uncovering a new function of vitamin A in redox signaling. Biochim Biophys Acta 1821:241–247
52. Hu Y, Kang C, Philp R, Li B (2007) PKCδ phosphorylates p52ShcA at Ser29 to regulate ERK activation in response to H2O2. Cell signal 19:410–418
53. Huang B, Gudi R, Wu P, Harris RA, Hamilton J, Popov KM (1998) Isoenzymes of pyruvate dehydrogenase phosphatase. DNA-derived amino acid sequences, expression, and regulation. J Biol Chem 273:17680–17688
54. Hurley JH, Newton AC, Parker PJ, Blumberg PM, Nishizuka Y (1997) Taxonomy and function of C1 protein kinase C homology domains. Protein Sci 6:477–480
55. Ilbert M, Graf PCF, Jakob U (2006) Zinc center as redox switch—new function for an old motif. Antioxid Redox Signal 8:835–846
56. Ilbert M, Horst J, Ahrens S, Winter J, Graf PC, Lilie H, Jakob U (2007) The redox-switch domain of Hsp33 functions as dual stress sensor. Nat Struct Mol Biol 14:556–563
57. Imam A, Hoyos B, Swenson C, Levi E, Chua R, Viriya E, Hammerling U (2001) Retinoids as ligands and coactivators of protein kinase C alpha. FASEB J 15:28–30
58. Jakob U, Muse W, Eser M, Bardwell JC (1999) Chaperone activity with a redox switch. Cell 96:341–352
59. Johansson H, Gandini S, Guerrieri-Gonzaga A, Iodice S, Ruscica M, Bonanni B, Gulisano M, Magni P, Formelli F, Decensi A (2008) Effect of fenretinide and low-dose tamoxifen on insulin sensitivity in premenopausal women at high risk for breast cancer. Cancer Res 68:9512–9518
60. Karpova T, Danchuk S, Kolobova E, Popov KM (2003) Characterization of the isozymes of pyruvate dehydrogenase phosphatase: implications for the regulation of pyruvate dehydrogenase activity. Biochim Biophys Acta 1652:126–135
61. Kazanietz MG, Bustelo XR, Barbacid M, Kolch W, Mischak H, Wong G, Pettit GR, Bruns JD, Blumberg PM (1994) Zinc finger domains and phorbol ester pharmacophore. Analysis of binding to mutated form of protein kinase C zeta and the vav and c-raf proto-oncogene products. J Biol Chem 269:11590–11594

62. Kazanietz MG, Wang S, Milne GW, Lewin NE, Liu HL, Blumberg PM (1995) Residues in the second cysteine-rich region of protein kinase C delta relevant to phorbol ester binding as revealed by site-directed mutagenesis. J Biol Chem 270:21852–21859
63. Kikkawa U, Matsuzaki H, Yamamoto T (2002) Protein kinase C delta (PKC delta): activation mechanisms and functions. J Biochem 132:831–839
64. Knapp LT, Klann E (2000) Superoxide-induced stimulation of protein kinase C via thiol modification and modulation of zinc content. J Biol Chem 275:24136–24145
65. Kolobova E, Tuganova A, Boulatnikov I, Popov KM (2001) Regulation of pyruvate dehydrogenase activity through phosphorylation at multiple sites. Biochem J 358:69–77
66. Konishi H, Tanaka M, Takemura Y, Matsuzaki H, Ono Y, Kikkawa U, Nishizuka Y (1997) Activation of protein kinase C by tyrosine phosphorylation in response to H2O2. Proc Natl Acad SAci 94:11233–11237
67. Korichneva I, Hoyos B, Chua R, Levi E, Hammerling U (2002) Zinc-release from protein kinase C as the common event during activation by lipid second messenger or reactive oxygen. J Biol Chem 277:44327–44331
68. Korichneva I, Waka J, Hammerling U (2003) Regulation of the cardiac mitochondrial membrane potential by retinoids. J Pharmacol Exp Ther 305:426–433
69. Kowalczyk JE, Beresewicz M, Gajkowska B, Zablocka B (2009) Association of protein kinase C delta and phospholipid scramblase 3 in hippocampal mitochondria correlates with neuronal vulnerability to brain ischemia. Neurochem Int 55:157–163
70. Kumsta C, Jakob U (2009) Redox-regulated chaperones. Biochemistry 48:4666–4676
71. Majumder PK, Neerad C, Mishra XS, Bharti A, Surender K, Satya S, Kufe D (2001) Targeting of protein kinase C δ to mitochondria in the oxidative stress response. Cell Growth Differ 12:465–470
72. Mayr M, Chung YL, Mayr U, McGregor E, Troy H, Baier G, Leitges M, Dunn MJ, Griffiths JR, Xu Q (2004) Loss of PKC-delta alters cardiac metabolism. Am J Physiol Heart Circ Physiol 287:H937–H945
73. Mayr M, Liem D, Zhang J, Li X, Avliyakulov NK, Yang JI, Young G, Vondriska TM, Ladroue C, Madhu B, Griffiths JR, Gomes A, Xu Q, Ping P (2009) Proteomic and metabolomic analysis of cardioprotection: interplay between protein kinase C epsilon and delta in regulating glucose metabolism of murine hearts. J Mol Cell Cardiol 46:268–277
74. McCollum E, Davis M (1913) The necessity of certain lipins in the diet during growth. J Biol Chem 15:167–175
75. Migliaccio E, Mele S, Salcini AE, Pelicci G, Lai KM, Superti-Furga G, Pawson T, Di Fiore PP, Lanfrancone L, Pelicci PG (1997) Opposite effects of the p52shc/p46shc and p66shc splicing isoforms on the EGF receptor-MAP kinase-fos signalling pathway. EMBO J 16:706–716
76. Migliaccio E, Giorgio M, Mele S, Pelicci G, Reboldi P, Pandolfi PP, Lanfrancone L, Pelicci PG (1999) The p66shc adaptor protein controls oxidative stress response and life span in mammals. Nature 402:309–313
77. Migliaccio E, Giorgio M, Pelicci PG (2006) Apoptosis and aging: role of p66Shc redox protein. Antioxid Redox Signal 8:600–608
78. Moraes-Vieira PM, Yore MM, Dwyer PM, Sayed I, Aryal P, Kahn BB (2014) RBP4 activates antigen-presenting cells, leading to adipose tissue inflammation and systemic insulin resistance. Cell Metab 19:512–526
79. Morita M, Matsuzaki H, Yamamoto T, Fukami Y, Kikkawa U (2008) Epidermal growth factor receptor phosphorylates protein kinase C {delta} at Tyr332 to form a trimeric complex with p66Shc in the H2O2-stimulated cells. J Biochem 143:31–38
80. Mott HR, Carpenter JW, Zhong S, Ghosh S, Bell RM, Campbell SL (1996) The solution structure of the Raf-1 cysteine-rich domain: a novel ras and phospholipid binding site. Proc Natl Acad Sci U S A 93:8312–8317

81. Nemoto S, Combs CA, French S, Ahn BH, Fergusson MM, Balaban RS, Finkel T (2006) The mammalian longevity-associated gene product p66shc regulates mitochondrial metabolism. J Biol Chem 281:10555–10560
82. Newton PM, Messing RO (2010) The substrates and binding partners of protein kinase Cepsilon. Biochem J 427:189–196
83. Nishizuka Y (1992) Intracellular signaling by hydrolysis of phospholipids and activation of protein kinase C. Science 258:607–614
84. Nutt DJ, Linnoila M (1988) Neuroreceptor science: a clarification of terms. J Clin Psychopharmacol 8:387–389
85. O'Connell M, Chua R, Hoyod B, Buck J, Chen CQ, Derguini F, Hammerling U (1996) Retro-retinoids in regulated cell growth and death. J Exp Med 184:549–555
86. Oridate N, Suzuki S, Higuchi M, Mitchell MF, Hong WK, Lotan R (1997) Involvement of reactive oxygen species in N-(4-hydroxyphenyl)retinamide-induced apoptosis in cervical carcinoma cells. J Natl Cancer Inst 89:1191–1198
87. Osborne TB, Mendel LB (1913) The relation of growth to the chemical constituents of diet. J Biol Chem 15:311–326
88. Patel MS, Korotchkina LG (2006) Regulation of the pyruvate dehydrogenase complex. Biochem Soc Trans 34:217–222
89. Pelicci G, Lanfrancone L, Grignani F, McGlade J, Cavallo F, Forni G, Nicoletti I, Pawson T, Pelicci PG (1992) A novel transforming protein (SHC) with an SH2 domain is implicated in mitogenic signal transduction. Cell 70:93–104
90. Petkovich M, Brand NJ, Krust A, Chambon P (1987) A human retinoic acid receptor which belongs to the family of nuclear receptors. Nature 330:444–450
91. Pinton P, Rimessi A, Marchi S, Orsini F, Migliaccio E, Giorgio M, Contursi C, Minucci S, Mantovani F, Wieckowski MR, Del Sal G, Pelicci PG, Rizzuto R (2007) Protein kinase C beta and prolyl isomerase 1 regulate mitochondrial effects of the life-span determinant p66Shc. Science 315:659–663
92. Preitner F, Mody N, Graham TE, Peroni OD, Kahn BB (2009) Long-term Fenretinide treatment prevents high-fat diet-induced obesity, insulin resistance, and hepatic steatosis. Am J Physiol Endocrinol Metab 297:E1420–E1429
93. Purdom S, Chen QM (2003) Linking oxidative stress and genetics of aging with p66Shc signaling and forkhead transcription factors. Biogerontology 4:181–191
94. Quadro L, Blaner WS, Hamberger L, Van Gelder RN, Vogel S, Piantedosi R, Gouras P, Colantuoni V, Gottesman ME (2002) Muscle expression of human retinol-binding protein (RBP). Suppression of the visual defect of RBP knockout mice. J Biol Chem 277:30191–30197
95. Ray WJ, Bain G, Yao M, Gottlieb DI (1997) CYP26, a novel mammalian cytochrome P450, is induced by retinoic acid and defines a new family. J Biol Chem 272:18702–18708
96. Reyland ME (2009) Protein kinase C isoforms: multi-functional regulators of cell life and death. Front Biosci (Landmark Ed) 14:2386–2399
97. Roche TE, Baker JC, Yan X, Hiromasa Y, Gong X, Peng T, Dong J, Turkan A, Kasten SA (2001) Distinct regulatory properties of pyruvate dehydrogenase kinase and phosphatase isoforms. Prog Nucleic Acid Res Mol Biol 70:33–75
98. Roffey J, Rosse C, Linch M, Hibbert A, McDonald NQ, Parker PJ (2009) Protein kinase C intervention: the state of play. Curr Opin Cell Biol 21:268–279
99. Ross A, Hammerling U (1994) Retinoids in the immune system. Raven, New York
100. Rosse C, Linch M, Kermorgant S, Cameron AJ, Boeckeler K, Parker PJ (2010) PKC and the control of localized signal dynamics. Nat Rev Mol Cell Biol 11:103–112
101. Rowles J, Scherer SW, Xi T, Majer M, Nickle DC, Rommens JM, Popov KM, Harris RA, Riebow NL, Xia J, Tsui LC, Bogardus C, Prochazka M (1996) Cloning and characterization of PDK4 on 7q21.3 encoding a fourth pyruvate dehydrogenase kinase isoenzyme in human. J Biol Chem 271:22376–22382

102. Sakai Y, Meno C, Fujii H, Nishino J, Shiratori H, Saijoh Y, Rossant J, Hamada H (2001) The retinoic acid-inactivating enzyme CYP26 is essential for establishing an uneven distribution of retinoic acid along the anterio-posterior axis within the mouse embryo. Genes Dev 15:213–225
103. Sugden MC, Holness MJ (2003) Recent advances in mechanisms regulating glucose oxidation at the level of the pyruvate dehydrogenase complex by PDKs. Am J Physiol Endocrinol Metab 284:E855–E862
104. Wald G (1968) The molecular basis of visual excitation. Nature 219:800–807
105. White JA, Beckett-Jones B, Guo YD, Dilworth FJ, Bonasoro J, Jones G, Petkovich M (1997) cDNA cloning of human retinoic acid-metabolizing enzyme (hP450RAI) identifies a novel family of cytochromes P450. J Biol Chem 272:18538–18541
106. Wolbach SB, Howe PR (1925) Tissue changes following deprivation of fat-soluble A vitamin. J Exp Med 42:753–777
107. Yang Q, Graham TE, Mody N, Preitner F, Peroni OD, Zabolotny JM, Kotani K, Quadro L, Kahn BB (2005) Serum retinol binding protein 4 contributes to insulin resistance in obesity and type 2 diabetes. Nature 436:356–362
108. Zhang G, Kazanietz MG, Blumberg PM, Hurley JH (1995) Crystal structure of the Cys2 activator-binding domain of protein kinase C8 in complex with phorbol ester. Cell 81:917–924
109. Zhao F, Ilbert M, Varadan R, Cremers CM, Hoyos B, Acin-Perez R, Vinogradov V, Cowburn D, Jakob U, Hammerling U (2011) Are zinc-finger domains of protein kinase C dynamic structures that unfold by lipid or redox activation? Antioxid Redox Signal 14:757–766

Chapter 9
Vitamin A and Vision

John C. Saari

Abstract Visual systems detect light by monitoring the effect of photoisomerization of a chromophore on the release of a neurotransmitter from sensory neurons, known as rod and cone photoreceptor cells in vertebrate retina. In all known visual systems, the chromophore is 11-*cis*-retinal complexed with a protein, called opsin, and photoisomerization produces all-*trans*-retinal. In mammals, regeneration of 11-*cis*-retinal following photoisomerization occurs by a thermally driven isomerization reaction. Additional reactions are required during regeneration to protect cells from the toxicity of aldehyde forms of vitamin A that are essential to the visual process. Photochemical and phototransduction reactions in rods and cones are identical; however, reactions of the rod and cone visual pigment regeneration cycles differ, and perplexingly, rod and cone regeneration cycles appear to use different mechanisms to overcome the energy barrier involved in converting all-*trans*- to 11-*cis*-retinoid. Abnormal processing of all-*trans*-retinal in the rod regeneration cycle leads to retinal degeneration, suggesting that excessive amounts of the retinoid itself or its derivatives are toxic. This line of reasoning led to the development of various approaches to modifying the activity of the rod visual cycle as a possible therapeutic approach to delay or prevent retinal degeneration in inherited retinal diseases and perhaps in the dry form of macular degeneration (geographic atrophy). In spite of great progress in understanding the functioning of rod and cone regeneration cycles at a molecular level, resolution of a number of remaining puzzling issues will offer insight into the amelioration of several blinding retinal diseases.

Keywords Vision • Retina • Rod • Cone • Regeneration • Visual cycle • CRALBP • RPE65 • 11-*cis*-retinol • 11-*cis* retinal • All-*trans*-retinol • All-*trans*-retinal

---

J.C. Saari (✉)
Departments of Ophthalmology and Biochemistry, School of Medicine,
University of Washington, 98195 Seattle, WA, USA

Department of Biochemistry, University of Washington, 357350, 98195 Seattle, WA, USA
e-mail: jsaari@u.washington.edu

M.A. Asson-Batres, C. Rochette-Egly (eds.), *The Biochemistry of Retinoid Signaling II*, Subcellular Biochemistry 81, DOI 10.1007/978-94-024-0945-1_9

Abbreviations

AAV	adeno-associated virus
ABCA4	ATP-binding cassette type 4
AMD	age-related macular degeneration
A2E	N-retinylidene-N-retinylethanolamine, a bisretinoid
At-Ral	all-*trans*-retinal
At-RE	all-*trans*-retinyl ester
At-Rol	all-*trans*-retinol
CRABP	cellular retinoic acid-binding protein
CRALBP	cellular retinal-binding protein
CRBP1	cellular retinol-binding protein type 1
DAPI	4',6-diamidino-2-phenylindole
IPM	interphotoreceptor matrix
IRBP	interphotoreceptor retinoid-binding protein
LCA	Leber's congenital amaurosis
MFAT	multifunctional O-acyltransferase
NHERF1	sodium hydrogen exchanger regulatory factor type 1
RDH	retinol dehydrogenase
11-Ral	11-*cis*-retinal
11-Rol	11-*cis*-retinol
11-RE	11-*cis*-retinyl ester
PA	phosphatidic acid
PC	phosphatidylcholine
PDZ	postsynaptic density 95, discs large, ZO1
PE	phosphatidylethanolamine
PI(3,4,5)P3	phosphatidylinositol (3,4,5) tris-phosphate
PS	phosphatidylserine
RalDi	retinal dimer
RAR	retinoic acid receptor
RBP	retinol binding protein
Retinene	earlier name for retinal
RGR-opsin	retinal g-protein receptor opsin
RIP1 or −3 kinases	receptor interacting protein kinases 1 or 3
ROS	rod outer segments
RPE	retinal pigment epithelial or epithelium
RPE65	retinal pigment epithelial protein 65 kDa, aka isomerohydrolase
RXR	retinoid X-receptor
SUV	small unilamellar vesicles
TNFα	tumor necrosis factor α
VCM	visual cycle modulation

Introduction

The vertebrate visual system responds to light over an approximate 10^{10} fold range of intensities [122, 142]. This amazing feat is accomplished by two sets of light receptors with overlapping sensitivities, known as rod and cone photoreceptor cells or more simply as rods and cones. Rods are exquisitely sensitive to low levels of illumination and will even signal the absorption of a single photon, but their range is limited because their response saturates as the intensity increases. Cones are less sensitive than rods but their response does not saturate over a physiologically relevant range of intensities. Overlap of the intensity response curves of the two classes of photoreceptor cells generates the impressive visual response mentioned above (duplicity theory of vision, [145]). Both rods and cones employ the same fundamental photochemistry and thus both require regeneration cycles to maintain functional amounts of light- absorbing visual pigment (chromophore).

Vision at its most basic in vertebrates, requires a chromophore or light-absorbing substance within a sensory neuron and a mechanism to couple information about the absorption of light to the output of a neurotransmitter. In all known visual systems, the chromophore is 11-*cis*-retinal or a closely related derivative (3-hydroxy or 4-hydroxy 11-*cis*-retinal, or 3,4-dehydro 11-*cis*-retinal derived from vitamin A2), covalently bound to a protein called an opsin (Fig. 9.1). Absorption of a photon causes 11-*cis*-retinal to isomerize to all-*trans*-retinal, which changes the shape of

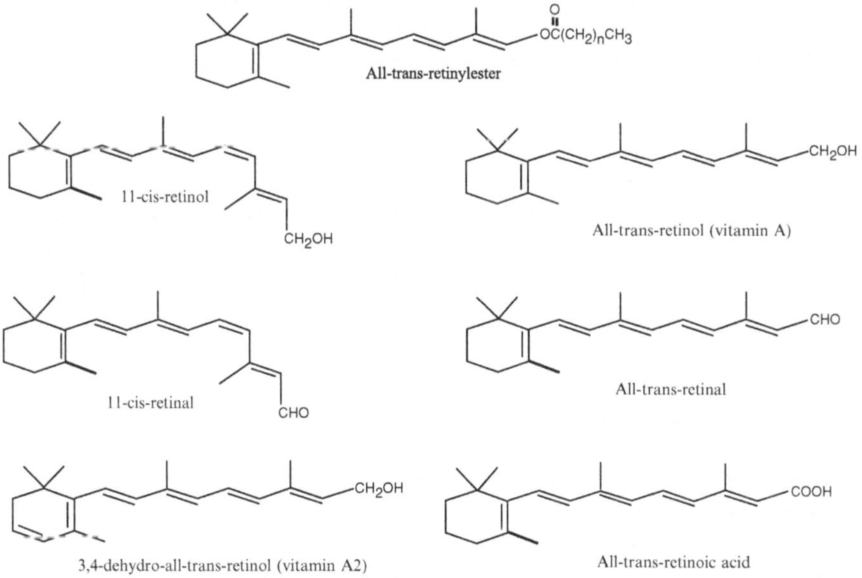

Fig. 9.1 Retinoid structures. Structures of retinoid intermediates in the rod and cone regeneration cycles are shown with all-*trans*-retinoic acid and vitamin A2 included for comparison. For all-*trans*-retinyl ester, n = 14 (most common) or 16

the opsin protein allowing it to interact with components downstream in a cascade and ultimately change the rate of release of glutamate from photoreceptor cells [77]. Our present understanding of the molecular details of this visual process was achieved by the intellectual effort of numerous vision scientists. The resulting mechanistic model serves as the prototype of signal transduction processes throughout biology.

Phototransduction, as sophisticated as it is, is only half the visual process because without a mechanism to reset the chromophore, rod and cone photoreceptor cells would eventually run out of functional visual pigment. This photoisomerization and resetting cycle was known originally as the visual cycle [164, 166] and is now also known as the retinoid or regeneration cycle to distinguish it from the phototransduction cycle. An understanding of the molecular functioning of the cone regeneration cycle lagged considerably behind that of the rod cycle but studies in recent years have opened our understanding of cone visual pigment regeneration, which differs considerably from that of rod visual pigment regeneration [90, 168–170].

Rod and cone regeneration cycles have turned out to be surprisingly complex. Whereas invertebrates rely on a photoreversal reaction that regenerates 11-*cis*-retinal from all-*trans*-retinal in a second photoisomerization reaction, all vertebrates utilize an enzymatically catalyzed isomerization for regeneration of 11-*cis*-retinal in the dark. In addition, this dark isomerase is situated not within the photoreceptor cells where the visual pigment is found, but within adjacent cells, retinal pigment epithelial (RPE) cells for rods and Müller cells for cones, necessitating mechanisms for translocation of the chromophore back and forth between the cell types. Additional complexity results from the chemistry involved in dealing with the retinoids. For instance, dark conversion of all-*trans*- to 11-*cis*-retinoid occurs at the oxidation level of the alcohol whereas visual pigments employ 11-*cis*-retinoid as the aldehyde, necessitating additional enzymology for interconversion of oxidation states. Finally, additional mechanisms are needed to protect photoreceptor cells from the toxic properties of the 11-*cis*-retinal and all-*trans*-retinal, the very molecules essential to vertebrate vision.

Humans commonly experience the rod regeneration cycle in the phenomenon of dark adaptation, which is the relatively slow gain in sensitivity of vision when we abruptly transition from a bright to a much darker environment. In an experimental setting following nearly complete photoisomerization of the 11-*cis*-retinal of our rods and cones, regain of visual sensitivity in the dark shows two distinct phases, an initial relatively rapid phase reflecting regain of cone function, and a slower phase reflecting regain of rod function [6, 76, 124]. Full regeneration of rod photoreceptor sensitivity takes approximately 40 min. Several active processes are responsible for quenching the signaling activity of opsin during this period including addition of 11-*cis*-retinal to regenerate the original visual pigment.

Several reviews of the rod visual cycle that have appeared recently provide more details and evidence supporting the roles of components of the cycle [71, 76, 77, 104, 105, 108, 125, 159, 161].

History

The modern view of vitamin A's role in the visual process developed when two independent lines of evidence merged in the early 1930s. One line of evidence resulted from observations that a fat-soluble factor could prevent a cluster of nutritional diseases including night blindness and xerophthalmia (corneal drying), diseases that we now associate with vitamin A deficiency. Eventual biochemical and structural studies led to the determination of the structure of fat soluble A, which was later named all-*trans*-retinol or vitamin A [62]. Several accounts of early events leading to the discovery of vitamin A have appeared recently [78, 123, 146, 150]. Wald later demonstrated that extraction of retinas yielded a substance then called retinene, which could be converted to vitamin A, thus providing chemical evidence linking vitamin A with the visual process [163, 164, 166]. Morton and colleagues later demonstrated that retinene was vitamin A aldehyde or retinal (reviewed in [96]). A second line of evidence led from photochemical studies with frog retina in the 1870s [14, 73], which noted that frog retina changed from purple to yellow and then colorless upon illumination. *Cis-trans* photoisomerization was tied to the visual process with the realization that chemically synthesized vitamin A (all-*trans*-retinol) did not yield a visual pigment upon oxidation to retinene (retinal) unless it had been irradiated with light and thereby isomerized [52]. The active retinene isomer was isolated and named neoretinene b [52]. However identification of its structure as 11-*cis*-retinal [103] required a great deal of experimentation in part because it was a "hindered" *cis*-form, which was believed to be unstable (summarized in [96]). Wald's recognition that dark isomerization of retinal could regenerate the original light absorbing substance complimented an early observation of Kühne who noted that proximity of the neural retina to the RPE was necessary for regeneration of the original purple pigment of the retina. By 1935, Wald had proposed a visual cycle that remarkably contained several of the key elements that we now associate with the set of reactions responsible for the photoisomerization and regeneration of rod visual pigment (rhodopsin) [164]. Dowling filled in many of the details of Wald's model in 1960s and his outline of the cycle has guided regeneration cycle research ever since [26].

Development of the Field

Anatomy of the Retina

Photoreceptor cells of vertebrates are positioned in the retina with their outer segments, the portions of the cells containing the light absorbing pigments, directed towards the back of the eye (Fig. 9.2). Thus, incoming light must pass through the layers of other neurons and the inner segments of photoreceptor cells before reaching the light absorbing visual pigment. Although seemingly a disadvantage, this

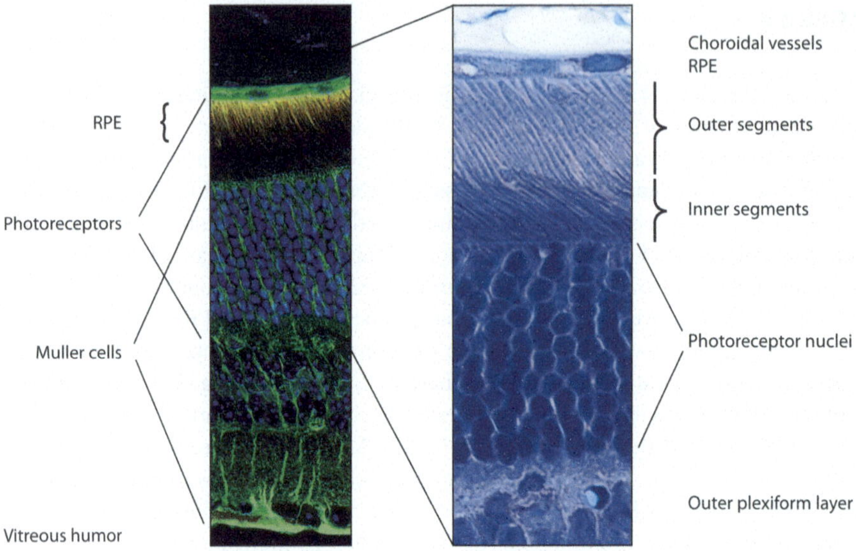

Fig. 9.2 Anatomy of the retina. (**a, left**) Fluorescence micrograph of a section of an albino mouse retina stained with fluorescent markers for RPE and Müller cells (*green*, anti-CRALBP), nuclei (*blue*, DAPI), and and RPE apical processes (*red*, anti-ezrin). The tips of rod and cone outer segments are interspersed between the RPE apical processes but are not visible in this micrograph. (**b, right**) Light micrograph of an albino mouse retina stained with *toludine blue*. Rod and cone outer segments are visible as indicated, RPE apical processes are not visible but are intercalated between the closely packed outer segments. Light enters the retina from the bottom in both images. *RPE* retinal pigment epithelium (The image in (**a**) is taken from [51], with modification)

arrangement places the photoreceptors cells and especially their outer segments (ROS) adjacent to the retinal pigment epithelium (RPE) whose basal surface is bathed by the choroidal circulation, the most active circulatory flux of vertebrates [154]. The basal surface of this single layer of polarized epithelial cells is highly infolded and contains receptors for retinol-binding protein (RBP4) and thus vitamin A (retinol) delivery [65, 155]. The apical surface elaborates long processes that extend centrally and wrap around the rod photoreceptor cell outer segments. The extracellular compartment between the photoreceptor tips and the RPE cells, called the interphotoreceptor matrix (IPM) by scientists and the sub-retinal space by clinicians, is further defined by the plasma membranes of Müller glial cells and the semipermeable junctions of the outer limiting membrane.

Vitamin A Chemistry, Metabolism, and Transport in the Eye

Figure 9.3 schematically illustrates our current model of vitamin A chemistry, metabolism and transport in the eye, as discussed in detail in the following sections.

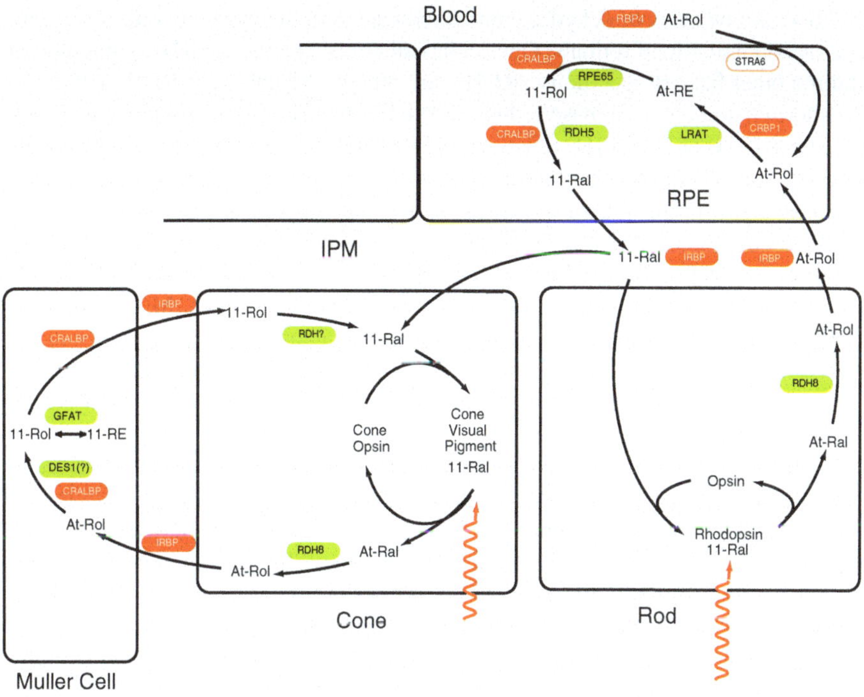

Fig. 9.3 Depiction of the reactions of rod and cone regeneration cycles. Schematic depicting retinoid signaling in retinal pigment epithelial cells (RPE) and photoreceptor (cone and rod) and Müller cells of the neural retina. Retinoid intermediates are abbreviated as indicated below. Enzymes and retinoid-binding proteins are shown in *green* and *red rounded-rectangles*, respectively. The receptor for RBP4 is shown in a clear rounded-rectangle. The rod regeneration cycle involves rod photoreceptor and RPE cells. The cone regeneration cycle involves cone photoreceptor, Müller cells, and RPE cells in mouse retina. *At-Ral* all-*trans*-retinal, *At-Rol* all-*trans*-retinol, *At-RE* all-*trans*-retinyl ester, *CRALBP* cellular retinal-binding protein, *CRBP1* cellular retinol-binding protein type 1, *RDH* retinol dehydrogenase, *DES1* dihydroceramide 4-desaturase type 1, *IPM* interphotoreceptor matrix, *IRBP* interphotoreceptor matrix retinoid-binding protein, *LRAT* lecithin:retinol acyltransferase, *RBP4* retinol-binding protein type 4, *RPE65* retinal pigment epithelial protein of 65 kDa, aka isomerohydrolase, *STRA6* stimulated by retinoic acid 6, aka RBP4 receptor, *11-Ral* 11-*cis*-retinal, *11-Rol* 11-*cis*-retinol, *11-RE* 11-*cis*-retinyl ester

Several features that appear in all systems involving retinoid metabolism are prominent in the visual regeneration systems and much of the complexity of the visual cycle likely results from the use of a toxic substance (11-*cis*-retinal) as the chromophore for the sensory system. In the visual system, retinoids alternate between the 11-*cis*- and all-*trans*-isomers of retinal and retinol. 11-*cis*-Retinal is an inverse agonist, binding to and reducing the weak intrinsic signaling activity of opsin and producing a light-sensitive receptor with amazingly low dark noise . Photoisomerization generates all-*trans*-retinal and a highly active signaling form of opsin.

There are two sets of dehydrogenases/reductases in the eye that control the generation and half-life of retinals. One set has all-*trans*-specificity (RDH8 and others) and the other has *cis*-specificity (RDH5 and others). Visual cycle RDHs belong to the short-chain dehydrogenase/reductase (SDR) super-family of proteins, and they utilize NAD(H) or NADP(H) as cofactors [70, 108]. In contrast to the single copies of genes specifying other components of the mammalian regeneration cycles, several genes specifying dehydrogenases of both all-*trans*- and *cis*-specificities are expressed in retina [70, 108]. Although the physiological substrate of each of these enzymes has yet to be established, perhaps they are present in multiple copies because of the need to control concentrations of toxic aldehydes. More generally, many short-chain reductases and medium chain alcohol dehydrogenases occur widely tissues with varying amounts of activity towards retinoids [36, 74, 98]. Determination of their relevancy for specific cellular processes remains an active area of research.

Both the rod and cone visual systems segregate dark isomerases (RPE65 and putative DES1, respectively) that convert all-*trans*-retinoids to 11-*cis*-retinoids, and some related regeneration reactions to different cells that are adjacent to rod or cone photoreceptor cells. Perhaps this is related to the possibility that photoisomerization and dark isomerization in the same cell would lead to futile cycling of the isomerization state of the retinoid. In addition, for rod cells this separation places the bulk of retinoid metabolism in a cell type adjacent to the blood supply (RPE cells) of the outer retina and thus ensures a continual supply of fresh retinol (vitamin A) as noted above.

The chemical properties and reactivities of retinoids demand mechanisms to increase their solubility in aqueous media, to protect them from inadvertent chemical reactions, and to direct their interactions with other components. Retina has a diverse set of retinoid-binding proteins that are tailored for these functions, including opsins, two cellular retinoid-binding proteins of different specificities (CRBP1 and CRALBP), and an extracellular retinoid-binding protein (IRPB) that mediates translocation of retinoids between cell types.

The chemically reactive aldehyde of retinal must be protected from non-specific reaction with amino groups of proteins and glycerophospholipids. Of particular importance is the necessity to prevent inappropriate oxidation of all-*trans*-retinal to all-*trans*-retinoic acid with its potent biological activity. Retinoid-binding proteins limit these reactivities because their affinities for retinoids are high and the aldehyde of retinal is protected from reaction when bound. In addition, Specific interactions of the retinoid-binding proteins with enzymes allow specific reactions to occur.

Retinoids must cross an extracellular compartment during the regeneration cycle because of the separation of photochemistry and dark isomerization as mentioned above. A retinoid-binding protein (IRBP) is present in the extracellular compartment in question.

The Rod Visual Cycle

In rhodopsin, the rod visual pigment, 11-*cis*-retinal is chemically bound to a lysine ε-amino group as a protonated Schiff base, which shifts its absorption maximum from 365 nm to 440 nm. Interactions of amino acid side chains of opsin with the chromophore further shifts its absorption maximum to ~500 nm, in the green range of visible light [107]. Absorption of a photon of the appropriate energy results in isomerization of 11-*cis*-retinal to all-*trans*-retinal. Photoactivated rhodopsin then decays through a series of spectroscopic intermediates (observable at low temperatures), reflecting relaxation of the opsin structure. Meta rhodopsin II (MetaII), the signaling state of photoactivated rhodopsin, consists of all-*trans*-retinal still linked to opsin via a protonated Schiff base [144]. The signaling half-life of MetaII is determined by several active processes including protein phosphorylation, binding of arrestin, and hydrolysis of the Schiff base with release of all-*trans*-retinal, all of which decrease signaling activity. The final quenching of signaling activity occurs when opsin binds 11-*cis*-retinal and rhodopsin is regenerated.

The life time of all-*trans*-retinal is determined by the activity of RDH8 [117] and other dehydrogenases of the rod outer segment [44, 70, 108], which reduce the aldehyde to the alcohol (all-*trans*-retinol), a less chemically reactive species. This reduction is one of the two slow steps in the rod regeneration cycle [26, 76, 135, 136] and involves NADPH [3, 20, 172].

ABCA4, a member of the ABC-transporter family [156] is also found in rod outer segment (ROS) membranes [156]. It was originally proposed to rescue all-*trans*-retinal that mislocated to the inner leaflet of ROS disk membranes by "flipping" its orientation to the cytosolic leaflet where it could be reduced to all-*trans*-retinol by RDH8 [9, 53]. Recent evidence suggests that it may also be involved in protecting the ROS from excess 11-*cis*-retinal [112].

All-*trans*-retinol leaves the ROS apparently in a passive process and traverses the interphotoreceptor matrix (IPM) chaperoned by a retinoid-binding protein (IRBP) in the IPM. IRBP is a water-soluble protein synthesized by photoreceptor cells and secreted into the IPM [2, 16, 41, 49, 75, 132]. IRBP binds 11-*cis*- and all-*trans*- retinols and retinals [101] with low affinity (Kd ~ 100 nM) compared to the other retinoid binding proteins. Evidence from various species demonstrateded that IRBP delivered all-*trans*-retinol to RPE more effectively than other retinoid-binding proteins [28, 29], suggesting the presence of an IRBP "receptor;" however no molecular characterization has been published. The highly asymmetric shape of IRBP (axial ratio 7:1) and its large size (MW ~ 144,000 Da) make it unlikely that the protein physically moves within the IPM to deliver its cargo of all-*trans*-retinol. It is more likely that a series of multiple binding equilibria are established that limit the amount of unbound retinoid [48].

Within RPE, all-*trans*-retinol binds to CRBP1 with high affinity ($K_d = 0.1$ nM) and is rapidly converted to all-*trans*-retinyl palmitate, stearate, or oleate in a reaction in which the fatty acyl group of the *sn1* position of a glycerophospholipid is transferred to the hydroxyl of retinol, catalyzed by lecithin:retinol acyltransferase

(LRAT) [8, 127, 128]. Studies with other tissues suggested that the role of CRBP1 in this reaction is to deliver all-*trans*-retinol to LRAT [110, 175]. The high affinity binding of retinol and its rapid esterification are likely to provide a concentration gradient that drives diffusion of all-*trans*-retinol from the ROS into the RPE [51, 137]. Retinyl esters are stored within RPE in lipid vesicles (retinosomes) that line the lateral interior perimeter of the generally hexagonal RPE cells [55]. Confocal microscopy demonstrated in mice that these vesicles fill with retinyl esters following exposure to light [54]. Control of the release of retinyl esters from retinosomes has been suggested to depend on photoisomerization of all-*trans*-retinal bound to RGR-opsin [114].

The mechanism for dark regeneration of the retinoid 11-*cis*-configuration puzzled investigators for decades and was finally resolved when Rando and associates demonstrated that enzymatic isomerization occurred at the oxidation state of retinol not retinal [11, 12]. Years later, RPE65, a prominent protein of RPE microsome preparations, was identified as the isomerase [59, 95, 118]. It is now apparent that RPE65 catalyzes the spontaneous hydrolysis and isomerizaton of all-*trans*-retinyl ester (an isomerohydrolase reaction) to generate palmitate and 11-*cis*-retinol. This reaction is the second slow step in the mouse rod regeneration cycle [79, 136, 137].

RPE contains CRALBP, a retinoid-binding protein with specificity directed to *cis*-retinols and *cis*-retinals. CRALBP is a water-soluble, globular protein that binds 11-*cis*-retinol (K_d = 60 nM) produced by RPE65 in these cells and relieves product inhibition of the isomerase [129, 153, 173]. CRALBP also facilitates oxidation of bound 11-*cis*-retinol to 11-*cis*-retinal, catalyzed by RDH5 and other *cis*-specific dehydrogenases in RPE [126, 133]. This observation was in striking contrast to esterification of 11-*cis*-retinol by LRAT, which was strongly retarded when the retinoid was bound to CRALBP [133]. Acceleration of oxidation and retardation of esterification suggested that CRALBP interacted with the dehydrogenase and that 11-*cis*-retinol was oxidized while it was bound to CRALBP. To verify this hypothesis, the oxidation of CRALBP-bound 11-*cis*-retinol was examined in the presence of O-methylhydroxylamine, a carbonyl reagent that reacts with free but not CRALBP-bound 11-*cis*-retinal. The results indicated that ~30 % of the 11-*cis*-retinal produced escaped reaction with the carbonyl reagent, suggesting that 11-*cis*-retinol may have been oxidized without being released from CRALBP [133]. However, one can not rule out the possibility that the rate of the reaction of the carbonyl agent was too slow to completely react with free 11-*cis*-retinal before rebinding to CRALBP.

These *in vitro* results suggested a model for the role of CRALBP in which the protein bound 11-*cis*-retinol generated by RPE65 and facilitated its oxidation to 11-*cis*-retinal by RDH5. This model was also supported by studies of the phenotype of CRALBP knockout mice [136, 174], which displayed delayed dark adaptation following a strong flash that isomerized 80 % of the rhodopsin and by examination of human patients with a mutation of the CRALBP gene (*RLBP1*), who also displayed delayed dark adaptation [17]. Examination of the kinetics of visual cycle retinoid intermediates during recovery from a flash demonstrated that all-*trans*-retinyl esters, the substrates for the isomerase, decayed much more slowly in the knockout, suggesting impairment in the isomerase step of the cycle in the absence of CRALBP [136].

Analysis of the distribution of visual cycle enzymes, retinoid-binding proteins, and retinoid-binding protein-associated cytoskeletal proteins in RPE cells provided evidence for the partition of visual cycle reactions to various cellular compartments [51]. CRALBP and CRBP1 were distributed throughout the RPE cell from the basal surface to the tips of the apical processes. Enzymes of the visual cycle (LRAT, RDH5, and RPE65) were found in the RPE cell body and were excluded from the apical processes. Finally, retinosomes were distributed within the RPE cell body, lining the lateral sides of the hexagonal cells [55]. These distributions led to a compartmentalization model of the rod visual cycle [51] in which CRBP1 and CRALBP facilitated the diffusion of retinoids throughout the cell, minimized free retinoid concentrations, directed retinoids to various enzymes, and facilitated their uptake and release.

However, this model postulated that 11-*cis*-retinal must be released from CRALBP and from RPE cells and passed on to IRBP, the chaperone from its traverse of the IPM. Attempts to demonstrate release *in vitro* were initially unsuccessful. The high affinity for 11-*cis*-retinal for CRALBP (Kd ~10 nM) and its much lower affinity for IRBP (Kd ~100 nM) indicated that net transfer of 11-*cis*-retinal from CRALBP to IRBP was forbidden. Indeed a slow exchange of retinoids between CRALBP and IRBP could be observed *in vitro* [48]; however, addition of crude RPE fractions, proteases, cAMP, sulhydryl agents, and ATP to the exchange assay failed to produce net transfer of retinoid to IRBP.

Interestingly, studies on the binding of CRALBP to lipids revealed a possible resolution of this problem. An overlay assay revealed that CRALBP bound strongly to acidic lipids with phosphatidic cid (PA), phosphatidyl inositol phosphates [PI(3,4,5)P_3], and phosphadidyl serine (PS), showing the strongest binding [100]. Moreover, addition of small unilamellar vesicles (SUVs) doped with acidic lipids, to CRALBP·9-*cis*-retinal (used as a surrogate for 11-*cis*-retinal) resulted in release of the retinoid for reaction with a carbonyl reagent [138]. The order of efficacy for this release matched that for the binding of CRALBP to lipid surfaces. This suggested that binding of an acidic lipid perturbed the structure of CRALBP and reduced its affinity for 9-*cis*-retinal. Analysis of the surface of CRALBP obtained first from a structural model [138] and then from an X-ray crystallographic structure [46] revealed a patch of basic residues that formed a recess. Though not demonstrated experimentally this basic recess could correspond to a binding site for the acidic lipids. These results suggested a mechanism for release of 11-*cis*-retinal in which CRALBP-11-*cis*-retinal bound to an acidic lipid in the plasma membrane inner leaflet and then released 11-*cis*-retinal into the IPM [138]. However, it must be stressed that the rate of release with PA-doped SUVs was very slow relative to turnover in the visual cycle. In addition, no extensive survey of acidic physiologic substances was carried out and it is possible that the release mechanism involves substances other than lipids. This area deserves more study.

Once 11-*cis*-retinal reaches the IPM, IRBP is available to chaperone the retinoid during diffusion to rod photoreceptor cells. However, it seems unlikely that the free retinoid with its chemically reactive aldehyde would be able to diffuse through the apical membrane of the RPE cell unassisted. A similar problem would appear to involve the mechanism for uptake of 11-*cis*-retinal by the rod outer segment, where another plasma membrane must be traversed. This problem does not appear to have been addressed experimentally. Association of 11-*cis*-retinal with opsin appears to occur spontaneously, completing the rod visual cycle.

The Cone Visual Cycle

The classical biochemical techniques that were so successful in deciphering the molecular details of rod phototransduction and rod pigment regeneration were not as useful for studying cone function because rods outnumber cones by about 20 to 1 in retinas that were commonly used for studies such as those from cow, mouse, and frog. Thus early, pre-genomic studies were limited to those employing electrophysiology, microspectrophotometry, and immunocytochemistry. These studies yielded many distinguishing features of cone physiology that only now are amenable to studies at the biochemical level. Electrophysiologists recognized that cones, in contrast to rods, could regenerate their visual pigments even when isolated from the RPE [39]. In addition, cones could regenerate their visual pigment from 11-*cis*-retinol as well as 11-*cis*-retinal whereas rods could only regenerate with 11-*cis*-retinal [5, 61]. Many electrophysiological studies demonstrated that visual pigment regeneration was faster in cones than rods [122], which was generally attributed to a more rapid rate of release and reduction of all-*trans*-retinal ([4]: [40]).

Strikingly, immunocytochemistry provided evidence for 11-*cis*- and all-*trans*-retinoid metabolism in neural retina, independent of RPE. CRALBP, which is specific for *cis*-retinols and retinal, was localized to RPE and Müller cells, the principle glial cells of the retina [16, 131]. CRALBP from neural retina (separated from RPE), purified with 11-*cis*-retinol and 11-*cis*-retinal as endogenous ligands, further implicating this glial cells in retinoid metabolism related to vision [130]. CRBP1 was also localized to Müller cells using immunocytochemistry [13, 16, 131] and mRNAs specific for CRALBP and CRABP were localized to Müller cells [134]. Application of biochemical studies to chicken retina, which has a rod cone ratio of about 1:2, yielded the important observation that cultured chicken Müller cells generated 11-*cis*-retinol following incubation with all-*trans*-retinol [25]. Subsequently several enzymatic activities were discovered in cone-enriched retinas, primarily chicken, and a plausible cycle of reactions was proposed occurring partly in Müller cells and partly in cones [90]. Recently, a combination of methods have been used in mice to demonstrate that 11-*cis*-retinoid produced in Müller cells was responsible for an early rapid phase of cone visual pigment regeneration whereas 11-*cis*-retinal from RPE completed regeneration in a slower phase [68, 169]. However, it should be pointed out that fundamental questions regarding the chemistry and enzymology of the cone regeneration cycle in various species remain to be determined.

A variety of cone visual pigments with varying spectral sensitivities and structures were found in vertebrates [58]. In cone visual pigments, 11-*cis*-retinal is bound to a lysine as a Schiff base as it is in the rod visual pigment and the fundamental photochemistry of cones is identical to that of rods in all vertebrates. Thus, it is reasonable to assume that the cone visual cycle for all types of cone photoreceptor cells will involve variations of a central theme. However, the species variations in dehydrogenases and isomerases noted below suggest that there may be differences in the enzymatic processes of retinoid during operation of the cycle. This issue remains to be resolved. The cone visual cycle as currently proposed for mouse involves the participation of primarily M-cones and Müller cells [66, 90], with contributions from 11-*cis*-retinal production in RPE.

Following photoisomerization, all-*trans*-retinal is released from cone visual pigments and reduced by NADPH within the cone outer segment. RDH8, the same dehydrogenase that is active in rods, catalyzes this reduction. In carp retina, an additional mechanism has been proposed in which reduction of all-*trans*-retinal is coupled to oxidation of 11-*cis*-retinol in NADP(H)-independent reactions [94]. The mechanism of this reaction has not been studied nor has it been determined if a similar process occurs in mammalian cone photoreceptors. Cones have been demonstrated to contain other dehydrogenases in addition to RDH8, RHD12 and RDH13 in mouse [44] and RDH8L2 and RDH13 in carp [94]. However, they are present in the cone inner segments and the identity of their physiological substrates has not been determined.

Following reduction to all-*trans*-retinol, the retinoid leaves cone photoreceptor cells and enters the IPM. As mentioned above, the IPM is bounded by the membranes of RPE, rod and cone outer segments, and Müller cells. IRBP is present throughout this compartment [16] and presumably chaperones retinoids during their diffusion to and from Müller cells just as it does for retinoids diffusing to and from RPE cells. It is worth noting that it is at this point that rod and cone regeneration cycles intersect.

Within Müller cells, all-*trans*-retinol is isomerized to 11-*cis*-retinol by an activity that has been called isomerase2 in mouse [66], isomerase II in chicken [97], and RPE65c in zebrafish [158]. According to recent studies, isomerase2 in mice is DES1 (dihydroceramide 4-desaturase-1), an enzyme of the membrane desaturase family, [66].

It must be stressed that the all-*trans*- to 11-*cis*-isomerization is endergonic. Indeed, the energy of 11-*cis*-retinoids is greater than that of all-*trans*-retinoids by ~ 4.1 kcal/mol [115] because of steric hindrance of the C-13-methyl with the C-10 hydrogen. In RPE, the energy for the isomerization catalyzed by RPE65 is obtained by coupling the hydrolysis of all-*trans*-retinyl esters ($\Delta G^{o\prime} \sim -5$ kcal/mole) to isomerization [115]. In contrast, in the cone visual cycle proposed for mice, all-*trans*-retinol is directly isomerized to 11-*cis*-retinol, which is immediately esterified or bound by CRALBP to drive the reaction in the direction of isomeriation [90]. DES1 has been demonstrated to interact with CRALBP [67], suggesting that a multiprotein complex is involved. Recently, the generation of the 11-*cis*-retinyl esters has been shown to involve a multifunctional O-acyltransferase (MFAT) [67].

A recent examination of the physiology of CRALBP-deficient mouse retina has clarified the role of the binding protein in retinal function [174]. M-cone dark adaptation was substantially delayed in CRALBP-deficient mouse retina supporting a role in chromophore recycling as previously demonstrated for rods [136]. M-cone sensitivity and amplification were also reduced in these animals. Restoration of CRALBP expression in either RPE or Müller cells improved M-cone dark adaptation indicating that both RPE and retinal visual cycles support cone function. However, restoration of CRALBP expression selectively in RPE cells did not restore dark-adapted M-cone sensitivity and amplification indicating that CRALBP in Muller cells is necessary for normal cone function.

The cone visual cycle has also been studied at the molecular level in zebrafish retina [22, 35, 161]. An RPE65 paralog, RPE65c, was detected in the zebrafish genome and its cDNA cloned, expressed, and characterized. The expressed enzyme was similar to RPE65 from mammalian species in its kinetic properties, dependence on iron, and expression in cells of the inner retina.

11-*cis*-Retinol, released from Müller cells, is taken up by nearby cone inner segments, and oxidized to 11-*cis*-retinal by an unidentified dehydrogenase. The ability of cone but not rod photoreceptor cells to use 11-*cis*-retinol for regeneration of their visual pigments [5, 61] may provide a mechanism by which the relatively small cone photoreceptor cell population can avoid competition for 11-*cis*-retinoid with the more numerous rod photoreceptors. In primates and birds however, the packing of cones into small regions of the retina (the macula and fovea) would also appear to serve this purpose. As mentioned above, in carp the oxidation of 11-*cis*-retinol to 11-*cis*-retinal in cone inner segments is apparently coupled to the reduction of all-*trans*-retinal to all-*trans*-retinol.

Association of 11-*cis*-retinal with cone opsin, which regenerates the original cone visual pigment, appears to occur spontaneously. However, this area has not been well studied. Cone visual pigment structure is more open than that of rhodopsin as judged by the greater accessibility of the Schiff base to water soluble carbonyl reagents [91], the greater rate of exchange of the chromophore with exogenous 9- or 11-*cis*-retinal [23], and the removal of 11-*cis*-retinal in the dark by apo-CRALBP [69]. Presumably, this open structure is one factor that facilitates the more rapid rate of cone visual pigment regeneration noted in electrophysiological studies [122].

Current State of the Field

Key Methods

Development of methods for inactivation of specific genes in mice permitted studies that completed our understanding of the roles of several enzymes and proteins whose functions previously could only be inferred from *in vitro* studies.

Methods for heterologous expression of genes also were instrumental in assigning enzymatic activities to specific gene products, a problem that plagued understanding the enzymology of the visual cycle for years. The parallel determination of the genome sequence of humans, mice, and numerous other species led to an appreciation of the conservation of genes and sequence motifs in visual processes. Confocal microscopy provided incomparable images of the cellular localizations of visual cycle enzymes and proteins and two-photon laser scanning confocal microscopy of living cells provided evidence for a functional role of retinosomes in the visual cycle. Methods for generating knockout and knockin mice have produced several animal models for human diseases [85, 171]. Naturally occurring mutations affecting the visual cycle in inbred dogs have provided large animals models for biochemical and physiological studies and subjects for initial trials with AAV delivery of corrective genes [10].

Controversies

From the beginning, the field of vision science has been fraught with controversies, competing theories, and seemingly insolvable problems. Controversies have frequently stimulated research into the relevant areas and have led to unexpected aspects of the underlying fundamental processes. An example is the early controversy regarding the identity of an "internal transmitter" in the phototransduction cascade, with investigators backing either cGMP or Ca^{2+}. The issue stimulated a great deal of research, which eventually revealed that both small molecules mediated important aspects of the visual response of rod and cone photoreceptors [7, 64, 139]. An additional example is the dark isomerase of the rod regeneration cycle, which was first detected in 1986 [11] but not associated with a protein structure until 2005 [59, 95, 118]. The mechanism of the isomerase reaction generated controversy with one group favoring a nucleophilic attack [33, 116] and another a carbocation intermediate [92]. Recent studies appear to have resolved the issue in favor of a radical cation mechanism [109, 119]. An understanding of the mechanism of this reaction is important because of recent efforts to decrease the activity of the visual cycle by the use of inhibitors of RPE65 (see discussion below). Isomerization in Müller cells continues to generate controversy with three different mechanisms proposed for isomerization in three different species. In addition, RPE65 has been detected in cones outer segments, where it has been postulated to play a non-enzymatic role, perhaps acting as a retinoid-binding protein (summarized in [159]). A more recent controversy relates to the relative importance of A2E, RalDi, and all-*trans*-retinal as toxic agents in progression of the form of macular degeneration known as geographic atrophy. This topic will be discussed in more detail in section "The future".

Relevance

The early symptoms of vitamin A deficiency including xerophthalmia and night blindness were keys to the discovery of fat soluble A (vitamin A) as mentioned in the introduction to this chapter. Retinoid research also stimulated the development of chemical methods for synthesis and characterization of vitamin A and carotenoids [57, 62]. Later studies of the design and synthesis of RAR agonists and antagonists relied, in part, on the chemistry developed during elucidation of the visual process.

The series of reactions by which a photon affects the release of transmitter from a neuron is arguably one of the best understood sensory signal transduction process in biology [77]. Numerous elaborations of this signaling mechanism, first discovered by investigators studying phototransduction, were later recognized in other G protein-coupled receptor cascades. In addition, the sequential cascade of phototransduction reactions and the derived amplification were used as a model for other activation cascades such as blood clotting [165]. The demonstration that retinal could be stably fixed to opsin by reduction of the Schiff base with $NaBH_4$ [15, 167], followed closely after the initial observation of reduction of a Schiff base in glycogen phosphorylase [32] and was one of the early applications of small molecule chemistry to proteins. Today the general concept of chemical modification of proteins and enzymes is a main-stay in the armentarium of protein chemists and enzymologists. Regeneration of a substance critical for receptor signaling in an adjacent cell type does not appear to be widely found in nature although examples are known such as the glutamate-glutamine cycle involving neurons and glial cells [47].

Vision would be impossible without regeneration of 11-*cis*-retinoids to serve as chromophores in visual pigments. However, retinoids are also needed for viability of rod and cone photoreceptor cells. One of the earliest detectable symptoms of vitamin A deficiency in adults is a deficit in night vision, which reflects the inability of the visual cycle to supply adequate amounts of 11-*cis*-retinal for regeneration of rhodopsin and elimination of the signaling activity of opsin. Manifestations of early deficiency were reversible but if the condition persists rod and cone photoreceptor cells degenerate to produce irreversible blindness [27]. Mechanisms for degeneration include toxicity resulting from continual activation of the signal transduction cascade by opsin in the absence of chromophore [30], the unfolded protein response resulting from the absence of chromophore during opsin gene translation [56, 80, 141], and proteosome clogging from continual opsin synthesis in the absence of chromophore [83, 176]. Thus, retinoids play several roles in maintaining photoreceptor health in addition to those of their classical function of in vision as chromophores.

Mutations in the genes for most of the rod visual cycle enzymes and associated proteins have been described in humans [160]. Nearly all caused delayed dark adaptation and many were associated with retinal degeneration. Patients with mutations in the gene encoding RPE65 (isomerase1) were treated with AAV carrying a correct copy of RPE65 gene [148]. The improvements in the vision of these patients were far from complete but-none-the-less represented one of the remarkable successful achievements of gene therapy.

The universal use of retinals as chromophores in all known visual systems obviously reflects the advantages of their unique chemical features of photoisomerization, reversible reaction with a lysine amino group within the active site of opsins, and facile adjustment of their absorption maxima by interaction of their conjugated chain of double bonds with polar groups of the opsin. However, safe use of these chemical features demanded the employment of several mechanisms to limit chemical reactivity and toxicity. The accumulation with age of fluorescent adducts derived from vitamin A, known in the visual system as lipofuscins, reveals that these mechanisms are not foolproof. The first lipofuscin fluorophore derived from retinal was isolated and identified in the 90s as the pyridinium bisretinoid A2E [120, 140]. Subsequently, a second fluorophore was identified as all-*trans*-retinal dimer (RalDi) [34, 151]. Considerable evidence testifies to the toxicity of these compounds *in vitro* and to their presence in elevated amounts in animal models of retinal disease such as Stargardt disease [89, 151]. However, the role of these substances in human geographic atrophy, a form of AMD, remains a matter of active research (see section "The future" below).

The notion of visual cycle modulation (VCM) as a therapeutic possibility had its origins in studies demonstrating that inhibition of the rod visual cycle slowed the rate of retinal degeneration in rodents resulting from exposure to light [24, 113, 136, 147]. It is likely that reduction of visual cycle activity limited the amount of all-*trans*-retinal in flux available for conversion to toxic *bis*-retinoids A2E and RalDi [140] or to exert other forms of toxicity [21]. The dry form of age-related macular degeneration (AMD), also known as geographic atrophy, leads to degeneration of the RPE and eventual blindness [81, 157]. Studies using retinal autofluorescence suggested that expanding areas of degenerating RPE were surrounded by a rim of hyperfluorescence attributable to *bis*-retinoids, suggesting that dry AMD involves defective processing of visual cycle retinoids and that the condition might be amenable to VCM [50]. Note however, that this view is not universally shared [1]. Four approaches to VCM that are currently undergoing investigation are outlined in the following sections (see [104]).

Inhibition of RPE65

All-*trans*-retinylamine is a potent inhibitor of rod visual cycle activity and its administration delayed retinal degeneration in several rodent models of retinal disease [38]. Preliminary results from clinical trials in human patients with dry AMD suggested that the delay in dark adaptation was proportional to the dose of administered drug (as N-retinylacetamide) [72]. Recent results indicate the efficacy of slow release of all-*trans*-retinylamine from nanoparticle delivery vehicles administered subcutaneously [111]. Another study demonstrated that non-retinoid isoprenoids delayed regeneration of rhodopsin by acting as antagonists of RPE65 and diminished the formation of A2E in a mouse model of Stargardt disease [88]. An inherent drawback to this therapeutic approach is delayed dark adaptation. However, this would appear to be an acceptable side effect if the drugs were effective in delaying the progression of geographic atrophy with its loss of cone function.

By-pass of RPE65

Patients with Leber congenital amaurosis (LCA) suffer from an early onset blindness, which results from inability to generate 11-*cis*-retinal for rhodopsin regeneration [160]. Mutations in these patients have been identified in the genes for RPE65 or LRAT, the enzyme responsible for synthesis of the substrate for RPE65. Studies with mouse and dog models of LCA established that gavage of the animals with 9-*cis*-retinol or its acetate ester restored vision as measured by the ERG response (mice) and behavioral studies (dogs) [37, 87, 162]. 9-*cis*-Retinal, which can be generated by oxidation of 9-*cis*-retinol *in vivo*, bound to rod opsin and forms a visual pigment with properties similar to those of rhodopsin [166]. Clinical trials are in progress to assess the efficacy of this approach in humans with LCA.

D_3-Vitamin A

Formation of A2E and RalDi involves a chemical step in which a proton is abstracted from the C-20 methyl group of the retinoid [63]. All-*trans*-retinol deuterated at the C-20 methyl group (C20-D_3-vitamin A) formed A2E and RalDi at rates approximately 7- and 12-fold slower, respectively, than those found with the parent compound due to a primary kinetic isotope effect. Administration of C20-D_3-vitamin A to ABCA4 knockout mice, a rodent model of human Stargardt disease, delayed retinal degeneration relative to the non-deuterated compound [84]. Subsequent studies revealed that the liver pools of retinol and retinyl esters of swine were 62 % and 57 % equilibrated, respectively, with C20-D_3-vitamin A 2 weeks after administration of the compound [93], suggesting that whole body equilibration of humans would be feasible. No recognized reactions of the visual cycle or systemic vitamin A function would appear to be affected by deuterium in the C-20 methyl group. Clinical trials of C20-D_3-vitamin A are planned with patients with Stargardt disease and geographic atrophy.

*Reversible Sequestration of All-*Trans*-Retinal*

Based on the premise that impaired processing of all-*trans*-retinal was responsible for retinal degeneration in dry AMD, a screen was devised to search for non-toxic amines that could reversibly sequester all-*trans*-retinal as a Schiff base [86] and thus reduce its free concentration. Pregabalin (Lyrica®), a drug used for the treatment of neuropathic pain, emerged as a lead compound because it reversibly formed Schiff bases with all-*trans*-retinal and did not inhibit any known reactions of the rod visual cycle [86]. Pregabalin is used clinically as a mixture of R- and S-enantiomers but only the S-enantiomer is biologically active for relief of neuropathic pain in the formulation. Studies with the R enantiomer demonstrated that the compound protected mice from light damage in a rodent model of retinal degeneration equally as well as the S-enantiomer. Major advantages of pregabalin include prior regulatory agency approval of the RS-mixture of enantiomers and the availability of the biologically inactive but chemically identical enantiomer, which should minimize or eliminate off-target effects of the drug. Clinical trials targeting Stargardt and geographic atrophy patients are planned.

The Future

Remaining Questions

A broad outline of the rod regeneration cycle has been in place for years, resulting from elaboration of the early studies of Wald and Dowling. The *enzymology* of most of the reactions of the rod cycle is now reasonably well understood with the exception of that of the dehydrogenases. For this class of enzymes, the non-specific nature of several dehydrogenases and their redundancy, have prevented a complete understanding of their physiological substrates and roles in retinoid processing.

The cone visual cycle remains more poorly characterized. For instance, three different cone visual cycle *isomerases* have been proposed resulting from studies in three different species, as was mentioned above. If these reports are substantiated, why would nature choose different solutions to the cone isomerase problem in different species when a single rod isomerase (RPE65) has been discovered in all vertebrate species examined? In addition, two of the proposed cone isomerase reactions involve different mechanisms than that involved in the reaction catalyzed by the rod isomerase.

Functional roles in the regeneration cycles for several proteins remain poorly characterized. *RGR-opsin* acts as a "reverse isomerase" *in vitro* with illumination resulting in an all-*trans*- to 11-*cis*-retinal photoisomerization [45]. Characterization of RGR-opsin knockout mice revealed that RGR was necessary for maintenance of a steady state level of 11-*cis*-retinal [19]. However, the effects were not pronounced and the role of this protein in visual physiology is not clear. Other studies suggested that RGR-opsin controls the hydrolysis of all-*trans*-retinyl esters and release of all-*trans*-retinol from retinosomes [114]. The ability of *IRBP* to bind retinols and retinals [18, 101], to carry these ligands *in vivo* [82, 102, 132], and its presence in the extracellular compartment that must be traversed by retinoids in the visual cycle all support its role as a chaperone for retinoids. However, the phenotype of knockout mice described in two studies did not support this function [106, 121]. Other studies suggested that results from the knockout studies might have been confounded by polymorphisms in the mouse RPE65 gene that affected rates of regeneration [60].

The issue of an IRBP receptor remains unresolved [28, 29, 102]. Recent results suggest that the multiple cysteines of IRBP are potent free radical quenchers and that IRBP is important in defining the redox environment of the IPM [42]. In addition, TNFα, RIP1 and 3, and RIP3 kinases were elevated in the IPM of IRBP-knockout mice, suggesting that enhanced signaling via the TNFα pathway could contribute to the toxicity observed in the absence of IRBP [143].

The roles of *A2E and RalDi* in human retinal disease and particularly in geographic atrophy (aka the dry form of macular degeneration) are currently a matter of contention [1, 149, 152, 159]. There is ample evidence to verify their accumulation in animal and human retinal diseases. However, their roles in developing retinal disease remain to be established. Some have argued that the real culprit associated with aberrant retinal processing is excess all-*trans*-retinal itself and not its conjugation products [21].

The role of CRALBP in mouse cone function was recently clarified by studies of CRALBP-deficient mice as described above [174]. However, the studies also revealed that M-cones in dark adapted CRALBP-deficient mice had a reduced response and a 20-fold lower sensitivity, in contrast to rods whose sensitivity was completely restored by dark adaptation [136, 174]. The authors suggest that the chronic chromophore deficiency observed in these animals resulted in reduced cone opsin levels and that minimizing light exposure might be a simple approach to reducing photopic vision loss in patients with CRALBP mutations [174].

The possibility of a *multi-protein complex* involving CRALBP via its PDZ motif, NHERF1, a PDZ domain protein that also binds the actin-binding protein Ezrin, and actin, was raised by *in vitro* studies and by the presence of all four proteins in the apical processes of RPE [99]. A complex of CRALBP and NHERF1 was isolated by gel filtration and partially characterized. The two proteins were present in approximately equal molar amounts and complex formation did not affect the absorption spectrum of 11-*cis*-retinal bound to CRALBP [100]. Immunoprecipitation studies provided additional support for the presence of a complex of CRALBP, RDH5, RPE65, RGR-opsin, and LRAT [43]. Studies in other laboratories provided evidence for a multiprotein complex consisting of RDH10, RPE65, and CRALBP [31]. Further experimentation is needed to determine the physiological relevance of these complexes.

A mechanism for release of 11-*cis*-retinal from CRALBP based on localization of a subfraction of the binding protein at the RPE apical membrane via a multiprotein complex has been proposed [100]. However, it relies on *in vitro* evidence for localization of CRALBP to actin of RPE apical processes via interaction with NHERF1 and ezrin and has not been substantiated by further work.

Acknowledgements The author acknowledges the support of the Departments of Biochemistry and Ophthalmology at the University of Washington during preparation of this Chapter.

References

1. Ablonczy Z, Higbee D, Anderson DM, Dahrouj M, Grey AC, Gutierrez D, Koutalos Y, Schey KL, Hanneken A, Crouch RK (2013) Lack of correlation between the spatial distribution of A2E and lipofuscin fluorescence in the human retinal pigment epithelium. Invest Ophthalmol Vis Sci 54:5535–5542
2. Adler AJ, Evans CD, Stafford WF III (1985) Molecular properties of bovine interphotoreceptor retinol-binding protein. J Biol Chem 260:4850–4855
3. Adler L, Chen C, Koutalos Y (2014) Mitochondria contribute to NADPH generation in mouse rod photoreceptors. J Biol Chem 289:1519–1528
4. Ala-Laurila P, Kolesnikov AV, Crouch RK, Tsina E, Shukolyukov SA, Govardovskii VI, Koutalos Y, Wiggert B, Estevez ME, Cornwall MC (2006) Visual cycle: dependence of retinol production and removal on photoproduct decay and cell morphology. J Gen Physiol 128:153–169
5. Ala-Laurila P, Cornwall MC, Crouch RK, Kono M (2009) The action of 11-*cis*-retinol on cone opsins and intact cone photoreceptors. J Biol Chem 284:16492–16500
6. Alpern M (1971) Rhodopsin kinetics in the human eye. J Physiol 217:447–471
7. Altman J (1985) New visions in photoreception. Nature 313:264–265
8. Barry RJ, Canada FJ, Rando RR (1989) Solubilization and partial purification of retinyl ester synthetase and retinoid isomerase from bovine ocular pigment epithelium. J Biol Chem 264:9231–9238
9. Beharry S, Zhong M, Molday RS (2004) N-retinylidene-phosphatidylethanolamine is the preferred retinoid substrate for the photoreceptor-specific ABC transporter ABCA4 (ABCR). J Biol Chem 279:53972–53979
10. Beltran WA, Cideciyan AV, Lewin AS, Iwabe S, Khanna H, Sumaroka A, Chiodo VA, Fajardo DS, Román AJ, Deng W-T, Swider M, Alemán TS, Boye SL, Genini S, Swaroop A, Hauswirth WW, Jacobson SG, Aguirre GD (2012) Gene therapy rescues photoreceptor blindness in dogs and paves the way for treating human X-linked retinitis pigmentosa. Proc Natl Acad Sci U S A 109:2132–2137
11. Bernstein PS, Rando RR (1986) In vivo isomerization of all-*trans*- to 11-*cis*-retinoids in the eye occurs at the alcohol oxidation state. Biochemistry 25:6473–6478
12. Bernstein PS, Law WC, Rando RR (1987) Isomerization of all-*trans*-retinoids to 11-*cis*-retinoids *in vitro*. Proc Natl Acad Sci U S A 84:1849–1853
13. Bok D, Ong DE, Chytil F (1984) Immunocytochemical localization of cellular retinol binding protein in the rat retina. Invest Ophthalmol Vis Sci 25:877–883
14. Boll F (1876–1877) Zur anatomie und physiologie der retina. Verhandlungen der R Accademia dei Lincei, Dritte Serie, Erster Theil, English translation by Hubbard On the anatomy and physiology of the retina. Vision Res. 1977, 17:1249–1265
15. Bownds D, Wald G (1965) Reaction of the rhodopsin chromophore with sodium borohydride. Nature 205:254–257
16. Bunt-Milam AH, Saari JC (1983) Immunocytochemical localization of two retinoid-binding proteins in vertebrate retina. J Cell Biol 97:703–712
17. Burstedt MS, Forsman-Semb K, Golovleva I, Janunger T, Wachtmeister L, Sandgren O (2001) Ocular phenotype of bothnia dystrophy, an autosomal recessive retinitis pigmentosa associated with an R234W mutation in the RLBP1 gene. Arch Ophthalmol 119:260–267
18. Chen Y, Saari JC, Noy N (1993) Interactions of all-*trans*-retinol and long-chain fatty acids with interphotoreceptor retinoid-binding protein. Biochemistry 32:11311–11318

19. Chen P, Hao W, Rife L, Wang XP, Shen D, Chen J, Ogden T, Van Boemel GB, Wu L, Yang M, Fong HK (2001) A photic visual cycle of rhodopsin regeneration is dependent on *Rgr*. Nat Genet 28:256–269
20. Chen C, Tsina E, Cornwall MC, Crouch RK, Vijayaraghavan S, Koutalos Y (2005) Reduction of all-*trans*-retinal to all-*trans*-retinol in the outer segments of frog and mouse rod photoreceptors. Biophys J 88:2278–2287
21. Chen Y, Okano K, Maeda T, Chauhan V, Golczak M, Maeda A, Palczewski K (2012) Mechanisms of all-*trans*-retinal toxicity with implications for Stargardt disease and age-related macular degeneration. J Biol Chem 287:5059–5069
22. Collery R, McLoughlin S, Vendrell V, Finnegan J, Crabb JW, Saari JC, Kennedy BN (2008) Duplication and divergence of zebrafish CRALBP genes uncovers novel role for RPE- and Müller-CRALBP in cone vision. Invest Ophthalmol Vis Sci 49:3812–3820
23. Crescitelli F (1984) The gecko visual pigment: the dark exchange of chromophore. Vis Res 24:1551–1553
24. Danciger M, Matthes MT, Yasamura D, Akhmedov NB, Rickabaugh T, Gentleman S, Redmond RM, La Vail MM, Farber DB (2000) A QTL on distal chromosome 3 that influences the severity of light-induced damage to mouse photoreceptors. Mamm Genome 11:422–427
25. Das SR, Bhardwau N, Kjeldbye H, Gouras P (1992) Müller cells of chicken retina synthesize 11-*cis*-retinol. Biochem J 285:907–913
26. Dowling JE (1960) Chemistry of visual adaptation in the rat. Nature 188:114118
27. Dowling JE, Wald G (1958) Vitamin A deficiency and night blindness. Proc Natl Acad Sci U S A 44:648–661
28. Duffy M, Sun Y, Wiggert B, Duncan T, Chader GJ, Ripps H (1993) Interphotoreceptor retinoid binding protein (IRBP) enhances rhodopsin regeneration in the experimentally detached retina. Exp Eye Res 57:771–782
29. Edwards RB, Adler AJ (2000) IRBP enhances removal of 11-*cis*-retinaldehyde from isolated RPE membranes. Exp Eye Res 70:235–245
30. Fain GL (2006) Why photoreceptors die (and why they don't). BioEssays 28:344–354
31. Farjo KM, Moiseyev G, Takahashi Y, Crouch RK, Ma JX (2009) The 11-*cis*-retinol dehydrogenase activity of RDH10 and its interaction with visual cycle proteins. Invest Ophthalmol Vis Sci 50:5089–5097
32. Fischer EH, Kent AB, Snyder ER, Krebs EG (1958) The reaction of sodium borohydride with muscle phosphorylase. J Am Chem Soc 80:2906
33. Fishkin N, Yefidoff R, Gollipalli DR, Rando RR (2005) On the mechanism of isomerization of all-*trans*-retinol to 11-*cis*-retinol in retinal pigment epithelial cells: 11-fluoro-all-*trans*-retinol as substrate/inhibitor in the visual cycle. Bioorg Med Chem 13:5189–5194
34. Fishkin NE, Sparrow JR, Allikmets R, Nakanishi K (2005) Isolation and characterization of a retinal pigment epithelial cell fluorophore: An all-*trans*-retinal dimer conjugate. Proc Natl Acad Sci U S A 102:7091–7096
35. Fleisch VC, Schonthaler HB, von Lintig J, Neuhauss SC (2008) Subfunctionalization of a retinoid-binding protein provides evidence for two parallel visual cycles in the cone-dominant zebrafish retina. J Neurosci 28:8208–8216
36. Gallego O, Belyaeva OV, Porté S, Ruiz FX, Stetsneko AV, Shabrova EV, Kostereva NV, Farrés J, Pares X, Kedishvili NY (2006) Comparative functional analysis of human medium-chain dehydrogenases, short-chain dehydrogenases/reductases and aldo-keto reductases with retinoids. Biochem J 399:101–109
37. Gearhart PM, Gearhart G, Thompson DH, Petersen-Jones SM (2010) Improvement of visual performance with intravitreal administration of 9-*cis*-retinal in *Rpe65*-mutant dogs. Arch Ophthalmol 128:1442–1448
38. Golczak M, Kuksa V, Maeda T, Moise AR, Palczewski K (2005) Positively charged retinoids are potent and selective inhibitors of the *trans-cis* isomerization in the retinoid (visual) cycle. Proc Natl Acad Sci U S A 102:8162–8167

39. Goldstein EM, Wolk BM (1973) Regeneration of the green-rod pigment in the isolated frog retina. Vis Res 13:527–534
40. Golobokova EY (2006) Govardovskii VI Late stages of visual pigment photolysis in situ: cones vs. rods. Vis Res 46:2287–2297
41. Gonzalez-Fernandez F, Gosh D (2008) Focus on molecules: interphotoreceptor retinoid-binding protein (IRBP). Exp Eye Res 86:169–170
42. Gonzalez-Fernandez F, Sung D, Haswell KM, Tsin A, Ghosh D (2014) Thiol-dependent anti-oxidant activity of interphororeceptor retinoid-binding protein. Exp Eye Res 129:167–174
43. Gu X, Crabb JS, Nawrot M, Saari JC, Crabb JW (2006) Quantitative mass spectrophometric analysis of visual cycle protein interactions. In: Proceedings of the 54 ASMS conference on mass spectrometry and allied topics. Seattle, May 28–June 1, 2006, Abstr citation no. A063296
44. Haeseleer F, Jang G-F, Imanishi Y, Driessen CAGG, Matsumura M, Nelson PS, Palczewski K (2002) Dual-substrate specificity short chain retinol dehydrogenases from the vertebrate retina. J Biol Chem 277:45537–45546
45. Hao W, Fong HK (1999) The endogenous chromophore of retinal G protein-coupled receptor opsin from the pigment epithelium. J Biol Chem 274:6085–6090
46. He X, Lobsiger J, Stocker A (2009) Bothnia dystrophy is caused by domino-like rearrangements in cellular retinaldehyde-binding protein mutant R234W. Proc Natl Acad Sci U S A 44:18545–18550
47. Hertz L, Dringen R, Schousboe A, Robinson SR (1999) Astrocytes: glutamate producers for neurons. J Neurosci Res 57:417–428
48. Ho M-TP, Massey JB, Pownall HJ, Anderson RE, Hollyfield JG (1989) Mechanism of vitamin A movement between rod outer segments, interphotoreceptor retinoid-binding protein, and liposomes. J Biol Chem 24:928–935
49. Hollyfield JG, Fliesler SJ, Rayborn ME, Fong S-L, Landers RA, Bridges CD (1985) Synthesis and secretion of interstitial retinol-binding protein by the human retina. Invest Ophthalmol Vis Sci 26:58–67
50. Holz FG, Bellman C, Staudt S, Schütt F, Völcker HE (2001) Fundus autofluorescence and development of geographic atrophy in age-related macular degeneration. Invest Ophthalmol Vis Sci 42:1051–1056
51. Huang J, Possin DE, Saari JC (2009) Localizations of visual cycle components in retinal pigment epithelium. Mol Vis 15:223–234
52. Hubbard R, Wald G (1952) *Cis-trans* isomers of vitamin A and retinene in the rhodopsin system. J Gen Physiol 36:269–315
53. Illing M, Molday LL, Molday RS (1997) The 220 kDa rim protein of retinal rod outer segments is a member of the ABC transporter superfamily. J Biol Chem 272:10303–10310
54. Imanishi Y, Palczewski K (2010) Visualization of retinoid storage and trafficking by two-photon microscopy. Methods Mol Biol 652:247–261
55. Imanishi Y, Batten ML, Piston DW, Baehr W, Palczewski K (2004) Noninvasive two-photon imaging reveals retinyl ester storage structures in the eye. J Cell Biol 164:373–383
56. Insinna C, Daniele LL, Davis JA, Larsen DD, Kuemmel C, Wang J, Nikonov SS, Knox BE, Pugh EN Jr (2012) An S-opsin knock-in mouse (P81Y) reveals a role for the native ligand 11-*cis*-retinal in cone opsin biosynthesis. J Neurosci 32:8094–8104
57. Isler O (1977) Progress in the field of fat-soluble vitamins and carotenoids. Experientia 33:555–573
58. Jacobs GH (2008) Primate color vision: a comparative perspective. Vis Neurosci 25:619–633
59. Jin M, Li S, Moghrabi WN, Sun H, Travis GH (2005) Rpe65 is the retinoid isomerase in bovine retinal pigment epithelium. Cell 122:449–459
60. Jin M, Li S, Nusinowitz S, Lloyd M, Hu J, Radu RA, Bok D, Travis GH (2009) The role of interphotoreceptor retinoid-binding protein on the translocation of visual retinoids and function of cone photoreceptors. J Neurosci 29:1486–1495

61. Jones GJ, Crouch RK, Wiggert B, Cornwall MC, Chader GJ (1989) Retinoid requirements for recovery of sensitivity after visual-pigment bleaching in isolated photoreceptors. Proc Natl Acad Sci U S A 86:9606–9610
62. Karrer P, Morf R, Schöpp K (1931) Zur kenntnis des vitamins A aus fischtranen. Helv Chim Acta 14:1431–1436, 1036–1040
63. Kaufman Y, Ma L, Washington I (2011) Deuterium enrichment of vitamin A at the C20 position slows the formation of deterimental vitamin A dimers in wild-type rodents. J Biol Chem 286:7958–7965
64. Kaupp UB, Koch K-W (1986) Mechanism of photoreception in vertebrate vision. Trends in Biochem Sci (TIBS) 11:43–47
65. Kawaguchi R, Yu J, Honda J, Hu J, Whitelegge J, Ping P, Wiita P, Bok D, Sun H (2007) A membrane receptor for retinol binding protein mediates cellular uptake of vitamin A. Science 315:820–825
66. Kaylor JJ, Yuan Q, Cook J, Sarfare S, Makshanoff J, Miu A, Kim A, Kim P, Habib S, Roybal CN, Xu T, Nusinowitz S, Travis GH (2013) Identification of DES1 as a vitamin A isomerase in Müller glial cells of the retina. Nat Chem Biol 9:30–36
67. Kaylor JJ, Cook JD, Makshanoff J, Bischoff N, Yong J, Travis GH (2014) Identification of the 11-*cis*-specific retinyl-ester synthase in retinal Müller cells as multifunctional O-acyltransferase (MFAT). Proc Natl Acad Sci U S A 111:7302–7307
68. Kefalov VJ (2012) Rod and cone visual pigments and phototransduction through pharmacological, genetic, and phpysiological approaches. J Biol Chem 287:1635–1641
69. Kefalov VJ, Estevez ME, Kono M, Goletz PW, Crouch RK, Cornwall MC, Yau K-W (2005) Breaking the covalent bond-A pigment property that contributes to desensitization in cones. Neuron 46:879–890
70. Kiser PD, Golczak M, Maeda A, Palczewski K (2012) Key enzymes of the retinoid (visual) cycle in vertebrate retina. Biochim Biophys Acta 1821:137–151
71. Kiser PD, Golczak M, Palczewski K (2014) Chemistry of the retinoid (visual) cycle. Chem Rev 114:194–232
72. Kubota R, Boman NL, David R, Mallikaarjun S, Patil S, Birch D (2012) Safety and effect on rod function of ACU-4429, a novel small molecule visual cycle modulator. Retina 32:183–188
73. Kühne W (1879) Chemische vorgänge in der netzhaut. In: Hofmann L (ed) Handbuch der Physiology, vol 3, part 1. Leipzig, FCW Vogel. English translation by Hubbard and Wald, Chemical processes in the retina. Vision Res 1977,17:1269–1316
74. Kumar S, Sandell LL, Trainor PA, Koentgen F, Duester G (2012) Alcohol and aldehyde dehydrogenases: retinoid metabolic effects in mouse knockout models. Biochim Biophys Acta 1821:198–205
75. Lai YL, Wiggert B, Liu YP, Chader GJ (1982) Interphotoreceptor retinol-binding protein: possible transport vehicles between compartments of the retina. Nature 298:848–849
76. Lamb TD, Pugh EN Jr (2004) Dark adaptation and the retinoid cycle of vision. Prog Retinal Eye Res 23:307–380
77. Lamb TD, Pugh EN Jr (2006) Phototransduction, dark adaptation, and rhodopsin regeneration. The proctor lecture. Invest Ophthalmol Vis Sci 47:5138–5152
78. Lanska DJ (2010) Chapter 29. Historical aspects of the major neurological vitamin deficiency disorders: overview and fat-soluble A. Handb Clin Neurol 95:435–444
79. Lee KA, Nawrot M, Garwin GG, Saari JC, Hurley JB (2010) Relationships among visual cycle retinoids, rhodopsin phosphorylation, and phototransduction in mouse eyes during light and dark adaptation. Biochemistry 49:2454–2463
80. Li S, Yang Z, Hu J, Gordon WC, Bazan NG, Haas AL, Bok D, Jin M (2013) Secretory defect and cytotoxicity: the potential disease mechanisms for the retinitis pigmentosa (RP)-associated interphotoreceptor retinoid-binding protein (IRBP). J Biol Chem 288:11395–11406

81. Lim LS, Mitchell P, Seddon JM, Holz FG, Wong TY (2012) Age-related macular degeneration. Lancet 379:1728–1738
82. Liou GI, Bridges CDB, Fong S-L, Alvarez RA, Gonzalez-Fernandez F (1982) Vitamin A transport between retina and pigment epithelium – an interstitial protein carrying endogenous retinol (interstitial retinol-binding protein). Vis Res 22:1457–1467
83. Lobanova ES, Finkelstein S, Skiba NP, Skiba NP, Arshavsky VY (2013) Proteosome overload is a common stress factor in multiple forms of inherited retinal degeneration. Proc Natl Acad Sci U S A 110:9986–9991
84. Ma L, Kaufman Y, Zhang J, Washington I (2011) C20-D3-vitamin A slows lipofuscin accumulation and electrophysiological retinal degeneration in a mouse model of Stargardt disease. J Biol Chem 286:7966–7974
85. Maeda A, Maeda T, Golczak M, Palczewski K (2008) Retinopathy in mice induced by disrupted all-*trans*-retinal clearance. J Biol Chem 283:26684–26693
86. Maeda A, Golczak M, Chen Y, Okano K, Kohno H, Shiose S, Ishikawa K, Harte W, Placzewska G, Maeda T, Palczewski K (2012) Primary amines protect against retinal degeneration in mouse models of retinopathies. Nat Chem Biol 8:170–178
87. Maeda T, Dong Z, Jin H, Sawada O, Gao S, Utkhede D, Monk W, Palczewska G, Palczewski K (2013) QLT091001, a 9-*cis*-retinal analog, is well tolerated by retinas of mice with impaired visual cycles. Invest Ophthalmol Vis Sci 54:455–466
88. Maiti P, Kong J, Kim SR, Sparrow JR, Alikmets R, Rando RR (2006) Small molecule RPE65 antagonists limit the visual cycle and prevent lipofuscin formation. Biochemistry 45:852–860
89. Mata NL, Weng J, Travis GH (2000) Biosynthesis of a major lipofuscin fluorophore in mice and humans with ABCR-mediated retinal and macular degeneration. Proc Natl Acad Sci U S A 97:7154–7159
90. Mata NL, Radu RA, Clemmons RS, Travis GH (2002) Isomerization and oxidation of vitamin A in cone-dominant retinas: a novel pathway for visual-pigment regeneration in daylight. Neuron 36:69–80
91. Matsumoto H, Tokunaga F, Yoshizama T (1975) Accessibility of the iodopsin chromophore. Biochim Biophys Acta 404:300–308
92. McBee JK, Palczewski K, Baehr W, Pepperberg DR (2001) Confronting complexity; the interlink of phototransduction and retinoid metabolism in the vertebrate retina. Prog Retin Eye Res 20:469–529
93. Mihai DM, Jiang H, Blaner WS, Romanov A, Washington I (2013) The retina rapidly incorporates ingested C20-D3-vitamin A in a swine model. Mol Vis 19:1677–1683
94. Miyazono S, Shimauchi-Matsukawa Y, Tachibanaki S, Kawamura S (2008) Highly efficient retinal metabolism in cones. Proc Natl Acad Sci U S A 105:16051–16056
95. Moiseyev G, Chen Y, Takahashi Y, Wu BX, Ma JX (2005) RPE65 is the isomerohydrolase in the retinoid visual cycle. Proc Natl Acad Sci U S A 102:12414–12418
96. Morton RA, Pitt GAJ (1957) Visual pigments. Fortschr Chem Org Naturst XIV:244–316
97. Muniz A, Betts BS, Trevino AR, Buddavarapu K, Roman R, Ma J-X, Tsin TC (2009) Evidence for two retinoid cycles in the cone-dominant chicken eye. Biochemistry 48:6854–6863
98. Napoli JL (2012) Physiological insights into all-*trans*-retinoic acid biosynthesis. Biochim Biophys Acta 1821:152–167
99. Nawrot M, West K, Huang J, Possin DE, Bretscher A, Crabb JW, Saari JC (2004) Cellular retinaldehyde-binding protein interacts with ERM-binding phosphoprotein 50 in retinal pigment epithelium. Invest Ophthalmol Vis Sci 45:393–401
100. Nawrot M, Liu T, Garwin GG, Crabb JW, Saari JC (2006) Scaffold proteins and the regeneration of visual pigments. Photochem Photobiol 82:1482–1488
101. Noy N (2000) Retinoid-binding proteins: mediators of retinoid action. Biochem J 348:481–495

102. Okajima T-IL, Pepperberg DR, Ripps H, Wiggert B, Chader GJ (1989) Interphotoreceptor retinoid-binding protein: role in delivery of retinol to the pigment epithelium. Exp Eye Res 49:629–644
103. Oroshnik W, Brown PK, Hubbard R, Wald G (1956) Hindered *cis*-isomers of vitamin A and retinene: the structure of the neo b-isomer. Proc Natl Acad Sci U S A 42:578–580
104. Palczewski K (2010) Retinoids for treatment of retinal diseases. Trends Pharmacol Sci 31:284–295
105. Palczewski K (2012) Chemistry and biology of vision. J Biol Chem 287:1612–1619
106. Palczewski K, Van Hooser JP, Garwin GG, Chen J, Liou GI, Saari JC (1999) Kinetics of visual pigment regeneration in excised mouse eyes and in mice with a targeted disruption of the gene encoding interphotoreceptor retinoid-binding protein or arrestin. Biochemistry 38:12012–12019
107. Palczewski K, Kumasaka T, Hori T, Behnke CA, Motoshima H, Fox BA, Le Trong I, Teller DC, Okada T, Stenkamp RE, Yamamoto M, Miyano M (2000) Crystal structure of rhodopsin: a G protein-coupled receptor. Science 289:739–745
108. Parker RO, Crouch RK (2010) Retinol dehydrogenases (RDHs) in the visual cycle. Exp Eye Res 91:788–792
109. Poliakov E, Parikh T, Ayele M, Kuo S, Chander P, Gentleman S, Redmond TM (2011) Aromatic lipophilic spin traps effectively inhibit RPE65 isomerohydrolase activity. Biochemistry 50:6739–6741
110. Posch KC, Boerman MHEM, Burns RD, Napoli JL (1991) Holocellular retinol binding protein as a substrate for microsomal retinal synthesis. Biochemistry 30:6224–6230
111. Puntel A, Maeda A, Golczak M, Gao S-Q, Yu G, Palczewski K, Lu Z-R (2015) Prolonged prevention of retinal degeneration with retinylamine loaded nanoparticles. Biomaterials 44:103–110
112. Quazi F, Molday RS (2014) ATP-binding cassette transporter ABCA4 and chemical isomerization protect photoreceptor cells from the toxic accumulation of excess 11-*cis*-retinal. Proc Natl Acad Sci U S A 111:5024–5029
113. Radu RA, Mata NL, Nusinowitz S, Liu X, Sieving PA, Travis GH (2003) Treatment with isotretinoin inhibits lipofuscin accumulation in a mouse model of recessive Stargardt's macular degeneration. Proc Natl Acad Sci U S A 100:4742–4747
114. Radu RA, Hu J, Peng J, Bok D, Mata NL, Travis GH (2008) Retinal pigment epithelium-retinal G protein receptor-opsin mediates light-dependent translocation of all-*trans*-retinyl esters for synthesis of visual chromophore in retinal pigment epithelial cells. J Biol Chem 283:19730–19738
115. Rando RR (1990) The chemistry of vitamin A and vision. Angew Chem Int Ed Engl 29:461–480
116. Rando RR (2001) The biochemistry of the visual cycle. Chem Rev 101:1881–1896
117. Rattner A, Smallwood PM, Nathans J (2000) Identification and characterization of all-*trans*-retinol dehydrogenase from photoreceptor outer segments, the visual cycle enzyme that reduces all-*trans*-retinal to all-*trans*-retinol. J Biol Chem 275:11034–11043
118. Redmond TM, Poliakov E, Yu S, Tsai J-Y, Lu Z, Gentleman S (2005) Mutation of key residues of RPE65 abolishes its enzymatic role as isomerohydrolase in the visual cycle. Proc Natl Acad Sci U S A 102:13658–13663
119. Redmond TM, Poliakov E, Kuo S, Chander P, Gentleman S (2010) RPE65, visual cycle retinol isomerase, is not inherently 11-*cis*-specific: support for a carbocation mechanism of retinol isomerization. J Biol Chem 285:1919–1927
120. Ren RF, Sakai N, Nakanishi K (1997) Total synthesis of the ocular age pigment 'A2E': a convergent pathway. J Am Chem Soc 119:3619–3620
121. Ripps H, Peachey NS, Xu X, Nozell SE, Smith SB, Liou GI (2000) The rhodopsin cycle is prserved in IRBP "knockout" mice despite abnormalities in retinal structure and function. Vis Neurosci 17:97–105
122. Rodieck RW (1998) The first steps in seeing. Sinauer, Sunderland

123. Rosenfeld L (1997) Vitamine-vitamin. The early years of discovery. Clin Chem 43:680–685
124. Rushton WAH (1965) The ferrier lecture, 1962. Visual adaptation. Proc Roy Soc Lond Series B 162:20–46
125. Saari JC (2012) Vitamin A, metabolism in rod and cone visual cycle. Annu Rev Nutr 32:125–145
126. Saari JC, Bredberg L (1982) Enzymatic reduction of 11-*cis*-retinal bound to cellular retinal-binding protein. Biochim Biophys Acta 716:266–272
127. Saari JC, Bredberg DL (1988) CoA- and non-CoA-dependent retinol esterification in retinal pigment epithelium. J Biol Chem 263:8084–8090
128. Saari JC, Bredberg DL (1989) Lecithin:retinol acyltransferase in retinal pigment microsomes. J Biol Chem 264:8636–8640
129. Saari JC, Crabb JW (2005) Focus on molecules: cellular retinaldehyde-binding protein (CRALBP). Exp Eye Res 81:245–246
130. Saari JC, Bredberg L, Garwin GG (1982) Identification of the endogenous retinoids associated with three cellular retinoid-binding proteins from bovine retina and retinal pigment epithelium. J Biol Chem 257:13329–13333
131. Saari JC, Bunt-Milam AH, Bredberg DL, Garwin GG (1984) Properties and immunocytochemical localization of three retinoid-binding proteins from bovine retina. Vis Res 24:1595–1603
132. Saari JC, Teller DC, Crabb JW, Bredberg L (1985) Properties of an interphotoreceptor retinoid-binding protein from bovine retina. J Biol Chem 260:195–201
133. Saari JC, Bredberg DL, Noy N (1994) Control of substrate flow at a branch in the visual cycle. Biochemistry 33:3106–3112
134. Saari JC, Huang J, Asson-Batres MA, Champer RJ, Garwin G, Crabb JW, Possin DE, Milam H (1995) Evidence of retinoid metabolism within cells of inner retina. Exp Eye Res 60:209–212
135. Saari JC, Garwin GG, Van Hooser JP, Palczewski K (1998) Reduction of all-*trans*-retinal limits regeneration of visual pigment in mice. Vis Res 38:1325–1333
136. Saari JC, Nawrot M, Kennedy BN, Garwin GG, Hurley JB, Huang J, Possin DE, Crabb JW (2001) Visual cycle impairment in cellular retinaldehyde-binding protein (CRALBP) knockout mice results in delayed dark adaptation. Neuron 29:739–748
137. Saari JC, Nawrot M, Garwin GG, Kennedy MJ, Hurley JB, Ghyselinck NB, Chambon P (2002) Analysis of the visual cycle in cellular retinol-binding protein type 1 (CRBP1) knockout mice. Invest Ophthalmol Vis Sci 43:1730–1735
138. Saari JC, Nawrot M, Stenkamp RE, Teller DC, Garwin GG (2009) Release of 11-*cis*-retinal from cellular retinaldehyde-binding protein by acidic lipids. Mol Vis 15:844–854
139. Saibil H (1982) Rival transmitters in visual transduction. Nature 297:106–107
140. Sakai N, Decatur J, Nakanishi K (1996) Ocular age pigment "A2-E", an unprecedented pyridinium bisretinoid. J Am Chem Soc 118:1559–1560
141. Sakami S, Maeda T, Bereta G, Okano K, Golczak M, Sumaroka A, Roman AJ, Cideciyan AV, Jacobson SG, Palczewski K (2011) Probing mechanisms of photoreceptor degeneration in a new mouse model of the common form of autosomal dominant retinitis pigmentosa due to P23H opsin mutations. J Biol Chem 286:10551–10567
142. Sakmar TP (2006) Timing is everything: direct measurement of retinol production in cones and rods. J Gen Physiol 128:147–148
143. Sato K, Li S, Gordon WC, He J, Liou GI, Hill JM, Travis GH, Bazan NG, Jin M (2013) Receptor interacting protein kinase-mediated necrosis contributes to cone and rod photoreceptor degeneration in the retina lacking interphotoreceptor retinoid-binding protein. J Neurosci 30:17458–17468
144. Scheerer P, Park JH, Hildebrand PW, Kim YJ, Krauss N, Choe H-W, Hofmann KP, Ernst OP (2008) Crystal structure of opsin in its G-protein-interacting conformation. Nature 455:496–502
145. Schultze M (1866) Zur anatomie und physiologie der retina. Arch Mikr Anat 2:176–286

146. Semba RD (2012) On the 'Discovery' of vitamin A. Ann Nutr Metab 61:192–198
147. Sieving PA, Chaudhry P, Kondo M, Provenzano M, Wu D, Carlson TJ, Bush RA, Thompson DA (2001) Inhibition of the visual cycle in vivo by 13-*cis*-retinoic acid protects from light damage and provides a mechanism for night blindness in isotretinoin therapy. Proc Natl Acad Sci U S A 98:1835–1840
148. Simonelli F, Maguire AM, Testa F, Pierce EA, Mingozzi F, Bennicelli JL, Rossi S, Marshall K, Banfi S, Surace EM, Sun J, Redmond TM, Zhu X, Shindler KS, Ying GS, Ziviello C, Acerra C, Wright JF, McDonnell JW, High KA, Bennett J, Auricchio A (2010) Gene therapy for Leber's congenital amaurosis is safe and effective through 1.5 years after vector administration. Mol Ther 18:643–650
149. Smith R, Bernstein PS, Curcio A (2013) Rethinking A2E. Invest Ophthalmol Vis Sci 54:5543
150. Sommer A (2008) Vitamin A, deficiency and clinical disease: an historical overview. J Nutr 138:1835–1839
151. Sparrow JR, Gregory-Roberts E, Yamamoto K, Blonska A, Ghosh SK, Ueda K, Zhou J (2012) The bisretinoids of retinal pigment epithelium. Prog Retin Eye Res 31:121–135
152. Sparrow JR, Dowling JE, Bok D (2013) Understanding RPE lipofuscin. Invest Ophthalmol Vis Sci 54:8325–8326
153. Stecher H, Gelb MH, Saari JC, Palczewski K (1999) Preferential release of 11-*cis*-retinol from retinal pigment epithelial cells in the presence of cellular retinaldehyde-binding protein. J Biol Chem 274:8577–8585
154. Strauss O (2006) The retinal pigment epithelium in visual function. Physiol Rev 85:845–881
155. Sun H (2012) Membrane receptors and transporters involved in the function and transport of vitamin A and its derivatives. Biochim Biophys Acta 1821:99–112
156. Sun H, Nathans J (1997) Stargardt's ABCR is localized to the disc membrane of retinal rod outr segments. Nat Genet 15:15–16
157. Swaroop A, Chem EY, Bowes Rickman C, Abecasis GR (2009) Unraveling a multifactorial late-onset disease: from genetic susceptibility to disease mechanisms for age-related macular degeneration. Annu Rev Genomics Hum Genet 10:19–43
158. Takahashi Y, Moiseyev G, Chen Y, Nikolaeva O, Ma J-X (2011) An alternative isomerohydrolase in the retinal Muller cells of a cone-dominant species. FEBS J 278:2913–2926
159. Tang PH, Kono M, Koutalos Y, Ablonczy Z, Crouch RK (2013) New insights into retinoid metabolism and cycling within the retina. Prog Retin Eye Res 32:48–63
160. Thompson DA, Gal A (2003) Vitamin A metabolism in the retinal pigment epithelium: genes, mutations, and diseases. Prog Retin Eye Res 22:683–703
161. Travis GH, Golczak M, Moise AR, Palczewski K (2006) Diseases caused by defects in the visual cycle: retinoids as potential therapeutic agents. Annu Rev Pharmacol 47:469–512
162. Van Hooser JP, Liang Y, Maeda T, Kuksa V, Jang G-F, He YG, Rieke F, Fong H, Detwiler PB, Palczewski K (2002) Recovery of visual functions in a mouse model of Leber congenital amaurosis. J Biol Chem 277:19173–19182
163. Wald G (1934–1934) Vitamin A in eye tissues. J Gen Physiol 18:905–915
164. Wald G (1935–1936) Carotenoids and the visual cycle. J Gen Physiol 19:351–371
165. Wald G (1965) Visual excitation and blood clotting. Science 150:1028–1030
166. Wald G (1968) Molecular basis of visual excitation. Science 162:230–239
167. Wald G, Hubbard R (1949) The reduction of retinene$_1$ to vitamin A$_1$ *in vitro*. J Gen Physiol 32:367–349
168. Wang J-S, Kefalov VJ (2009) An alternative pathway mediates the mouse and human cone visual cycle. Curr Biol 19:1665–1669
169. Wang J-S, Kefalov VJ (2011) The cone-specific visual cycle. Prog Retin Eye Res 30:115–128
170. Wang J-S, Estevez ME, Cornwall MC, Kefalov VJ (2009) Intra-retinal visual cycle required for rapid cone dark adaptation. Nat Neurosci 12:295–302

171. Weng J, Mata NL, Azarian SM, Tzekov RT, Birch DG, Travis GH (1999) Insights into the function of rim protein in photoreceptors and etiology of Stargardt's disease from the phenotype in *abcr* knockout mice. Cell 98:13–23
172. Winkler BS, DeSantis N, Solomon F (1986) Multiple NADPH-producing pathways control glutathione (GSH) content in retina. Exp Eye Res 43:829–847
173. Winston A, Rando RR (1998) Regulation of isomerohydrolase activity in the visual cycle. Biochemistry 37:2044–2050
174. Xue Y, Shen SQ, Jui J, Rupp AC, Byrne LC, Hattar S, Flannery JG, Corbo JC, Kefalov VJ (2015) CRALBP supports the mammalian retinal visual cycle and cone vision. J Clin Invest 125:727–738
175. Yost RW, Harrison EH, Ross AC (1988) Esterification by rat liver microsomes of retinol bound to cellular retinol-binding protein. J Biol Chem 263:18693–18701
176. Zhang T, Enemchukwu NO, Jones A, Wang S, Dennis E, Watt CB, Pugh EN Jr, Fu Y (2015) Genetic deletion of S-opsin [prevents rapid cone degeneration in a mouse model of Leber congenital amaurosis. Hum Molec Gen 24:1755–1763

Index

M.A. Asson-Batres, C. Rochette-Egly (eds.), *The Biochemistry of Retinoid Signaling II*, Subcellular Biochemistry 81, DOI 10.1007/978-94-024-0945-1

The manufacturer's authorised representative in the EU is Springer Nature Customer Service Centre GmbH, Europaplatz 3, 69115 Heidelberg, Germany. If you have any concerns regarding our products, please contact ProductSafety@springernature.com

Printed and bound by CPI Group (UK) Ltd, Croydon, CR0 4YY
30/09/2026
02225446-0006